RICE SCIENCE
AND DEVELOPMENT POLITICS

RICE SCIENCE AND DEVELOPMENT POLITICS

Research Strategies and IRRI's Technologies
Confront Asian Diversity (1950–1980)

*Robert S. Anderson, Edwin Levy,
and Barrie M. Morrison*

CLARENDON PRESS · OXFORD
1991

Oxford University Press, Walton Street, Oxford OX2 6DP
Oxford New York Toronto
Delhi Bombay Calcutta Madras Karachi
Petaling Jaya Singapore Hong Kong Tokyo
Nairobi Dar es Salaam Cape Town
Melbourne Auckland
and associated companies in
Berlin Ibadan

Oxford is a trade mark of Oxford University Press

Published in the United States
by Oxford University Press, New York

© Robert Anderson, Edwin Levy, Barrie Morrison 1991

The publication of this book was supported
by a grant from Simon Fraser University.

All rights reserved. No part of this publication may be reproduced,
stored in a retrieval system, or transmitted, in any form or by any means,
electronic, mechanical, photocopying, recording, or otherwise, without
the prior permission of Oxford University Press

British Library Cataloguing in Publication Data
Data available

Library of Congress Cataloging in Publication Data
Anderson, Robert S., 1951–
Rice science and development politics: research strategies and
IRRI's technologies confront Asian diversity, 1950–1980/Robert S.
Anderson, Edwin Levy, and Barrie M. Morrison.
p.   cm.
Includes bibliographical references and index.
1. International Rice Research Institute—History. 2. Rice—
Research—International cooperation—History. 3. Rice—Research—
Sri Lanka—History. 4. Rice—Research—Bangladesh—History.
5. Agriculture—Sri Lanka—Technology transfer. 6. Agriculture—
Bangladesh—Technology transfer. I. Levy, Edwin. II. Morrison,
Barrie M. III. Title.
SB191.R5A7599 1991   633.1'8'072—dc20   90-48926
ISBN 0-19-828341-5

Typeset by Best-set Typesetter Ltd.

Printed in Great Britain by
Bookcraft (Bath) Ltd,
Midsomer Norton, Avon

To

Lewis S. Anderson (1910–1971)
and Louis B. Mezei (1906–1990)

Ed Levy's Family

Elizabeth R. Morrison

Preface

THIS book appears after years of research collaboration during which the issues it addresses have remained fresh and important. It was during the 1974 American Association for the Advancement of Science meetings in Boston that Bob Anderson began to discuss this project with AAAS officers and eventually the US National Science Foundation. Robert Baum, Rachelle Hollander, William Blanpied, and Arthur Norberg of the NSF encouraged the development of a collaborative project on the Ethical and Human Value Implications of Science and Technology among Bob Anderson, Ed Levy, and Barrie Morrison at the Institute of Asian Research at the University of British Columbia, Vancouver, and Paul Brass at the University of Washington, Seattle. The result was Foundation project OSS 7684625 A04. Anderson, Brass, Levy, and Morrison collaborated on an earlier book, *Science, Politics, and the Agricultural Revolution in Asia* (1982), a book which began to articulate some of the ideas offered here.

During the course of the project each of the authors visited IRRI and surrounding areas of the Philippines and carried out research in Asia. Morrison returned to Sri Lanka again after 1980 and completed the case study offered here in the course of doing other work. Anderson did the same in Bangladesh and in India. He and Levy spent extensive time with researchers in the southern US rice research community while working on the project. Levy visited IRRI and India to discuss technology-transfer issues. All this activity gave rise to numerous discussions about rice, IRRI, and Asia, so that the point of view offered here is a collective one. In addition the authors had the benefit of discussions with Paul Brass about his research in India on these questions and read his reports and manuscripts. The voluminous Indian material, which is not included here, provided an additional and valuable comparison. Anderson travelled to rice research centres at Chinsurah, Kharagpur, Cuttack, Hyderabad, and Madras to interview Indian officials and scientists and to collect evidence on the mutual influences of Indian research and international rice research. Finally, members of the group travelled in China, Thailand, and Burma to augment their field work and to establish comparisons.

Each part of the manuscript was written by a specific author: Levy wrote Chapters 1 and 10, with additions by Morrison and Anderson; Anderson wrote Chapters 2, 3, and 4 on IRRI, with an addition by Levy; Morrison wrote Chapters 5 and 6 on Sri Lanka; and Anderson wrote

Chapters 7, 8, and 9 on Bangladesh. Chapter 11 was written jointly. Each author has made significant contributions at every stage of the book.

In so lengthy and so far-flung a process many people were our guides, supporters, and benefactors. We all thank Nyle Brady, director of IRRI and researchers at Los Banos who generously discussed and corresponded with us regarding these issues; the insights of Randolph Barker, Gelia Castillo, Lloyd Evans, Grace Goodell, Thomas Hargrove, Richard Harwood, Robert Herdt, Peter Kenmore, Edmund Oasa, and Hubert Zandstra were invaluable.

Morrison is grateful to C. Narayamaswami, director of the Agrarian Research and Training Institute in Sri Lanka for an attractive research base and accommodation. Numerous officers and former officers of the Sri Lankan Department of Agriculture gave advice, information, and hospitality, most particularly C. R. Pannabokke, E. Abeyratne, and D. Senadhira.

Anderson is grateful to S. M. H. Zaman, former director of the Bangladesh Rice Research Institute, and to numerous researchers, particularly Ekramul Ahsan, for conversations and correspondence; also to A. Quasem of the Bangladesh Institute of Development Studies for research assistance. Anderson wishes to thank the staff of the Rockefeller Archives Center and the Ford Foundation Archives for guiding him in the search for and correct use of relevant documents, although responsibility for interpretation of this material is solely his. Their archivists are remarkable for their patient assistance to scholars, particularly those at a long distance.

In Vancouver we are most appreciative of the excellent editorial advice of Marjorie Sinel and Lynne Hissey. Additional research assistance is gratefully acknowledged from Alice Kidd, James Pratt, Amparo Cadavid, Shafiqur Rahman, and Greg Durward for his contribution to Chapter 3 with Levy. We are thankful to Simon Fraser University for supporting publication of this book, and to Keith Griffin for his encouragement to publish it. Andrew Schuller, Anna Zaranko, and Michael Belson of Oxford University Press helped us at every turn.

Finally, we acknowledge the long and patient support of our families, friends, and colleagues. The joke that 'rice is nice' became nearly legendary, and we are glad these efforts have finally been harvested.

R.A., E.L., and B.M.

Vancouver
December 1990

About the Authors

Robert Anderson is professor in the Department of Communication at Simon Fraser University, Burnaby, Canada. He began studying in Asia in 1961, and has written on the role of science and technology in international development. In particular, he wrote *Saha and Bhabna: Building Scientific Institutions in India* (1975) and *The Hour of the Fox: Tropical Forests, the World Bank, and the Indigenous People of Central India* (1988). He also has had experience in managing and evaluating rural development projects in Bangladesh and other countries, beginning in 1972.

Edwin Levy is Vice-President of Quadra Logic Technologies Inc., Vancouver, Canada where he is responsible for regulatory affairs. He was previously professor in the Department of Philosophy at the University of British Columbia in Vancouver. He is co-author of *Mandated Science* (1988), which is about science and scientists in official bodies (such as expert committees, courts) with a mandate to make policy decisions.

Barrie Morrison is professor in the Department of Asian studies at the University of British Columbia, Vancouver, Canada. He began studying in Asia in 1963 and has written extensively about the changing structure of rural society in Asia. He wrote *Political Centres and Culture Regions in Early Bangladesh* (1970) and *The Disintegrating Village: Social Change in Rural Sri Lanka* (1979).

Contents

List of Maps and Figure	xii
List of Tables	xiii
List of Abbreviations	xiv

1. Introduction 1

   1.1 Our Original Assumptions and their Limitations 5
   1.2 Two Theses for Evaluating Research Strategies
       and the Transfer of Technology 11
   Notes 18

2. American Interests, Foundation Planning, and IRRI's Origins ... 22

   2.1 Foreign Policy, the Food Problem, and the Rockefeller
       Foundation 26
   2.2 Rice Politics and Research Planning 33
   2.3 An International Rice Research Institute? 40
   2.4 A Single Centre: The Process of Specification 47
   2.5 Specifying Research Priorities for IRRI: Yields Again . 50
   Notes 57

3. IRRI's Mission: To Develop Technology and Test Methodology ... 62

   3.1 New Plant Types and the First Variety 64
   3.2 Beyond IR 8 70
   3.3 Shifts in IRRI's Organization 73
   3.4 Methodological Problems at IRRI 78
   3.5 IRRI's Early Efforts to Gauge the Constraints and Consequences
       of New Rice Technology for Large vs. Small Farmers 94
   3.6 Traumatic Change and Continuity in Approach 107
   Notes 110

4. Communicating IRRI's Mission 117

   4.1 IRRI's Concept of Research 119
   4.2 An Office of Communication 122
   4.3 The Outreach System and Asian Diversity 127
   4.4 IRRI's Mission Criticized, and the Underwriters' Response . 136
   Notes 144

5. Sri Lanka: Success and Stagnation in an Independent
   Research Tradition 149

5.1 The Different Physical Environments and Social Organization
of Rice-Growing ... 152
5.2 The Foundations and Beginnings of Rice Research to 1931 ... 159
5.3 The Politicians' Concern for the Rice Grower and the Place of
Research, 1931–1950 ... 161
5.4 The Successes and Limitations of the Rice-Breeding Strategy,
1950–1971 ... 165
Notes ... 183

6. Sri Lanka: The Quest for a New Research Strategy, 1971–1980 ... 186

6.1 The Central Agricultural Research Institute (CARI) in the Mid-
Country and Highlands Region of the Wet Zone ... 188
6.2 Maha Illuppallama and the Northern Dry Zone ... 192
6.3 The Central Rice-Breeding Station, Batalagoda, and the
Mid-Country ... 198
6.4 The Rice Research Station at Bombuwela and the Low-Lying Wet
Zone ... 206
6.5 The Relevance of IRRI to Sri Lankan Agricultural Research in
1977–1978 ... 211
6.6 New Research Strategies Emerge from Earlier Experience ... 214
Notes ... 219

7. Bangladesh: A Research Tradition Becomes Absorbed
in the Green Revolution ... 222

7.1 Agricultural Diversity, Social Complexity ... 225
7.2 From Historic Surplus to Rice Deficits ... 232
7.3 Rice Research and Agricultural Development
in Bengal before 1947 ... 237
7.4 The Continued Neglect of Rice Research 1948–1960 ... 242
7.5 Forces for the Green Revolution ... 248
7.6 Researchers' Values and Green Revolution Practices ... 256
7.7 Chronology of the Green Revolution ... 273
Notes ... 275

8. Bangladesh: Independent Nation, Dependent Research ... 280

8.1 The Politics of Self-Sufficiency and the Weakness of Rice
Research ... 284
8.2 Changing the Forces of Rice Production:
More Seeds and Their Requirements ... 300
Notes ... 326

9. The Limitations of BRRI's Objectives, Organization, and Donors ... 330

9.1 BRRI's Objectives and Organization ... 334
9.2 Relations with Rice Cultivators ... 346
9.3 The Responsibilities of BRRI's Donors ... 348
Notes ... 356

Contents

10. Conclusions and Reflections on Research Strategies ... 360
    - 10.1 The Foundations' Strategy and IRRI's Strategy ... 360
    - 10.2 The National Research Strategies ... 365
    - 10.3 The Natural and Social Scientific Paradigms in Research Strategies ... 368
    - 10.4 Integration: A Fundamental Problem for Science and Technology ... 370
    - 10.5 The Underspecification Thesis Reconsidered ... 372
    - 10.6 The Mango Thesis Reconsidered ... 374
    - 10.7 Responsibility of Science and Technology ... 375
    - Notes ... 379

11. IRRI in the 1980s ... 381
    - Notes ... 385

Glossary ... 386

Index ... 387

Maps

1	Sri Lanka provinces and districts	150
2	Bangladesh districts	223
3	Bangladesh: HYV rice cultivation, 1969–1970	270
4	Bangladesh: HYV rice cultivation, 1973–1974	281
5	Bangladesh: HYV rice cultivation, 1977–1978	282

Figure

3.1	Factors in adoption of modern rice varieties	102

Tables

3.1 Relationship of farm size to labour, capital, and return 96
3.2 Farmers benefiting from modern varieties 103
3.3 Estimates of the differential impacts of technical progress 105
3.4 IRRI allocation of core budget research funds 106
5.1 Rice imports, Sri Lanka, 1886–1977 151
5.2 Cost of the Sri Lanka food subsidy programme 151
5.3 Distribution of agricultural land in Sri Lanka, 1962 153
5.4 Paddy smallholdings in up-country Sri Lanka, 1946 and 1962 154
5.5 Landlessness and tenancy in up-country Sri Lanka, 1946 155
5.6 Paddy smallholdings in Sri Lanka dry zone, 1946 and 1962 156
5.7 Landlessness and tenancy in Sri Lanka dry zone, 1946 157
5.8 Paddy smallholdings in Sri Lanka low country, wet zone, 1946 and 1962 158
5.9 Landlessness and tenancy in Sri Lanka low country, wet zone, 1946 158
5.10 Adoption of H4 and sister lines by district of Sri Lanka, 1966–1967 168
5.11 Change in area and cultivation practices in Sri Lanka, 1957–1968 170
5.12 Sri Lanka government credit for paddy production 172
7.1 Landless Bangladesh agricultural households, 1976–1977 230
7.2 Big-farmer agricultural households, 1976–1977 231
7.3 Self-sufficiency in rice of Bengal districts, 1938 235
7.4 Agricultural research budgets, Pakistan, 1964–1965 260
8.1 Estimates of 1972–1973 aman rice crop, Bangladesh 296
8.2 Food sufficiency plan and actual grain imports, Bangladesh, 1973–1978 298
8.3 Import of HYV rice seed to Bangladesh, 1966–1976 301
8.4 Annual allocation/sale of chemical fertilizer, East Pakistan/Bangladesh, 1966–1980 312
8.5 Percentage of aman rice acreage planted to HYV, 1974–1977 323
9.1 Aspects of BRRI finances 349

Abbreviations

A & M	Texas Agricultural and Mechanical University	CIAT	Centro Internacional de Agricultura Tropical, Colombia
ADB	Asian Development Bank	CIDA	Canadian International Development Agency
ADC	Agricultural Development Council	CIMMYT	Centro Internacional de Mejormiento de Maiz y Trigo, Mexico
AI	Agricultural Instructor		
AICRIP	All India Crop Rice Improvement Program	CFR	Council on Foreign Relations, New York
AID	(Village) Agricultural and Industrial scheme (Pakistan)	CGIAR	Consultative Group on International Agricultural Research, World Bank
AR	*Annual Reports*, IRRI, including Agricultural Economics Department	CRRI	Central Rice Research Institute (India)
ARI	Agricultural Research Institute (East Pakistan)	CRVT	Coordinated Rice Varietal Testing programme
ARTI	Agrarian Research and Training Institute	DAEO	District Agricultural Extension Officer
		EFT	Extension Field Trials
BADC	Bangladesh Agricultural Development Corporation	EPARRI	East Pakistan Accelerated Rice Research Institute
BLB	bacterial leaf blight	FAO	UN Food and Agriculture Organization
BPH	brown plant hopper		
BRRI	Bangladesh Rice Research Institute	FF	Ford Foundation
		FFA	Ford Foundation Archives
CARE	Committee for American Relief Everywhere	GEU	genetic evaluation and utilization programme, IRRI
CARI	Central Agricultural Research Institute	HYVs	high yielding varieties

Abbreviations

IADP	Intensive Agricultural Development Program (India, FF)		Party, Bangladesh
		NASA	National Aeronautics and Space Administration (US)
IADS	International Agricultural Development Service (created by some CGIAR donors, 1973)	NSF	National Science Foundation (US)
		PCARR	Philippines Council for Agricultural Research
ICRISAT	International Crop Research Institute for the Semi-Arid Tropics, India	RAC	Rockefeller Archive Center (see Chapter 2, nn. 1 and 34)
IDRC	International Development Research Center, Ottawa	RF	Rockefeller Foundation
		S and T	science and technology/scientists and technologists
IITA	International Institute of Tropical Agriculture, Nigeria	SLFP	Sri Lanka Freedom Party
INFER	International Network for Fertilizer Efficiency	TAC	Technical Advisory Committee, CGIAR
IRC	International Rice Commission	UBC	University of British Columbia
IRRI	International Rice Research Institute	UNDP	UN Development Program
IRTP	International Rice Testing Program	UNP	United National Party (Sri Lanka)
ISI	Indian Statistical Institute	UNROD	UN Relief Operation Dhaka
JVP	Janatha Vimukthi Peramuna, Sri Lanka militant Maoist party	UPI	United Press International
		UPLB	University of the Philippines at Los Banos
KVSs	Krushi Karma Viapthi Sevakas (farm or secondary school-educated extension staff, Sri Lanka)	USAID	United States Agency for International Development
LRPC	Long Range Planning Committee (IRRI)	USDA	US Department of Agriculture
LRPCR	*Long Range Planning Committee Report*	WAPDA	Water and Power Development Authority (East Pakistan and Bangladesh)
MPCS	Multi Purpose Co-operative Society		
NAP	National Awami		

WBD	Water Development Board (East Pakistan and Bangladesh)

All weights and measures have been converted to metric values. These are not the values found in the original documents.

1
Introduction

IN 1960 the International Rice Research Institute (IRRI) was opened near Manila in the Philippines. IRRI was conceived and financed by the Rockefeller and Ford Foundations, whose goal was to help solve the world food problems in such a way that the foundations' economic and political concerns about Asia would be allayed. The foundations began to persuade Asian governments to adopt a style of rice research and conception of agriculture which were congruent with IRRI's objectives. Our book is about IRRI's planning and development, the efforts to persuade Asian governments to try to increase food production by following IRRI's advice, and the interplay of IRRI's technology, rice scientists, and national research strategies. In order to show the outcomes of this interplay, we focus on conditions in Sri Lanka and Bangladesh: experience in those countries was, by 1980, part of the impetus for change at IRRI itself. Our book testifies to the necessity that the fundamental diversity of rice agriculture should be the organizing principle for an adequate research strategy and shows how strategies which are based on other organizing principles (indeed, on other conceptions or models of rice agriculture) have met continuous difficulties and have ultimately been revised or abandoned. We think this story has implications for science and technology research on questions far outside agriculture and the world food problem.

The mission of IRRI's scientists and technologists was to develop by genetic manipulation varieties of rice that would, according to IRRI's first Annual Report, yield several times what current types did and that would do so 'almost any time and anywhere in the torrid zone'. Although the foundations' planners could not foresee *in detail* the effects of introducing new rice strains into the Asian social and political ecology, they understood full well that the changes that could be wrought by these technological innovations could be vast and cascading. IRRI's efforts contributed to what was called the 'Green Revolution'—'green' because of agriculture and in contrast to political shades of red. Although science and technology have obviously produced many large-scale effects in societies, the Green Revolution ranks with railroad-building in North America and electrification in the Soviet Union as being among the most ambitious, *deliberate* efforts to employ science and technology to effect rapid structural change. Of course the view that science and technology were integral to modernization was not limited to the foundations. Many other Western institutions and a number of Asian governments held similar views.[1] But what was unique was the foundations' effort to make science and technology the engine of societal change by focusing research and development on the main agricultural crop of a region.[2]

The strategy of employing plant genetics to produce large-scale changes in agriculture was not new to the Rockefeller Foundation (RF). Beginning in the 1940s the RF had sponsored Norman Borlaug's research team in Mexico which had developed new strains of wheat resistant to local diseases. Indeed in Asia adapting and distributing wheat varieties constituted another arm of the Green Revolution, an arm that operated independently of IRRI. But the focus of this book is almost exclusively on rice.[3]

Many questions have been asked about the Green Revolution, and the answers have been hotly debated. Was there really a revolution? Whose lot was improved—and whose was diminished—as a result of the Green Revolution: cultivators? landlords? moneylenders? the urban population? particular political parties? particular agencies? specific nations? Western capitalism?[4] These questions are undoubtedly important, but the debate about outcomes tends to leave obscure the processes by which the effects were achieved. In contrast we set out to study the assumptions, beliefs, values, theories, and means by which science and technology were to be intentionally employed as an engine of social change.

Thus our purpose in this book is not to decide whether the Green Revolution was a good or a bad thing—although our studies of IRRI itself and of its relations with rice research in Sri Lanka and Bangladesh indeed contain negative and positive comments about some features of the Green Revolution. Rather our intention is to reveal and to evaluate some of the most fundamental principles involved in the link between science and development. It is unlikely that there will be another undertaking exactly like the Green Revolution, but we are convinced that the bonds between science and development are not transient. The transfer of technology and the communication of science are now permanent features of international development. We hope that the conclusions we reach here will be useful in illuminating and improving future efforts to apply science and technology as components of the solution to social problems.

The sustained attempt to foment a Green Revolution involved a double transfer. Not only were the products of modern science and technology—such as the new rice varieties—to be transferred to the very different contexts of Asian agriculture, but the means for developing the

Introduction

new technology were also to be transferred and developed in Asia. Thus the planners and agents of the transfer—the personnel at the foundations, the scientists and technologists, and the government officials— had to employ either implicitly or explicitly conceptions and models of the desired products, of the research system that would achieve them, of the methods for implementing the newly developed technologies, and of the Asian cultures and agricultures that were to embrace them.[5]

The central scientific products were to be high yielding varieties (HYVs), that is, types of rice that would respond positively to the application of chemical or natural fertilizers and thus yield more per hectare.[6] In an important respect response to fertlizer was less of a problem than the strength of the plant after response. Several traditional tropical varieties would respond so well to the application of additional nutrients that their heavy heads would topple over into the waters of the paddy fields, and no gain or even a loss would be incurred. Japanese, Taiwanese, and American scientists had solved this problem for the temperate zones by developing semidwarf varieties that had short, stiff stalks. However, these plants were of the *japonica* type, unsuitable to the tropical regions where *indica* varieties were the rule. Because the foundations were addressing what they called 'the problem of tropical rice' they concluded that a new type of *indica* rice plant was required. This is one reason, among others, why plant-breeding and plant genetics dominated the attention of planners.

These transfers were to be accomplished by means of a centralized scientific institution, IRRI, that would develop varieties of rice suitable throughout the tropics.[7] As we shall see, IRRI represents not only an institutional narrowing or focusing of effort, but also pivotal conceptual reductions. The food problems of Asia were subsumed under the problem of low yield per hectare, and questions about rice were seen largely as isolable from other components of agricultural and cultural systems. Acceptable solutions were sought almost exclusively in the area of plant genetics, so that eventually other sciences became involved primarily as a result of problems posed by plant-breeding. Because of these reductions the potential recipients of IRRI's products shrank from the set of all Asian cultivators to those who tilled irrigated land, that is 20–30 per cent of the total.

Two major, interrelated issues are at the centre of our enquiry: (1) the transferability of agricultural research and implementation strategies; and (2) the roles and responsibilities of scientists and technologists in the Green Revolution. In sections 1.1 and 1.2 below we provide a framework for addressing these issues. Section 1.1, which is devoted mainly to the transferability question, chronicles our changing views about the conceptual foundations of the Green Revolution. We include descriptions of abandoned positions not because they were once ours, but because we believe that these views were held by others and still have influence.

In section 1.2 we set the stage for considering the roles and responsibilities of scientists and technologists. We regard this area as a vital one that is often entirely neglected or, much worse, addressed superficially. We attempt to avoid these difficulties by posing and defending two major claims. According to the Underspecification Thesis the tasks undertaken by scientists and technologists are conventionally underspecified. Thus, technologists usually make a significant contribution to characterizing both their tasks and the nature of acceptable solutions; as such they bear responsibility for their products. According to the Mango Thesis although what is sent from laboratories to the field (for example new types of rice seeds) may have a technological core that is in some sense neutral, what is usually transferred and implemented is a great deal more than the core technology. And there is no plausible way to maintain that this wider entity is neutral.[8] That is, one can always maintain the view that technology is neutral by employing a narrow, restrictive definition of what counts as technology; then any differences that occur after implementation are due to factors external to 'the technology' rather than due to 'the technology' itself. Using this ploy one can, for example, argue that neutrality is demonstrated by cases in which the same type of seed is planted in very different agricultural systems and any different results are due to the physical or social systems and not to the seed technology itself. The Underspecification and Mango Theses, however, emphasize these factors: usually the entity implemented is not a narrowly construed technological object; what is implemented should be understood as consisting of a broad spectrum of technological features plus some features very closely linked to the scientific and technological enterprise. Thus what is implemented is not neutral.

The plan of the book, which focuses on the period between 1950 and 1980, treats the American conceptions of a research strategy first, showing how these conceptions were simplified and put into practice through IRRI. The interplay of agricultural development thinking and American foreign policy is evident here. IRRI's first twenty years of operation are then examined in detail, including its varied attempts to transfer both its products and its methodology to other countries. Some of the effects at IRRI of reactions from these other countries are discussed here. Then we contrast the research strategies of Sri Lanka and Bangladesh. On the one hand, Sri Lanka maintained autonomy and, resisting IRRI strategy, exercised a largely decentralized, locally responsive research activity (despite internal contests over what the proper course was). Bangladesh, on the other hand, saw its own research efforts absorbed in IRRI's Green Revolution strategy and when it emerged as a new nation in 1972, centralized its research, making it more dependent financially upon IRRI. The two theses described above, the Mango Thesis and the Underspecification Thesis, are helpful in explaining the various effects which result when research strategies meet agricultural diversity.

Our focus is on the strategies of foundations, IRRI, governments, development banks, and other official institutions connected with the operation of the state and the market. We show how these strategies have met the local diversity of rice agriculture and what the results were for the planners and agencies of the strategy. But we do not examine in detail farmers' 'research' strategies that are part of the practices which contribute to agricultural diversity. Innovation and empirical testing by farmers is legendary, including their selection and adaptation of rice seeds. Fortunately, there are detailed studies not only of the meeting of official research strategies with farmers' practices, but also of the intricacies of farmers' testing and innovation—the research strategy of farmers.[9] Our object here is to show whether and how farmers' strategies are taken into account in official research strategies—at IRRI, in Sri Lanka, and in Bangladesh. What becomes clear is that although each of the strategies sought increased production of rice, the cultivator's world formed only a small part of the model on which official strategies were based. Other interests and other conceptions usually prevailed.

1.1 Our Original Assumptions and Their Limitations

The original goal of this study was to examine the ethical and value implications of the transfer of American rice science and technology to South Asia. We initially assumed that a seed-driven expansion of production sustained by innovative plant-breeding and supported by basic research was characteristic of American rice technology and science. We assumed that the scientific and technological innovations in the United States were in the forefront of the world's understanding of how to increase rice production, that there was a transfer of this understanding downward and outward from the American centres of research, through IRRI, to the national research stations of Asia and from there to the fields of the millions of paddy growers, and that consequently the Americans had an ethical responsibility for this transfer. Our eight years of research persuaded us that the original goal was inadequate and some of the assumptions mistaken.

The model which initially guided our researches was not an implausible one in the mid-1970s. As early as 1958 it was widely known that the wheat crops of Sonora province in Mexico had nearly doubled from their 1950–1 levels.[10] It was argued that the pivotal contribution to the increase had been the plant-breeding conducted by Mexicans and Americans under the joint sponsorship of the Mexican government and the Rockefeller Foundation. This plant-breeding research drew upon the work of the American land grant colleges and federal research stations.[11] The leading role of agricultural research in increasing food grain production was popularized among agricultural economists in the United States by the writings of Theodore Schultz, one of the leading American scholars in agricultural economics. Schultz argued that the key to transforming a traditional agricultural sector into a source of economic growth is investment to make modern 'high pay-off inputs' available to farmers in poor countries.[12] The high pay-off inputs were new technical knowledge, new manufactured inputs such as petroleum-based fertilizers, and a new capacity of the farmers to use the knowledge and inputs effectively. Such views were widespread by the late 1960s. Paralleling the published arguments were the internal analyses of such agencies as the USAID, for whom John W. Mellor, Professor of Agricultural Economics at Cornell, had written in 1969, the 'agricultural research breakthrough symbolized by the new cereal varieties offers an opportunity to turn away from defeatist agricultural development policies directed toward the race to keep food supplies in balance with population growth and famine relief and to turn toward a positive role for agriculture, which places it at the leading edge of the total development process'.[13] Mellor continued by claiming that 'an effective research system is probably the most important institution-building job to be performed in the development process'.[14]

We knew that from its inception the International Rice Research Institute had argued that the route to improving rice yields in monsoon Asia was to follow the research and technology model that appeared to have worked so well in Mexico and in the United States. IRRI was staffed and guided largely by Americans who were trained in the American agricultural system, but was increasingly joined by Asian researchers, some of whom were trained in Asia. By 1980 its staff was largely Asian, although many senior scientists and its director were American. Its early annual reports were uniformly optimistic that rice research would solve the food problems of Asia by implementing such a plan. However, by 1981, when we had completed our field work and began writing up our findings, three persistent problems, which we will address in turn, challenged the initial model: (1) the physical and cultural diversity of Asian agriculture; (2) the importance of lateral as well as vertical linkages in Asian agricultural systems; and (3) the interplay of various goals held by the multitude of actors and agencies concerned with the food problems of Asia. Suffice it to say that the food problems of rice-growing countries of Asia in 1980, at the end of our study period, were not less complex and in some cases not much less precarious than in 1950.

Physical and Cultural Diversity

Each of us was impressed by how varied rice agriculture was not only across countries but within countries. National departments of agriculture had earlier identified many different agro-ecological regions within their own countries. In Sri Lanka alone there were twenty-two major agro-ecological regions distinguished within the 65,600 sq. km. of the island. In Bangladesh thirteen agro-ecological regions were identified within the 142,500 sq. km. of the country. This complexity was repeated in every country whose rice research we reviewed. These regions were defined by topography, soil types, and water-climatic regimes. It was recognized that the crops, landholding size, farming practices and even the socio-economic stratification were related to the physical characteristics of the regions. Experienced officials claimed that they could predict the crops, landholdings, and stratification if they were told the topography and soil type. However, it was recognized that the broad classification into agro-ecological regions was only a first step; a finer, more discriminating system of classification was established to facilitate realistic targeted research and effective extension programmes. So we learned what was already obvious to farmers and experienced agricultural officers, that rice cultivation in Asia was not simply specific to agro-ecological regions but varied from valley to valley, or within a few miles, from one side of a hill or river to another.[15]

While it was common among the experienced observers to acknowledge the ways in which the physical environment defined the forms of rice cultivation, it was less common for them to recognize that there were large historical and cultural components in the practices of farming. Not only were farmers influenced by their physical context, they were also affected by a cultural context which worked on at least two different levels, the local or microlevel and the national or macrolevel. At microlevel the community's accumulated understanding, technological resources, and local infrastructure formed a pool unique to that community. Even villages with similiar physical endowments of topography, soil type, and rainfall varied in their cultivating practices and in the social organization of agriculture. This fact underscored the point that new technology designed to promote development actually displaces older, traditional technology; the choice was not between new technology and none.

Related to this local microlevel is an analytically subordinate question. There had been an intense debate among rural sociologists and anthropologists as to the economic rationality of the rice farmer. Did the farmer calculate the costs and benefits to be expected from managing his fields in different ways or was he caught in the lock-step of tradition with little opportunity to innovate? Field researchers, such as W. David Hopper, had little doubt that the former was correct. He pointed for example to the long record of innovation in the Punjab.[16] Nevertheless it was recognized throughout Asia that the rationality was not a narrow market-oriented rationality but one which took into account the family's social status and cultural aspirations. In many communities the higher-status families would not let their women transplant paddy or let their sons work for other households, even though it would increase the household's disposable income. This more complex reckoning of how to use the household labour and resources was clearly related not only to the local cultivating practices but also to the enveloping local culture which guided the understanding of sex roles, generational behaviour, social stratification, and, perhaps, support for a 'moral economy of the peasant'.[17] Thus the location specificity of the way rice was cultivated was influenced by the local culture, not simply by the established practices of farming.

A second level concerns what has been called the macrostructure: the laws of property-holding, taxation, and inheritance; the market for land, inputs, and products; the transportation services into rural areas; the relative power of the farming communities in the political system; etc. Some researchers recognized that the specific form of rice cultivation practised depended on the intersection of at least three major components: the physical environment, the local cultivating culture, and the macrostructure of agriculture. The location-specificity of rice farming challenged the initial model of a seed-led change supported by an American style of applied and basic research spreading downward and outwards to bring improved yields to Asian farmers. As the growing of rice is carried out within a specific local context which is shaped by the convergence of the physical environment, of the local culture which is more extensive than simply the practices of cultivation, and of the macrostructure of agriculture, then the technology would need to be tested against these distinctive local requirements if it was to be taken up by the villagers. Of course, we recognize the forcing role of technology and other specific changes whether originating from the environment, the cultivating community, or the macrostructure; a single initial change may trigger sets of linked changes throughout the local rice-cultivation system. Yet in our view when changes are observed they are not solely due to a single exogenous innovation but rather to the interaction of the exogenous with other permitting factors within a specific setting.[18]

Lateral Linkages

A second and related set of observations which challenged the initial model was centred on our discovery that although there were vertical connections, for example, from the regional seed farm to the cultivating household, lateral linkages were both more numerous and more important than vertical connections in the rural areas that we knew about. New seeds, information, labour, tools, and credit were transferred, exchanged, borrowed, and given from neighbour to neighbour and kinsman to kinsman far more often than any transfer from the district agricultural officer to the extension worker to the farmer. This is not to deny that external inputs such as new varieties of seed were introduced into the local agricultural economy from outside. Rather it is to make two observations: first, exogenous inputs reach the cultivator along multiple pathways some of which are outside the control of the national governments and, second, once introduced, the more common suppliers are local people with whom the cultivator has many-sided relations, not simply as a one-time supplier of a needed input. There is almost always some kind of unofficial system, an informal but important network of supply within which the individual has some social standing, sources for whom there is some trust and within which he prefers to carry out his exchanges.

To illustrate the point about multiple pathways, Grace Goodell, an anthropologist employed at IRRI, conducted a survey in a nearby rice-growing area in Laguna province in the Philippines.[19] She systematically enquired of each farmer what variety of rice he was planting and where he had obtained it. She then collected a sample of the seed paddy which she brought back to the plant breeders at IRRI. While the farmers were reporting the names of some of the more popular IRRI varieties, the actual seeds that they gave her, which were identified by the breeders, were largely made up of breeding stock or varieties that had not been released. The seeds were not what the farmers said they were. She found that the Filippino labourers working on the IRRI fields were carrying seed off the station, selling or giving it to their relatives and friends. There was a 'grey market' in seed which paralleled the official channels of distribution of the approved seeds.[20]

Moreover, as we reviewed the actual historical routes of the diffusion of the high yielding rice varieties, we found the same type of plural pathways and lateral exchanges. This raised the possibility that the accumulated rice science and technology and even some cultivating practices of the United States were largely irrelevant to the changes in cultivation practices in Asia where there was already a multitude of sources of genetic materials, input supplies, and agronomic techniques.[21] The genetic materials and design of plant morphology of interest to IRRI were based upon the earlier work of the Japanese in Taiwan and the Dutch in Indonesia. The combining of the elements into the stiff-strawed, low-statured, heavily headed plant was being carried out at different research centres in Asia, including of course IRRI. The decision to concentrate on breeding varieties that would do well under conditions of reliable irrigation was not novel: it took up the fifty-year-old judgement of colonial departments of agriculture and generations-deep preference of cultivators. In addition, the most technologically advanced agronomic practices of rice farmers in Louisiana, Texas, and California, with their intensive application of fertilizer and their airborne spraying of pesticides, had little relevance to the paddy farmers of countries such as Bangladesh and Sri Lanka. It was these practices which were expected to be transferred and modified in Asia by strategy planners of the 1950s.

Competing Goals

A close examination revealed that there were really a number of different definitions of food problems in Asia and that a collage of competing goals, interests, and perceptions infused the proposed solutions. Comparing notes from our interviews at IRRI, at the foundations, and with government officials, scientists, and farmers in several countries, we came to the conclusion that rice is not simply rice; that is, the hope of introducing new high yielding varieties of rice into Asia meant very different things to different people.

Rice, in the context of the Rockefeller Foundation's discussions in the 1950s, involved identifying a high-profile project which would help ward off economic stagnation, political instability, and the spread of communism in Asia. For the Ford Foundation, it offered the chance to join a prestigious major agrarian undertaking in Asia after very unsatisfactory experiences there in the early 1950s. Rice for the new research institution, IRRI, was a magnificent opportunity to display the scientific and technological virtuosity of American or American-trained plant breeders, geneticists, and agronomists and for the individuals concerned to advance their professional careers dramatically. Rice for the national governments of South Asia was a means of supplying cheap food to the cities and of reducing their expensive and politically damaging dependence on foreign governments' supply of food grains—though from the earliest discussions of technological solutions the possibility was raised that new dependencies would result from the need for required inputs which had to be transferred with the new seed. Rice for the national departments of agriculture was a new means of building up their standing in the bureaucratic infighting, of lining up new sets of allies among the politicians from the rice-growing areas, and

Introduction

of distinguishing them from advocates of other crops like jute or tea. And rice for the small cultivators meant different things depending on how significant rice-growing was in the household income, in marshalling resources for the support of maturing children and in providing for old-age security. In short, even if all agreed that more rice would be desirable, each of these interest groups wanted to ensure that additional rice would be provided in such a way that its values were realized at the same time. Rice production and distribution were thus clearly linked to values among each of these different groups.

If rice does not mean the same thing to each cluster in the hierarchy which was to promote the diffusion of technical innovations; if American rice science is not seen as the sole source of innovation but rather that innovation moves along plural pathways with lateral exchanges being of great practical importance; and if smallholder rice cultivation in Bangladesh and Sri Lanka is location-specific as defined by the intersection of the physical environment as well as the micro- and macrostructures of agriculture—then not much was left of the initial definition of the research problem.

These three persistent problems, and the additional questions they raised, became the central concerns of the study. Why did the Rockefeller Foundation come to be concerned about rice production in Asia, and why did it think that seed-led innovations would improve production? Why was the choice made for a centralized institution that would combine both plant-breeding and basic scientific research? Why did the RF, the FF, and IRRI assume that they were able to generate both seeds and science that would be universal in their applications throughout Asia? What impact did the export of the IRRI model have upon the national agricultural research programmes in Bangladesh and Sri Lanka? How appropriate was the IRRI model and the national variants for the rice farmers of Bangladesh and Sri Lanka? Finally, what were the issues surrounding the roles and responsibilities of scientists and technologists?

We should make very clear that we do not attempt to propose and defend a general theory of agricultural development in this book. Indeed, the criticisms of our initial research problem given here strongly suggest that, given the local diversity evident in smallholder cultivation in Asia, efforts to devise such a generalized theory are efforts wasted.

1.2 Two Theses for Evaluating Research Strategies and the Transfer of Technology

We approach issues about the responsibilities of scientists for research strategies by asking whether the following speech would have been appropriate if given in Stockholm by the agricultural scientist who was awarded the 1970 Nobel Peace Prize.

I am extremely sorry, but as a matter of principle, I cannot accept this award. The reason is that a necessary precondition for praising (or blaming) someone for his or her actions is that (s)he can be held responsible for the foreseeable consequences of those actions. Although I acknowledge that I am indeed responsible for some modest achievements in plant genetics—keeping in mind that all scientists stand on the shoulders of their predecessors—I cannot accept the Peace Prize, since it is awarded primarily for the *moral and political consequences of scientific discoveries, not for the discoveries themselves.*

Concerning moral consequences, I share the view of Anna J. Harrison, a past president of the American Chemical Society. Dr Harrison maintains that 'Sciences are amoral. The *use* of science and technology may be judged by society to be moral, amoral, or immoral.' Thus according to this view moral considerations are correctly ascribed to those who decide to use science and technology, not to scientists.

Similarly, science is politically neutral. It can be used to achieve either beneficial or destructive ends. Scientists and technologists are merely in the business of making options available to individuals and to political authorities. Choices about use or non-use are in the hands of the implementors, and thus they deserve the credit or blame for the consequences of technical implementation.

Needless to say, Dr Borlaug did *not* make this apocryphal rejection speech—nor should he have done so.[22] However this excursion into fantasy makes an extremely important point: people who *blame* scientists and technologists for the consequences of their work are often offered retorts like the ones above. Our current purpose is neither to heap praise nor to pour scorn on scientific and technological activities and their consequences; rather the goal is to clarify a more fundamental point, namely the ways in which science or scientists and technology or technologists (hereafter referred to as S and T) may be held responsible for their activities, especially when those activities are related to development. We shall explore the conditions under which the S and T enterprise can be correctly held at least partially responsible for some of the sociopolitical consequences of research. It should be clear from the preceding that this question of responsibility is conceptually prior to and distinct from questions of praise or blame. Lest there be residual confusion, we emphasize that our position is neither antiscience nor romantic; our point is that scientists and technologists are much more responsible for the consequences of their actions than is usually thought, and to the degree that they are responsible the question of praise or blame cannot be turned aside by shallow arguments like those contained in the apocryphal rejection speech.

In our view one of the major factors confounding attempts to assess

Introduction

the roles and responsibilities of scientists and technologists is that a much too restrictive concept of technology is employed. We shall not attempt to offer an exhaustive account of science and technology; instead we present two theses which we believe are required for an adequate evaluation of the role of science and technology in this period of the Green Revolution (1950–80). The first, the Underspecification Thesis, is a claim about the significant degree to which S and T participate in the characterization of problems and solutions; the second, the Mango Thesis, argues that what is implemented or transferred is usually much more than some narrow, isolated piece of technology and that the S and T enterprise bears some responsibility for the broad aspects of transfers. We consider these theses to have wide but not universal applicability, so we shall first present them briefly here in their general form.[23]

The Underspecification Thesis

Our thesis is this:

Although the broad characterizations of projects are often provided by bodies external to S and T, these characterizations are inherently underspecified and are supplemented by the S and T enterprise. This provision of additional specifications is thus a reason why the S and T enterprise bears considerable responsibility for implementation.

The external agencies mentioned in the thesis usually stand in an underwriter relation to S and T; they provide or manage the funds on which the S and T establishment depends. The very broad characterizations provided by these agencies are a general framework for S and T pursuits, but the framework must be supplemented by additional, often subsidiary but nevertheless significant priorities and specifications. These additional factors may be wittingly or unwittingly added by the S and T enterprise, that is, by individual scientists or by scientific groups. We will show that the provision of these factors is a major source of the responsibility attributable to S and T.

There are at least two ways in which projects may be *underspecified*. First, the characterization may include a qualitative goal—for example 'higher yield' in the rice case—without specifying a quantitative aim: how much higher must the yield be? Second, broad characterizations may omit some qualitative features altogether. This second sense of underspecification also highlights a possible source of confusion. Both the term 'underspecification' and the remarks about omitting to specify some qualitative features may suggest that we consider underspecification itself a bad thing. On the contrary, there are good reasons why projects are and should be underspecified. It is both practically and theoretically impossible to specify fully all the characteristics of an undertaking (especially when the project involves scientific and technological research the results of which cannot be fully known in advance).

If we consider, for example, the case of a contractor building a structure according to an elaborately worked out set of specifications, surely one of the most extreme cases of specifying a project, we see that even here there will be features which are not fully covered by the initial instructions or by the blueprint. Our point then is not to argue that 'underspecification' is a failing, but to investigate the complex ways in which S and T activities start out being partially specified and end up being relatively completed projects. Our claim in the case of rice examined here is that scientists and technologists are not merely passively following orders. Instead they are contributing to the conception of what they are or should be doing. Such contributions are expected, and they are a source of the responsibility attributable to S and T.

The distinction between broad characterizations and subsidiary priorities is not sharp. In the rice case the grey area contains, for example, the issue whether to seek varieties which would produce large, dramatic yield increases or those which would involve small, incremental increases. This lack of a sharp distinction offers a kind of strategic ambiguity for the interplay of different interests. The trade-off here has to do with risk; varieties which differ only slightly from traditional ones are less likely to involve major risks but are also less likely to produce large yield increases. As to what body made the decision, it does appear that IRRI scientists—especially plant breeders—did debate these strategies, but it is less clear to us whether the decision to concentrate on larger increases was made primarily by the scientific establishment or by governing bodies.[24]

Again it is important for us to emphasize that these examples are intended to demonstrate not that the S and T establishment or its members made wrong decisions with respect to the subsidiary priorities, but that they made such decisions at all. The scientists and technologists at IRRI are, for the most part engaged in mission-oriented research; our thesis is that the precise mission is underspecified, giving the S and T enterprise substantial latitude in the way in which it attempts to fulfil that mission. The presence of such latitude means that S and T make choices which affect the acceptability and utility of their results. Some of these decisions may be taken after prolonged debate, some after informal discussions, and some merely by continuing 'standard' practices after little or no conscious examination of alternatives, for example by applying priorities and procedures which have been inculcated by education, by training, or by cultural heritage. In each of these modes of decision-making, but especially in the last mentioned, the value structures of scientists and technologists play an effective role. There are, of course, significant differences among ways in which choices are made and in the relative importance of the effects flowing from such decisions; individual cases must be examined. Nevertheless the presence of such decisions is one factor which makes it meaningful to ask about the degree to which scientists and technologists share the responsibility for the effects of their research. If every aspect of the S and T mission were specified in advance by a legitimate overriding institution, or if each decision taken by researchers were submitted individually to a legitimate overriding body for approval, then in this regard the responsibility of scientists and technologists would indeed be small. We shall show that at IRRI significant decisions were made by members of the S and T enterprise and that this same responsibility holds for the operations of the Sri Lanka and Bangladesh rice research establishments.

The Mango Thesis

What exactly is transferred when science and technology are employed in development? Our answer involves a simile: what usually is implemented can be likened to a mango, a fruit which contains a solid seed pit surrounded by soft, juicy flesh. The mango seed pit, however, has a distinctive feature in that its surface is covered by a network of hairy fibres which extends into the flesh. Thus it is exceedingly difficult to separate completely the pulp from the pit of a freshly cut mango.[25] In the case of the Green Revolution the analogy runs this way: the mango pit is comparable to a scientific/technological core of the research strategy, namely the HYV strains of rice; the fibrous covering of the pit represents a network of requirements, generated by implementing the core, which includes fertilizer, irrigation, and pest control; and the mango flesh is analogous to a mass of values which suggest how requirements generated by the core should be met, for example the overwhelming emphasis placed on 'quick' adjustments and technical fixes rather than long-term structural solutions.

Many of the disputes about the possible 'neutrality' of technology are traceable to different views about what constitutes the technology in question. If one is speaking merely about the core technology, for instance the HYV strains, then one has some grounds for claiming a sort of social, political, and economic neutrality. After all, the performance of the *seeds* themselves is directly a function of water, soil, and other environmental conditions. However, even if the *core* technology of a transfer is essentially scale-neutral, one must still ask about the social, political, and economic effects of the *entire* transfer. In the case of rice, for example, it was the whole mango, and not merely the core that was transferred. Thus if those debating the neutrality of technology mean to be referring to what is transferred—namely the whole mango—then the anti-neutrality position win hands down. In short the Mango Thesis consists of this claim:

> *The mango metaphor correctly characterizes most technology transfers and as such shows ways in which technology is not neutral.*

The additional components of the transfer are, at least, values and information which suggest how requirements generated by the core are to be met and, at most, actual social, political, and economic strategies for meeting these requirements.

The Mango Thesis gives us a response to the argument that because the same technology can be introduced into radically different sociopolitical matrices resulting in radically different ramifications, technology must be regarded as being sociopolitically neutral. Neutrality can be achieved—or at least aspired to—either if one has succeeded in 'cleansing' the technology of the values and dynamic which adhere to it or if one has taken steps to neutralize those features which still adhere. Such 'cleansing' operations usually demand from recipient governments or societies foresight, understanding, resources, and, in the short term, sacrifices. In this case, there were claims for the neutrality of the transfer, but the evidence we assembled left us sceptical of those claims. If, though, what is transferred is a technological system plus a number of values—most of the mango—and no countervailing measures are taken, then the widespread adoption of this object will set off a predictable and relatively unvarying series of events.

Virtually from the beginning the Green Revolution in rice was regarded as involving the transfer of a 'package'. A minimal package would consist of new HRV seed, fertilizer to which the seed would respond, and new cultivating practices involving irrigation, fertilizer, weeding, and pest control. A maximal package would include all of the above, plus new irrigation facilities; new machinery for land preparation, harvesting, and post-harvest processing; new credit arrangements; adjustments in rice prices, taxes, and import-duty structures; revision of land-tenure arrangements; enlarged storage, milling, transportation, and communication facilities; development of indigenous capability to produce the inputs for the realization of the full potential of the new varieties—for example units to multiply the new seed and factories to produce the fertilizer; and increased extension and research systems to adapt, deliver, and maintain the package.

If the Mango Thesis is construed simply as the claim that the technology transferred should be understood as extending into and producing effects in all of the areas mentioned in the maximal package, then the thesis merely directs us to examine the results of the Green

Revolution in the same manner as innumerable investigators have done. There is, however, another interpretation of the thesis. As enumerated, the items in the maximal package are comparable to the fibrous covering of the pit, for they are merely areas in which the technology generates requirements; nothing has been said yet about the particular measures which were adopted. The point is that where the Mango Thesis holds it is obvious both that the values of those engaged in technology transfer are likely to play a significant role in decisions about exactly how their requirements are to be met and that the values of the S and T enterprise are an important component of what is transferred. We believe, for example, that the following values, among others, were transferred to Asia along with the HYV seeds and that these factors were at least reinforced, if not originated, by the S and T enterprise, which, in turn, specified them in its own way:

1. the goal of and priority given to increased production and the consequent subordination of distribution and consumption concerns;
2. an approach whose flow of influence and information is overwhelmingly from the top down;
3. the view that the fundamental and higher order problems in the agricultural sector are solvable primarily by Western technology.

This third value is of central importance here and in the application of the Underspecification Thesis as well. Beliefs such as this are what underpin 'technological fixes', that is, the approach that largely ignores the social context of problems and focuses on some technological features. In the case of the Green Revolution the planners at the foundations saw the technological approach, the development of new types of rice, as being addressable in relative isolation from most social and political factors. However, they saw the technological solutions producing cascading effects, the end result of which would be newly structured agrarian sectors.

Scientists and technologists at IRRI certainly shared the view of the planners that Western technology could fix the problem of low rice yield, but the degree to which S and T were aware of the intended large-scale social effects is more debatable. Still, the discourse of scientists and technologists in the following chapters is shot through with this value and the others mentioned above. It is clear to us that these values and priorities of scientists and technologists, taken either individually or collectively, did significantly affect what was transferred and thus what consequences the Green Revolution has had. Furthermore the S and T enterprise is not exonerated of responsibility for bad as well as good effects simply because its value structure is coincident with those of underwriting foundations or of other actors or agencies. Silent partners are still partners.

Finally, in this introduction it is necessary to clarify the intricate relationship between individual persons and the institutions in which and for which they work. This relationship must be kept in mind throughout a reading of this book. Complicated corporate institutions like governments and foundations prefer to suggest that the world can be seen through a single window. Their official positions on various issues attest to this preference. But those who work there do not see the world all through the same window. Most institutional decisions and actions are not the simple mechanical expression of established principles, but are the outcome of a contest based on significant internal differences of view and interest. Definitions of problems, in this case, tend to settle the limits of this contest. But institutions are, nevertheless, the site of this contest as much as, say, families. At the same time, persons who work for very long in these institutions tend to become infused with the collective values of the place. Some become 'organization men', others actually lead and manage these individuals and direct the policies and practices of their institutions. But this identity of persons and institutions is not to be found in every case. Thus when we say a foundation or government or American agricultural scientist said or did or thought something, we must remember that these terms stand as simplifications of the more complex relation of individuals and institutions. In fact this study was in part a long process by which we could separate a plethora of evidence from these more complex relations and characterize them into a meaningful institutional picture. It is in this sense that we say that the values and priorities of scientists and technologists, individually or collectively, did significantly affect what was transferred and thus what consequences the Green Revolution had.

NOTES

1. For a discussion of some Asian governments' views of the role of science and technology in Asia at the end of World War II, see Christopher Thorn, *The Issue of War*, (London: Hamish Hamilton, 1985), esp. pp. 307 ff.
2. A good example of the development of the technology and its very large-scale effects is the monumental work of Thomas P. Hughes, *Networks of Power: Electrification in Western Society 1880–1930*, (Baltimore: Johns Hopkins University Press, 1983).
3. After IRRI began operation, the Rockefeller and Ford Foundations also created CIMMYT, an international institute for research on wheat, simply by renaming an existing research centre in Mexico in 1966.
4. A large literature exists on the Green Revolution and its effects. Among those works in which summaries of much of the research can be found are:

B. Bayliss-Smith and Sudhir Wanmali (eds.), *Understanding Green Revolutions: Agrarian Change and Development Planning in South Asia* (Cambridge: Cambridge University Press, 1984); I. S. Friere de Sousa, E. G. Singer, and W. L. Finn, 'Sociopolitical Forces and Technology: Critical Reflections on the Green Revolution', in J. C. Super and T. C. Wright, (eds.), *Food, Politics, and Society in Latin America* (Lincoln, Neb.: University of Nebraska Press, 1985), pp. 228–45.

5. These agencies make their own evaluations of research strategies: see Consultative Group on International Agricultural Research (CGIAR), *International Agricultural Centres: A Study of Achievements and Potential*, (Armidale, Australia: University of New England, 1987). For an example of the relation of international centres and one nation, see Barry Nestle, *Indonesia and the CGIAR Centres: A Study in Their Collaboration in Agricultural Research*, (Washington, DC: World Bank (CGIAR), 1985). For a contrasting account of national systems of agricultural research from the point of view of research managers, see Douglas Daniels (ed.), *Evaluation in National Agricultural Research*, (Ottawa: IDRC, 1987). Although our analysis and interpretation are sometimes at variance with these studies, we have greatly benefited from their observations and insight.

6. The new rice seeds are sometimes called 'highly responsive varieties' or 'HRVs' in order to emphasize the fact that generally dramatic results occur only when significantly increased inputs are also used. We think 'HRV' is more appropriate, but we use 'HYV' because we draw upon sources that do.

7. There are studies of rice which have helped us enormously. V. Wickizer and M. Bennett, *The Political Economy of Monsoon Asia* (Stanford, Calif.: Stanford University Press, 1941). More recently, see Randolph Barker, Robert Herdt, and Beth Rose, *The Rice Economy of Asia*, Washington, DC: Resources for the Future, 1985; Irene Norlund, Sven Cederroth, and Ingela Gerdin (eds.), *Rice Societies* (London: Curzon Press, 1986); Francesca Bray, *The Rice Economies: Technology and Development in Asian Societies*, (Oxford: Blackwell, 1986). For a recent summary of the status of rice research, see M. S. Swaminathan, 'Rice', *Scientific American*, January 1984, 80–93.

8. We acknowledge that seeds may have been seen to have political qualities: for a complete review of this issue, see Jack R. Kloppenburg, *Seeds and Sovereignty: The Use and Control of Plant Genetic Resources*, Durham, NC: Duke University Press, 1988. See also Donald L. Plucknett et al., *Gene Banks and the World's Food*, Princeton, Princeton University Press, 1987, especially, 'A Case Study in Rice Germplasm: IR 36', pp. 171–85.

9. Gerald G. Marten (ed.), *Traditional Agriculture in Southeast Asia*, Boulder, Colo.: Westview Press, 1986; Grace Goodell, 'Bugs, Bunds, Banks and Bottlenecks: Organizational Contradictions in the New Rice Technology', *Economic Development and Cultural Change*, 33, 1 (October 1984), 22–41; Paul Richards, *Coping with Hunger: Hazard and Experiment in an African Rice Farming System*, London: Allen and Unwin, 1986 (see esp. ch. 8, 'Rice Varieties and Farmer-Experiments'). See also Donald H. Lambert, *Swamp Rice Farming: The Indigenous Pahang Malay Agricultural System*, Boulder, Colo.: Westview Press, 1985.

10. Norman Borlaug, 'The Impact of Agricultural Research on Mexican Wheat

Production', *Transactions of the New York Academy of Sciences*, Series II, 20, 3 (1958).
11. Cynthia Hewitt de Alcantara, *Modernizing Mexican Agriculture: Socioeconomic Implications of Technological Change 1940–1970*, Geneva: United Nations Research Institute for Social Development, 1978. See also Bruce Jennings, *Foundations of International Agricultural Research: Science and Politics in Mexican Agriculture*, Boulder, Colo.: Westview Press, 1988.
12. Theodore W. Schultz, *Transforming Traditional Agriculture*, New Haven: Yale University Press, 1964, pp. 145–61.
13. John W. Mellor, *The Role of Government and New Agricultural Technologies*, reprinted in *Food Policy Statement*, Washington, DC: International Food Policy Research Institute, November 1985.
14. Ibid.
15. For a lucid discussion of the intrinsic value of local complexity in ecological systems, see Bryan G. Norton, *Why Preserve Natural Variety?*, Princeton: Princeton University Press, 1987.
16. W. David Hopper, 'Mainsprings of Agricultural Growth in India', *Indian Journal of Agricultural Science*, 35 (June 1965), iii–xxvii.
17. James C. Scott, *The Moral Economy of the Peasant*, New Haven: Yale University Press, 1976. Compare Samuel L. Popkin, *The Rational Peasant*, Berkeley: University of California Press, 1979. And Charles F. Keyes (ed.), 'Peasant Strategies in Asian Societies: Moral and Rational Economic Approaches—A Symposium', *Journal of Asian Studies*, 52, 4 (August 1983), 753–868. By adopting the term "rational" we do not suggest the rice farmers understood their interests perfectly or did not make mistakes in their own research strategies.
18. See Richard R. Harwood, 'Centralized Research and the Complexity of Social Agriculture', in Robert S. Anderson *et al.*, (eds.), *Science, Politics, and the Agricultural Revolution in Asia*, Boulder, Colo.: Westview Press, 1982, pp. 299–322; Richard R. Harwood, *Development: Understanding and Improving Farming Systems in the Humid Tropics*, Boulder, Colo.: Westview Press, 1979. See also Larry Burmeister, *Research, Realpolitik, and Development in Korea: The State and the Green Revolution*, Boulder, Colo.: Westview Press, 1988.
19. Private communication, 2 May 1978.
20. Similarly, Morrison found in his work in a village in Sri Lanka that most of the sampled farmers were getting their new seeds from relatives and other farmers and not from the District Agricultural Officer and the KVS. Compare Willem van Schendel, *Peasant Mobility: The Odds of Life in Rural Bangladesh*, Assen: Van Gorcum, 1981, p. 149: 'In the early 1970s a new type of aman rice was introduced by extension workers of the Thana Agricultural Office in Khetlal but hardly any cultivators grew it. In 1975 and 1976 the first high-yielding varieties of rice started arriving in the area, "spontaneously" spreading from village to village in the way new varieties have been diffused for centuries.'
21. Robert S. Anderson, 'Rice Revolution in the Southern United States, 1865–1945: The Technical and Political Ingredients of Agrarian Transformation', for Canadian Conference on Agrarian Systems, February 1979.
22. For a summary of his recent views, see Norman Borlaug, 'Accelerating

Agricultural Research and Production in the Third World', *Agriculture and Human Values*, Summer 1986, 5–14.
23. Although these theses are widely applicable, they are not needed to assess responsibility either at the pure extreme of S and T or at the applied extreme. (Of course individuals are responsible for being, say, mathematicians or computer assemblers rather than social workers or politicians, but that is a different matter.) However, if the embodiment of scientists' or technologists' work is either totally unpredictable or quite predictable, then assessing responsibility does not require subtle analysis. The responsibility is negligible in the first instance (totally unpredictable) and heavy in the second (totally predictable). Our research has shown us that little of the work at IRRI lies at either extreme.
24. The actual procedures involved in plant-breeding should be kept in mind. Breeders make dozens, or even hundreds, of crosses, and then evaluate the progeny for a very large number of properties. Of course the breeder cannot be certain in advance just which plant properties will show up in crosses. Thus to suggest that breeders 'concentrated' on some rather than other properties primarily means that some rather than other varieties were chosen for further testing, multiplication, and, eventually, distribution. The point in the text is that decisions about which varieties are so chosen probably involved a complex interplay among breeders, administrators of the institute, and governing bodies. See also Thomas Hargrove, *Scientific Communication among Rice Researchers*, IRRI Research Paper 13, 1978.
25. James Cameron, the distinguished British journalist, wrote, 'I met [M. K.] Gandhi on the train ... I had hoped for a few minutes of the eternal verities; all I got was valuable instruction on how to eat a mango.... "Be patient, young man; how can I tell what is in store for us? Now you cut it thus."' *Guardian*, 19 December 1982.

2
American Interests, Foundation Planning, and IRRI's Origins

> The struggle of the 'East' versus the 'West' in Asia is in part a race for production, and rice is the symbol and substance of it.
> 'Rice Politics', *Foreign Affairs*, 1953

TEN years after the government of the Philippines invited the RF in 1950 to improve agriculture in that country, through application of scientific research, the RF, in collaboration with the FF, concluded plans for a major rice research institute located near Manila. Examining the interplay of the RF's interests and capabilities, as well as the larger picture of American solutions to the problems of underdevelopment in Asia during the 1950s, reveals the variety of forces and calculations which brought IRRI into being.[1] The close links between the foundation, members of the Rockefeller family, and US planning for world economic and political development help to explain, in part, why American foundations undertook to change rice cultivation in Asia.

In 1950 American foreign policy and development officials were reviewing the reasons for the loss of China to the communists in 1949. Their reflections were tempered by the recriminations of the McCarthyites and by the commencement of the Korean War. At this time President Truman appointed Nelson Rockefeller as chairman of the International Development Advisory Board, whose task was to expand the Marshall Plan and Point Four activities on a global scale and to recommend the design of a government agency to carry this expansion out.

In 1951 Rockefeller wrote in *Foreign Affairs* that the biggest problem in a 'world economic policy and increased investment' was the problem of 'underdevelopment'.[2] Unless the United States could solve the problem of underdevelopment in certain nations and, specifically, could help them to increase their food production, they could not, said Rockefeller, become trading partners for the free world. The correct response should be a 'widening of the boundaries of US national interest' and the first objective of US policy should be 'a drive to increase food production in the underdeveloped areas by 25 per cent, which would bring them barely above the minimum need for health'.[3] This drive should be followed by raw material development and extraction and, finally, by increased export of manufactured goods to those areas from the US and Europe. These were the only ways, he said, to increase private investment 'in frontier areas'. The problem with underdeveloped areas was 'that they are largely agricultural, with low yields per person employed and per acre, so they seldom produce sufficient foodstuffs to provide a satisfactory diet; and their manufacturing and industrial activity is generally rudimentary'.[4]

Rockefeller warned that 'any reckless handling of this problem [of world trade and underdevelopment] can create such chaos in the underdeveloped areas that our present imports of raw materials from them may be completely upset, and future imports lost as a result of their being thrown into the closed economic orbit of the enemy'.[5]

The rudimentary character of manufacturing and industry in the underdeveloped world explained why the first objective was to be a drive to increase food production. Of course this would be costly, but he pointed to the annual US military budget of $60 billion: the cost of conquering hunger would be lower, he said, than the cost of military control, and his strategy might even reduce military budgets. Doubters would do well 'to consider the volume of market demand if the billion population of the underdeveloped free-world areas could raise per capita incomes from the present average of $80 per year to the $473 level of Western Europe or to the $1,453 level of the United States'.[6] If this were done, Rockefeller said, 'with due regard to the interests of the entire free world, the tempo of economic advance in the areas in question would be, in truth, revolutionized'.[7]

Rockefeller's statements represented a dominant view within the American economic and political establishment. After serving as Assistant Secretary of State, working for Truman, and serving as Governor of New York, Rockefeller became subsequently the Vice-President of the United States. He presented the food problem in its global context in relation to other economic and political problems as they were perceived by a leading member of the American political élite. In the same year as he advised the President (1951) his family's foundation was creating a new research and funding division to define more clearly the food problem and its solution. Although John D. Rockefeller had more continuing interest in the foundation's work in agriculture and Asia, Nelson's views are important for their very public character and joint government and private-sector nature.

Into the wider context of underdevelopment, analysts and planners who followed Rockefeller inserted the problem of revolutionary Asia, which combined elements of the problems of underdevelopment and the threat of world communism. Because of Asia's underdevelopment, dissatisfied peasants, students facing unemployment, and underpaid urban workers were listening to the campaigners for communist movements and parties throughout Asia. According to government officials and analysts for the influential Council on Foreign Relations, weak Asian governments could not long hold communist movements at bay unless the underlying causes were realistically treated by the US.

Harvard historian John Fairbank argued that a preponderantly military effort against communism in Asia could defeat itself: 'may we again win the war [with Korea] but lose the peace to follow?'[8] The US had supported Chiang's 'nationalist' forces in China, which had been first and foremost anticommunist; the Nationalists' interest in social, political, and economic reform was minuscule and was expressed, under pressure, only at the end of their regime. The US thus had failed, said Fairbank, to understand social conditions in Asia and had lost its paramount position in mainland China. He said that though Americans had 'learned to fight in the rice paddies', reform was 'the more fundamental force' to help 'an insecure peasantry and frustrated intelligentsia' which were ready for communism. The Communist rise to power in China, he argued, 'deserves our closest study', because the US had previously failed to understand it. Fairbank suggested that the US, instead of militarily challenging the Chinese revolution and the communist movements which could follow it, must support the forces of social change and learn from the lessons of the past:

> our agricultural specialists can do much to raise Asia's productivity. But this is a political as well as an economic problem. One lesson learned from our private Assistance to China's Rural Construction Movement in the 1930s—such as support the Rockefeller Foundation gave to Jimmy Chen's mass education movement at Ting-Hsien in North China—was that the upbuilding of peasant life through programs of literacy, health, technical improvement and co-operative effort sooner or later threatens the established order of tenant–landlord relationships.
>
> In short, it is plain that American specialists, backed by American technology and some supplies, can help carry through genuine rural reform programs; but must have the active political backing of the local regime.[9]

He warned American specialists in rural reform 'that the technology alone, which can be prostituted for political purposes, is no bar to totalitarianism of the right or left'.[10]

There should be two approaches by Americans to Asia, said Fairbank: first, private citizens and groups should build a network of strong and well-informed contacts in Asia; this should include universities and professors, the Rotary, the YMCA, and the Associated Press, and the like. Until the American public understood the nature and interests of Asian society, private investment should wait. Second, the US government should create well-informed and helpful relations with Asian goverments; given the increasing strength of the USIS and Voice of

America which promoted American views, Fairbank said, the need was to improve knowledge about Asia in America to guide the formation of more effective policies toward Asia. 'The great private sector of American education should be enlisted in this ideological effort at home, and mobilized for greater activity abroad', said Fairbank.

American planners' views were reinforced by those from other countries; for example, the Minister of External Affairs for Australia, P. C. Spender, wrote at the same time that:

International Communism is not passive: its agents move among the Asian people preaching doctrines of national independence, of reform, of social equality, of economic development and the elimination of evils of landlord–tenant relations and the other material burdens that weigh heavily upon the people. These are objectives the Asian people want. They are objectives which only the non-Communist world can help the Asian people to gain.[11]

The former Minister of Overseas Affairs in France, Jacques Soustelle, reminded the United States that although the Korean War had 'tied down all the American forces available in the Pacific ... the entire strategy of the West must be conceived as a whole and it would be foolish to consider Korea and Indo-China separately'.[12] Indo-China was important to the political stability of non-communist Asia, he said, and should receive American military aid. Among other reasons, it was one of the rice granaries of the Far East, having experienced dramatic increases in production and export under French rule since the end of the nineteenth century.

What solutions were offered by these writers for the problem of impoverished Asia? Spender said that the Commonwealth Conference in 1950 which resulted in the Colombo Plan was a good means of forming a partnership to seek an alternative to the promises of communist agents. The time had come for the United States to join in the Colombo Plan or to assist it financially because 'the struggle for an independent Asia in which the reformist movements are expected to bring about relief from poverty and to advance stable government and democracy will make demands upon the economic resources of the non-Asian countries'.[13] US analysts agreed but felt that the Colombo Plan was too small for the massive task. At this stage there was no general agreement on the specific forms which US involvement should take.

Part of the responsibility of the US in the Philippines, Albert Ravenholt (former UPI correspondent in the Far East) said, lay 'in our policy of emphasizing agricultural research and education while failing to ensure extension of improved methods to ordinary farmers'.[14] He pointed out that Hukbalahap guerrillas had not only forced the US Army to supply Clark Field airbase in guarded convoys from nearby

Manila, but had also set aside government authority in some of the richest rice-growing areas on the mainland of Luzon. It was a backward agriculture, he said, with no technical help in sight, which made peasants desperate.[15] Economist Daniel Bell had already reported in 1950 that: 'the Philippines farmer is between two grindstones. On the top is the landlord, who often exacts an unjust share of the crop in spite of ineffective legal restrictions to the contrary. Beneath is the deplorably low productivity of the land he works. The farmer cannot see any avenue of escape'.[16]

In November 1950, within a month of the Bell mission's recommendation of $250 million in US grants and loans to the Philippines in return for economic and social reform under US supervision, the two countries signed an agreement which provided for massive US aid. This included appointment of a land reform adviser who had previously worked with the US-led land reform in Japan and a director of the agricultural division of the US Special Mission.

2.1 Foreign Policy, the Food Problem, and the Rockefeller Foundation

Defining the food problem and situating it among other problems, like revolutionary Asia, was the background against which the RF worked every day. From the many voices offering advice and counsel foundation officials were fully aware of the variety of practical solutions being considered by the US government, of the increased military expenditures planned for forceful intervention in Asia, of the mobilization of a vast non-military apparatus, and of the debate about identifying and backing the right forces of social change. The foundation was interested in widening US national interests and in investing in frontier areas which would modernize economic advancement in Asia. These are the reasons why the problem of tropical rice became so significant to the RF and why the foundation's officers began to plan for its solution.

What was distinctive about the RF, when compared with other major American foundations, is the critical role played by family members in its thinking and practice. From John Rockefeller onward, the family had a stable kinship pattern with the father or eldest son playing a dominant role in the hierarchy. The family encouraged specialization of its various members in various enterprises. John D. Rockefeller, jun., took a great interest in the foundation which had been established by his father in 1913 following three years of public controversy about the implications of allowing such a wealthy 'robber baron' to establish a foundation. There was clear continuity when four of his five sons—John D. III, Laurence, Nelson, and David—began to act as trustees in the 1940s. Three generations used these family funds, added to them, and employed the most prestigious technical advisers, committee members, and skilled field workers which money could buy.

Apart from the data and guidance obtained through RF activities in education, medicine, and agriculture, the trustees of the foundation kept informed through the elaborate network of the Council on Foreign Relations (CFR). From time to time both trustees and Rockefeller family members also served in official positions for the government. Through close ties among the RF, the CFR, and the Department of the State, the trustees of the RF, in their various roles, came to be very well informed and influential on the foreign policies of the United States. As recent analysts noted, it is in this sense that 'the foundations sustain the complex nerve and guidance mechanisms for a whole system of institutional power'.[17] The foundation was, however, sometimes at odds with official U.S. policy.

In 1939 two-thirds of the trustees of the RF were members of the Council on Foreign Relations; in addition the RF financed the massive and influential 'War and Peace Studies' project of the Council. During World War II the Council lobbied strongly for a new set of US-based international institutions like the United Nations and the World Bank. In 1941–2 the planning groups of the CFR and the State Department met together to discuss post-war US strategy, largely because of the relative superiority of the CFR's research capacities. Members of the CFR were also members of the 'informal political agenda group' which acted as advisers to President Roosevelt. CFR personnel overlapped with Department of Treasury planners to help set up the World Bank and the International Monetary Fund. The Council was a strong proponent of the formation of the United Nations, and the Rockefellers took an active role in its establishment in New York, including donating some of the land for construction of the UN building.[18]

The influence of the Rockefeller family and foundation with regard to US policy is seen in a variety of situations. In 1946 David Rockefeller was secretary of a CFR study group on reconstruction in Western Europe (the Spofford group): he advised the RF to make a $50,000 grant in 1948 to the Aid to Europe study group which changed its name to the Marshall Plan Study Group when its ideas were adopted by Secretary of State Marshall. In 1951 John Rockefeller was appointed special consultant to John Foster Dulles, who was head of the US delegation to Japan to negotiate a final treaty between the two countries, and in the same year Nelson Rockefeller was made chairman of the International Development Advisory Board by President Truman. Dean Rusk moved from the State Department in 1952 to become President of the RF, where he stayed until being appointed US Secretary of State in 1960. John Rockefeller was instrumental in having the RF support the new Russian Institute at Columbia University in 1952; at the same time he was active in the CFR study group, with Rusk and Bundy, on US–Soviet policies.

George Franklin, friend and former Harvard roommate of David Rockefeller, was made the executive director of the CFR, and David Rockefeller became its vice-president in 1952. The importance of RF trustees for US foreign policy thinking is shown when, in 1960, President Kennedy's list of possible candidates for Secretary of State included five trustees of the Rockefeller Foundation, of whom Dean Rusk was chosen.[19] Douglas Dillon, also a trustee of the Rockefeller Foundation, was appointed Secretary of the Treasury at the same time.

The mutual supportive and reinforcing relationships between the RF and the American foreign policy establishment is seen most clearly in the work of Dean Rusk, president of the foundation from 1951 to 1960. His previous experience and his continuing relations with US policy formulation while he was the president of the RF explain why the foundation's international undertakings may be understood as expressions of American policy. When Rusk was invited by John Foster Dulles to stand for the RF's presidency in October 1951, Rusk had already been a foundation trustee for over three years. In 1948, when Rusk (then thirty-nine) had responsibilities in the State Department for UN Affairs, Dulles had invited Rusk to meet the incumbent president of the foundation. Soon he was appointed a foundation trustee, and within a year was also promoted to be deputy undersecretary of state, responsible for liaison between the State Department and the Defense Department: 'as the new positions evolved (in 1949) Rusk became one of the triumvirate who had de facto control of American policy'.[20] In March 1950, with the State Department under heavy fire from Joseph McCarthy, Rusk volunteered for the post of assistant secretary for Far Eastern Affairs. No stranger to the region (he was stationed in New Delhi in World War II and became deputy chief of Stillwell's staff for the China–Burma–India War theatre), Rusk co-ordinated foreign policy during the Korean War. At this time Rusk guided legislation for the congressional appropriation of 'the general area of China funds', intended by Congress to contain the spread of communism from mainland China. He immediately directed $50 million of that appropriation to the support of the French in Indo-China: the idea of 'reserving' Indo-China and South-East Asia from further communist encroachment was most vigorously articulated by Rusk. 'Rusk's activities in 1950 and 1951 left no doubt that he and Dulles and Jessup were building a little containment line across the Pacific'.[21] Containment would not succeed, he argued, through force of arms alone or through Marshall Plan-like reconstruction. It would be necessary to build new institutions in Asia and to orient both American and UN policies to do so.

Anticipating the election in 1952 of General Eisenhower as President, and knowing that a Republican administration would appoint new officials in the State Department, Rusk responded enthusiastically to the RF's invitation in 1951. He knew the foundation well as its trustee. Someone in the selection committee, however, feared that Rusk's appointment might expose the foundation to an attack by McCarthy. Dulles successfully argued Rusk's case to the foundation's trustees, and his appointment as president was confirmed late in 1951. Beyond his immediate task of testifying for the RF before a congressional committee investigating tax-exempt foundations, he was a key participant in the CFR's study of American–Soviet relations. For example, Secretary of State Dulles asked him to make a study and report on 'the colonial question' at the time of the Bandung Conference in 1955, his views were frequently solicited as a leading authority on UN affairs, and a close study of his own diary and records shows he had 'involvement in virtually every question of significance in the 1950s'.[22] It is, however, Rusk's development of foundation priorities and specific projects that interests us here. To understand the choices which shaped IRRI, it is essential to understand the RF's existing traditions and values, many of which were incorporated in IRRI.

In addition to working within the context of the objectives of American foreign policy, the people at the RF who planned IRRI worked on a daily basis with the traditions and practices of the foundation. Over the years of experience in the American South, Latin America, and China, where in each case the major goal was to integrate the region with the mainstream US economy, the foundation had developed a set of guiding principles which were subsequently applied in the discussions leading to IRRI. These were building on strength, employing technology as the leading factor, managing scientific research, and working through governments.

From the earliest days of the RF, the trustees decided to 'restrict the foundation's direct operations to specific areas such as public health, medicine and agriculture'.[23] These areas were largely the problems of frontier regions of the world like Latin America and China where economic activities were sluggish and where American political presence and economic investment was inhibited by the consequences of poor health and backward agriculture.[24] Thus while types of activities were carefully delimited, vast areas of the globe were considered the proper arena in which its activities should be carried out. The selection of specific locations in which to demonstrate the efficacy of its approaches and techniques was typically done to take advantage of any favourable conditions. In this sense, the foundation was committed to building on strength. For example, in China, the RF concentrated on working with successful farmers who already employed labour and experimented with new techniques.

Beginning in the American South and paralleling its medical activities, the RF pursued a programme of agricultural modernization and improvement. When the RF moved its operations beyond the US, it continued to concentrate on public health and agriculture. Following its construction in 1921 of the 'Johns Hopkins of China'—the Peking Union Medical College—the RF signed a contract with Cornell University to begin research and extension in Chinese agriculture from 1924. This included supporting the Nanking Agricultural University, as well as the rural mass education movement organized by James Chen at Ting-hsien in North China during the 1930s. Following his visit to Mexico in 1940, US Secretary of Agriculture Wallace urged the RF to undertake a joint research programme on agriculture with the government of Mexico, which it began in 1941. Again Cornell University personnel were involved; the leader was a plant pathologist named George Harrar, who became president of the RF in 1961.[25] Excited with the potential of their work in Mexico, impressed with the possible consequences of reform in post-war Japan and Taiwan, the RF began to investigate agriculture in the rest of Asia. For example, small grants were made in newly independent India in 1948. Chastened by the failure of American policy in China and the communist victory in 1949, the RF began to take the South and South-East Asian region more seriously. There was already a separate Far Eastern division of the RF for these purposes. The RF was building on its experiences in the American South, China, and Mexico when it transferred its attention to the rest of Asia.[26] In this manner the RF relied primarily on strengths; rather than trying to reform weakness, it intervened in those parts of the world where American interests were emerging.

The identification of health and agriculture occurred in conjunction with a belief in the universal applicability of science and technology. It was firmly believed that the only effective solutions to problems would emerge in the form of new technology produced by research and that the application of new technology would head off the direct structural transformations advocated by the communists. If there was not enough existing knowledge available to deal with a problem, such knowledge could be paid for by the foundation and produced by scientists. More and more the RF was coming to the conclusion that in public health and agriculture the knowledge relevant to solutions of a problem could be embodied in the technologies, for example, new forms of malaria control chemicals or new higher yielding food plants and changed modes of production in agriculture. There was an attempt to define problems to suit this concept of solutions. Foundation records refer frequently to 'tractable' problems, those which will yield to the application of science and technology. Also these problems offered leverage in dealing with governments and public opinion; work on disease and hunger was called 'morally unassailable' by RF officials.

The RF's increasingly technology-first approach was expressed in the firm belief that new technology is the key leading factor in the process of desired social change because technology is also the locomotive of economic growth. RF officials understood the importance of demonstrating the efficacy of technology in the field, visible to governments and publics. But their approach was always in a measured relation with their attempt to reform existing institutions or to create new ones. In the RF's immediate environment, there was also a compelling attraction to the prospect of a 'behavioural science', a kind of science which would permit social engineering comparable to biological engineering being proposed for agriculture and which actually was being tested in the RF's project in Mexico. Even if advocates of this behavioural science did not enjoy the prestige conferred on the natural sciences and biological engineering, some social sciences were part of the thinking within the RF about the problems of world food, revolutionary Asia, and tropical rice. The creation of the Population Council and of the Agricultural Development Council in 1952 and 1953 is evidence of this thinking.

But despite this interest in social engineering there was the perception that eventually new technologies could be applied to the solution of the problem of insufficient food production with little reference to the less tractable problem of food maldistribution or the thorny issue of land reform. Technology would be the leading factor in the profound changes which governments appeared to desire and which the RF believed to be inevitable. The strength of their belief was mirrored in the communist belief in new forms of social organization and new technology. If new knowledge was to be produced by scientists, research would have to be paid for and given proper direction and would probably have to take place in new institutions. Scientific research and scientists were more influential in the RF than in other private US foundations.[27] Beginning in the 1920s with the appointment as president of the RF of mathematical physicist Max Mason, then president of the University of Chicago, scientists had as strong a role in the RF as lawyers, businessmen, or economists, and they were listened to seriously by the trustees, who by and large were not scientists. In 1932 a mathematician from the University of Wisconsin, Warren Weaver, was appointed to direct the Division of Natural Sciences. This division strongly supported new basic research in genetics, biophysics, and biochemistry and in nuclear research which contributed to the development of nuclear weapons, for example at Lawrence Radiation Laboratory at Berkeley.[28]

Weaver, who coined the term 'molecular biology' in 1938, urged able younger physicists to study biological problems, and funded the application of new physical and chemical techniques to biology in the 1930s and 1940s. Equally important, Weaver articulated a new role of managers of science, 'an executive role which did not imply detailed supervision'.[29] Weaver's work in this division was so successful that the expanded RF investment in agriculture in the 1950s was channelled through it and its name changed to the Divison of Natural Sciences and Agriculture. Consistent with the RF's conviction that practical solutions lay in applications of science and technology, it was conceived that the new agriculture should be the result of the application of basic biological knowledge. As an RF advisory committee said in 1951, 'Agriculture is nothing more than the application of the principles of biology and other natural sciences to the art of growing food'.[30]

The RF had come to believe that most existing institutions were not really adequate for the discovery, production, or management of technological solutions. A new research institution could be the source of applying new science and technology, if given specific missions. For example, the Rockefeller medical research institutes in New York, which later became Rockefeller University, informed and gave direction to dozens of public health projects around the world. The Rockefeller-supported malaria research institute in Italy was the means to carry out the mission to eradicate malaria. To build new institutions the RF made a commitment to train and retrain locally influential personnel not only at universities in the United States but also at universities in Peking, Delhi, or Los Banos. Of its work in Asia, one analyst has said that the RF and those co-operating with it were building through training 'a human infrastructure' which could lobby with its own governments for new investments in agricultural development programs.[31] This is precisely what had been done in the field of public health in many countries.

Building these new research institutions thus established the idea of 'mission-oriented research'. Giving direction to research, guiding the curiosity of scientists towards the problems of frontier areas, and pressing for institutional changes which might allow fuller utilization of research results were all values which guided RF officials during the 1950s. They already had experience in creating new institutions in politically unstable conditions, for example, in the 1930s in Germany. Macrakis shows how the RF insisted in 1935 on releasing \$650,000 for the establishment of the Institut für Physik five years after the money was committed by vote in 1930, despite the fact that political conditions had demonstrably deteriorated. Four years later, in 1939, the Institut was commandeered for the study of 'military uses of nuclear fission'. Before releasing the money to the German government, an internal memorandum was prepared by the RF stating (in part) 'We go where there is the largest opportunity of advancing human welfare. We are not deterred by the political and economic complexion of nations except as it may handicap what we desire to do'.[32]

Throughout its work, officials and trustees of the RF seemed to have consistently believed in the superiority of private initiatives over government undertakings, not simply because such initiatives were a more direct expression of their interests but also because private initiatives were more efficient and less easily distracted from their objectives. But, on the other hand, foundations like the RF understood the importance of priming the pump, of developing the solutions of a problem to the point at which governments, with their greater resources and long-term commitments, could take over on a routine basis, thus allowing the foundation to withdraw and consider, define, and solve new problems, in a role analogous to venture capitalists and their relentless identification of new, often risky, projects.

Both in its international work and in the US, the foundation's officials believed they maintained a low profile, avoiding the spotlight—as in the agricultural collaboration with the Mexican government following its nationalization of the Rockefeller oil company in 1940. Many foundation officials and trustees had, it should be remembered, alternated between the foundations and work in US government positions. They did not have a high regard for government work and capabilities. All the RF's international projects had to have the support of governments in underdeveloped areas, and RF officials were unimpressed with the standards of most of those governments. Dean Rusk cautioned RF officials that they had to take pains to distinguish their work from that of the US government. But the need to work through and lead governments in specific directions required that the foundation maintain both distance and excellent relations with governments. These complex relations were essential to the work of the foundation and to the planning of IRRI.

2.2 Rice Politics and Research Planning

It was with the four guiding principles of building on strength, employing technology as the leading factor, managing new research, and working through governments that Dean Rusk, the president, and John D. Rockefeller, the chairman of the board of trustees, worked. Their views were harmonious, despite the disparity between Rusk's origins in the poor South and Rockefeller's. Rusk guided the RF away from the health and disease problems, because he believed that governments could now handle them, and pressed the RF towards getting agriculture moving in order to conquer hunger. Rockefeller was also thinking of new directions of investment in population policy and in agricultural policy. Both Rusk and Rockefeller had developed a very strong interest in the problems of Asia, and they worked hard to increase the work of the new Far Eastern division. Within a year of his appointment, in 1951, Rusk asked the trustees for an expanded programme with an additional $5 million annually for the next ten years, specially for the work in Asia, followed by lesser concentration on the Middle East and Africa. Rusk warned trustees that the vigorous, growing populations of Asia and Africa would not much longer tolerate the unequal distribution of resources and benefits. While the planning of IRRI was going on in 1958, Rusk was chairing a large Rockefeller Brothers Fund study called 'the Mid Century Challenge to US Foreign Policy'. In it he stressed the power of the rising expectations in poor new nations and warned that American corporations, private institutions, as well as the US government, would have to understand correctly the origins and thrust of these expectations in order to deal successfully with them.

Meanwhile John D. Rockefeller was creating new organizations in the 1950s to further the RF's work in specific ways. As early as 1934, when he was twenty-eight years old, John Rockefeller had written to his father that population 'is the field in which I will be interested—as I feel it is so fundamental and underlying'.[33] He toured Asia in 1948 with a special RF team of social scientists and physicians to survey the interrelationships between demography and public health. In 1952 he founded a separate Population Council with RF funds and was made its first president. It quickly became the most influential American institution in the movement for family planning and population control, with an annual budget by 1959 of $1 million. John Rockefeller remained sensitive to the population and food supply equation throughout his life; he expected much from a technology-first approach. The family planning approach developed since the 1950s had essentially consisted of developing and delivering contraceptive technologies to countries with high population growth rate, making them available at low cost or no cost to women of childbearing age who agreed to adopt them as a regular practice, thereby lowering, it was expected, the birth rates. The parallel to the HYV rice strategy is obvious: plant technologies were developed at IRRI and delivered to countries with low yields per acre and low food productivity, made available to farmers who would adopt them as a regular practice, thereby raising the yields and increasing the production of food. The evidence shows the difficulty of this approach in both cases.

In 1950–1, the RF contemplated establishing a major new Agricultural Sciences Division, building on its experiences in the US South, China, and Mexico. Presumably as a consequence of informal discussions, P. L. Mapa, the Secretary of Agriculture and Natural Resources of the Philippines, wrote to John D. Rockefeller III inviting him to have foundation officers 'look into conditions here in our country'.[34] Mapa cited the RF achievements in agricultural research and development in Mexico which had raised 'the standard of living of the masses'. In the view of Filippino and US agricultural specialists, Mapa said, increased rice and corn production would be the key to economic stability, but varieties of seeds then available did not yield as much as those planted in other countries. Raising the standard of living was important because the Philippines was a prime example of democracy in Asia, and it was crucial for democracies to achieve economic stability.

The president of the RF, Chester Barnard, replied that the foundation was reviewing broad types of operation in public health and agriculture and that it would soon discuss Mapa's request.[35] In the foundation's recommendations to its board of trustees in 1951, entitled 'The World Food Problem, Agriculture, and the Rockefeller Foundation', the Advisory Committee stated that there was a special problem in the Philippines in regard to the relation of hungry people and the growth of communism. This committee also said that there was perhaps a special responsibility on the part of the US to do something about agriculture in the Philippines.[36] Officers of the RF and the government of the Philippines knew that a new agricultural research and education programme at the University of the Philippines at Los Banos was being developed by Cornell University and paid for by the US government under the Mutual Security Agreements. It would thus be possible to build on the strengths, not only of the new Cornell project and large government funding, but also of the historic American presence in Philippines agricultural development since the begining of the twentieth century.

The RF's trustees decided to establish a new division of agriculture, partly on the basis of their advisory committee's report in mid-1951. Harrar, Mangelsdorf, and Weaver were sent to Asia, particularly to India, to evaluate the foundation's potential role. They reported early in 1952 that:

We would suggest that there are two types of activity which make sense: first, activities which explicitly face up to the complex and interrelated problems of ignorance and tradition, and to seek to attack these problems; and second, isolable technical problems which are so important that their solution would find acceptance and application even under present circumstances.[37]

The idea of isolable problems and the acceptability of solutions even in present circumstances were to become ideological keys to understanding the planning of IRRI.

Asia presented many agricultural problems, but, the report concluded, 'of much larger potential importance for India and for the whole of the Orient, is the problem of breeding improved hybrid varieties of rice'. The report, acknowledging that excellent work had been done in Japan and that an FAO rice-breeding project was underway in India, insisted that important opportunities continued to exist for the RF. Thus the authors proposed to start by sending 'an expert plant breeder, acquainted with rice problems, to Japan, to India and probably to other parts of the Orient' to make an intensive study of what was already being done. 'It is estimated that varietal improvement and disease and pest control could result in significant improvement in yields and contribute markedly to relieving Asia's perpetual problem of insufficient rice production'.[38]

Scientists in the RF were not alone in underlining the importance of the problem of tropical rice. It was common to American strategic thinking about Asia in the early 1950s that food supplies were fundamental to US interests and to regional stability. Just at this time an agricultural economist from the University of Virginia, John King, wrote 'Rice Politics' for *Foreign Affairs*. In this article, King called specific attention to the political significance in Asia of a country's ability to produce a rice surplus to feed all its people and to trade rice on the world market. He said that the reputed success of mainland China in food production was being received enthusiastically throughout Asia. Urging rapid action on the problem of rice production by the United States and by the United Nations, he presented the relationship of rice to the political situation thus: 'South and South East Asia must be made to realize that increased production and a higher standard of living are possible in their own countries without adoption of totalitarian methods. The struggle of the "East" versus the "West" in Asia is in part a race for production, and rice is the symbol and substance of it'.[39]

Having stated that 'the practical course is to seek to expand Asian rice production to meet the needs of the region', King reviewed the long list of constraints on rice production in South and South-East Asia. He said that though China experienced the very same constraints before 1949, China was trading rice in 1952 and 1953 despite previous rice shortages. King suggested how this was possible: 'Refugees making their way out of China bring constant reports that experimental farms established and financed by the United Nations and the United States have been taken over by Communists, the fruits of their experiments accepted, and their teachings forced upon Chinese farmers'.[40] Relying on studies in Asia by the FAO, he said that 'output could be greatly increased, on land now under cultivation, by improvement in seed, plant-breeding, irrigation and use of fertilizers, and by better methods of milling, storage, and transport'. Studies by the UN's Economic Commission for Asia and the Far East, King said, showed that 'the minimum needs of South and southeast Asia could be met by a 10 per cent increase in rice production in 1953, and some experts are convinced that a 50 per cent increase in rice production is possible on land now under production'.[41] He then cautioned: 'Economic and sociological changes are the preconditions for greater production, and they cannot be achieved quickly by non-totalitarian methods ... a long road stretches between scientific investigation and the lessons of the experimental farms, and the practices of the rice farmer.'[42]

King said that while an FAO International Rice Conference in 1953 had wrestled with this problem, 'the gap between the possible and the actual can be bridged only by putting into practice the countless long-range plans now languishing in countless desk drawers. The fruits of experimental research must be passed on quickly to the men who ultimately implement the plans and programs—the rice farmers.'[43] What is significant about King's article is that rice, the key to Asian food supplies, was placed in the context of regional security and US relations with Asia. King advised that the required increases in production could be brought about by scientific investigation on experimental farms, by new technologies, and by providing sufficient incentives to the farmer. These were all the ingredients of the Green Revolution; thus 'Rice Politics' made a timely announcement of the feasibility and necessity of vigorous promotion of an integrated agricultural development strategy.

King also suggested international shipments of rice to deficit countries could be increased and would not interfere with efforts to increase rice production, a suggestion that spoke directly to officials and politicians with farm constituencies worried about mounting stocks of US surplus food commodities. Within a year, in 1954, Congress passed PL 480, which simultaneously aimed at disposing of surplus food, stabilizing the US dollar, and developing influence in recipient nations. Eventually very large local currency funds were generated by the sale of this food in countries like India, Pakistan, and Sri Lanka, funds partially at the disposal of US agencies for agricultural development and research in those nations. For example, the large-scale rice research programme initiated by the RF in India in 1964–5 was largely supported by PL 480 rupee grants for the following eight years. Agricultural research expanded while international food shipments increased steadily, although some American food producers worried about what the potential self-sufficiency of these countries would do to their own business.

Within the RF the exploration for a suitable rice-related project in Asia was continuing. John D. Rockefeller III toured Asia in 1953 to view his petroleum refinery and distribution establishment and to examine philanthropic investment prospects. He was accompanied by one of the RF Board of Trustees, William Myers, former Dean of Agriculture at Cornell University. Rockefeller and Myers had already toured together in the US South in 1941 and in Mexico in 1946 and 1951. They were met in Asia by RF officers like Harrar and Weaver, who were following up recommendations made on the basis of their previous tour of India in 1952. Harrar was specifically considering the possibility of a project in the Philippines, a country which he said had great potential, largely because of its cheap manpower.[44] He noted that while Americans and Filipinos seem to get along well together, the present political conditions were too uncertain to warrant a major RF commitment. 'The threat of dissident [Huk] action makes it impossible to travel freely and develop broad plans for long-range agricultural programs'. He was also surprised to find that 'there is even some sentiment against greatly increased agricultural production'. But he noted the possibility of a change of government in the forthcoming election; 'many persons believe that if this does happen the new government will be much more progressive, democratic and conscious of popular needs'.

In December 1953 the board of trustees met to consider the RF's plans to expand agricultural activities to reflect 'our interests in the basic problems of food'. The 'atmosphere of the discussion was clearly favorable, and the NSA Division was authorized to proceed to get personnel to start activating the program'.[45] Subsequently, in October 1954, Weaver and Harrar wrote an important memorandum–manifesto called 'Research on Rice' in which they sought and gained approval from the board of an annual expenditure of $5 million for the period between 1955 and 1960.

'Research on Rice' began with a description of the position of rice in the world's food system. Acknowledging that there were more hungry people in the Orient than anywhere else in the world, Weaver and Harrar, partly relying on Lousiana State University economist J. Norman Efferson's book *The Production and Marketing of Rice* (1952), stated that with respect to volume of production, the most important food crop in the world was rice. Most of the rice crop was not marketed and was thus the basis of subsistence of a large proportion of Asia's population. In fact, they said, 85 per cent of the world's population obtained 90 per cent of its calories from carbohydrates in the form of grains like rice, wheat, and corn. The race between food production and population was 'so grim that starvation was a constant threat'.[46] And speaking to John D. Rockefeller III's personal concerns about population, they said that only if the presumed upper limit on food production and limits on yield were raised could an answer be given to neo-Malthusians who predicted 'inescapable doom unless population increases are rather promptly curbed'.

Despite the importance of the rice plant, they said, there was little known scientifically about rice because rice was important in those countries 'where science has not progressed very rapidly'.[47] The rice plant was vigorous, adaptive, resistant to disease, and would 'produce a tolerable crop under almost any circumstances'. There had therefore been 'little incentive to study this marvelous plant'. And that is precisely what was changing.

The report then announced that the priority should be to learn how to increase yields. Yields were to be the Archimedean point on which the RF could stand and move the world food problem. In probably the most significant expression of the values of American agricultural scientists, Weaver and Harrar proposed a broad and imaginative attack to discover the factors which combined to influence yields. Since no others seemed prepared to set up or finance such a study of rice yields, the RF was being offered a unique and most significant opportunity. 'It is considered high time that such a study be made for one (or more) of the great food plants of the world: and it is considered well within possibility that such a study would reveal yield potentialities not now viewed as possible'.[48] The range of normal yields for rice around the world was stated as 1,200 lb. per acre to 6,000 lb. per acre. But this range should not be seen as an upper limit. Though the efficiency of plant conversion of solar energy was, they believed, 20 per cent, this limit was not nearly reached by rice. In fact present levels of photosynthesis could be multiplied by a factor of 120 before reaching this limit. 'Thus when one thinks of the ultimate gains that might be made in crop yields there is no reason, if one considers only the matter of available energy, why yields could not be increased not 50%, or 100%, but by a factor of 100 or more'.[49] These very high expectations persisted at IRRI until the early 1970s and certainly affected the thinking of scientists and allocation of rice research budgets. They constituted the basic premiss of the Green Revolution.

Four subsidiary technical questions were raised by 'Research on Rice' in regard to rice: (1) Does the phenomenon of 'hybrid rice' affect only unimportant aspects of the rice plant, rather than yield and uniformity, as in the case of corn? Or does hybrid vigour not occur at all when pure lines of rice are crossed? If hybrid vigour does occur in rice, 'is there any practical breeding program which could exploit it?' (2) When two types of rice are crossed 'lethality' occurs in a high percentage of the progeny. Lethality also occurs in 'other important crop plants, such as corn'. 'Why? No one knows.' (3) Why does the *japonica* type of rice respond so well to fertilizers while the *indica* type does not? (4) Could the secret of the rice plant's resistance to root rot be transferred to other cereal grains which suffer from root rot? Answers to questions like these would be very significant because contributions to a basic understanding of this plant would directly benefit millions of persons—in South America, Africa, and other parts of the world in addition to Asia.[50]

The problem of tropical rice was thus seen as a classic case of low productivity, that is, low yields per acre. Low yields were judged, in turn, an 'isolable technical problem' amenable to investment and scientific research. A technical solution could be introduced to Asia under present circumstances and accepted there without prerequisite changes in the social relations of agricultural production, in infrastructure, or in government. No matter what the environment, the RF asserted, transferring the technical solution would result in desirable social changes without additional political effort. At this stage the problem was defined so the key variable, yields per acre, seemed to be a dependent variable and new technology the independent variable. That is what was meant by an isolable technical problem, although it seems clear in retrospect that such isolation is a mental construct. This approach to tropical rice was consistent with the RF's earlier conviction that agriculture is nothing more than the application of the principles of biology and other natural sciences to the art of growing food.

But in the foundation's eyes the solution had to be capable of embodiment in an institution which could produce new technologies. How was a basic scientific understanding to be achieved? Where could scientists solve the problem of tropical rice?

2.3 An International Rice Research Institute?

The heart of 'Research on Rice' considered arguments for and against a single definitive centre for rice research in Asia. The advantages of a single centre were easily apparent to the foundation, they said. The international friendships and understanding based on co-operation in science would contribute to a pattern of global living. More important, 'the really fundamental physiological, biochemical, and genetic problems are essentially independent of geography and are certainly independent of political boundaries; so that these problems could effectively and efficiently be attacked in one central institute'. In addition there should be financial savings and efficiencies by eliminating unnecessary duplications; expensive instruments could be located here, such as phytotrons, electron microscopes, mass spectroscopes, and the like. It would be possible to 'concentrate a high-powered and efficient international team of experts' forming a more effective group than any one country could hope to assemble. There would be unique facilities for training, a single depository for research knowledge, and a centre for multilingual translation of research knowledge.

In a talk some years earlier Harrar had spoken about the clear need to make agricultural research 'truly internationalized'; differences of language, culture, race, creed, colour, and tradition 'must be rendered unimportant'.[51] Rendering these differences unimportant, presumably in some central place and by some external and powerful agency, would be the only way to 'develop broad plans for long-range agricultural problems'. Yet every informed person with whom Harrar and Weaver had spoken was of the opinion that it would be extremely difficult if not impossible to get international co-operation for an international rice research institute. The rice-raising countries of the Orient were in a nationalistic frame of mind, still too sensitive to the bitter experience of World War II and their struggles for independence to co-operate in supporting a single research centre.[52] No one believed relevant countries would pledge funds to sustain such a venture. A country only expressed interest when it seemed likely that the institute would be located there. Countries were reported preoccupied with the problem of using for their own purposes, and not lending, the few trained personnel they had. What is more, Weaver and Harrar found that many 'scientists cannot use English effectively, if at all; and from some points of view it would be unfortunate to force an occidental language on an eastern institute'. The distance from Japan to Pakistan, that is, the sweep of rice-growing Asia, is also very great, and a large part of a budget would have to be spent on air travel.

But the preceding difficulties were minor in comparison to other disadvantages outlined in the report. Those arose directly out of the local and particular nature of rice cultivation in each country. Because of their significance, these disadvantages are quoted here in full:

Although certain basic questions are universal in character, the more applied, and hence more pressing and more easily appreciated, problems tend to be pretty local in character. How should one raise rice here, with our soil, our climate, our water conditions, our cultural preferences? These conditions, even though admittedly of limited validity, tend to make each country interested in a specific local program, but rather cool toward a general program located elsewhere.

It is surely true that, generally speaking, indigenous developments, rooted in the local soil, the local culture, the local problems, have a natural stability and a natural promise of growth which can never be matched by an activity which is to a large extent foreign in character.

On the one hand it makes a country-by-country development the easier, and doubtless the more realistic, alternative. On the other hand, one can sensibly argue that a basic over-all program of improving rice yields and quality would, in the long run but inevitably, have a sound and desirable influence on the scientific factors.[53]

Here were the central contradictions faced by the RF's planners: a centralized way of doing science versus the extremely local character of rice, commitment to problems independent of geography and politics versus applied problems more pressing in the perception of national governments, the whole world as the proper arena for work versus a more realistic, country-by-country alternative which had a natural promise of growth and stability.

Weaver and Harrar concluded that the costs for ten years' operation of a rice research institute would be $5 million.

It is by no means too large a sum to devote to a program of rice improvement. But it is too much money to wager unless the terms of the bet are good. It is our present judgement that it would be unwise to wager $5 million on such an institute, located in either Japan or India (the locations of strength) or in Siam; Indo-China or Indonesia (the more centrally and neutrally located positions of weakness).[54]

They recommended, therefore, that a single centre was premature and that the RF 'sponsor rice research activities in several different locations in several countries', planning each so 'they add up to a significant whole'. One million dollars would be spent immediately in rice research in Japan. Ministry of Agriculture officials there were already meeting with the RF to indicate 'their strong desire to work with us on a rice program'. In India, at the Central Rice Research Institute in Cuttack, 'we could profitably invest up to $500,000'; 'this is now, in our opinion, the best rice research program, group, and facility in the world'. These grants were to involve basic genetic physiology, biochemistry research, as well as the training of indigenous personnel. In addition, other funds might be used for the 'creation of new facilities . . . primarily aimed at rice culture in the Orient, but of value to rice culture in Latin America, southern Europe, Africa and elsewhere'.

In 1954 the RF announced the appointment of Richard Bradfield, head of the Department of Agronomy at Cornell University and one of the original group of three who had explored and planned an initial agricultural operating activity in Mexico.[55] Bradfield was assigned to work in Asia for the next twelve to eighteen months. He and Chandler toured Asia together in 1955, though neither had any previous experience in Asia. (Both were to become very important in IRRI, Chandler as its first director (1960–72) and Bradfield as tireless promoter of its multiple cropping research.) Following approval by the board at the beginning of December, Bradfield met with RF officials to clarify his authority while in Asia. There was a discussion of the extent of involvment of US universities. 'Research on Rice' had recommended 'some limited assistance to places in the United States and to one or two places in Latin America, where men can be trained for international service in connection with rice'. At this December 1954 meeting the role of US universities 'was broadened to include the possibility of aiding in the establishment of a training center in the USA where Orientals and others might receive basic training in agricultural science, with particular reference to rice improvement. LSU [Louisiana State University] was mentioned as one strong possibility'.[56] In fact the challenge to the RF officials was to learn what to do. Except for a few with experience in China, they were not old Asia hands and they were quite unfamiliar with conditions and scientific resources in the American rice-growing states.

Bradfield and Chandler toured Asia between February and August 1955 and reported back in New York in August, Bradfield returning again to Asia from September 1955 to May 1956. Bradfield, clearly the senior partner in the RF's eyes, made his final report on the feasibility of a rice research institute in July 1956. During 1955 and 1956 the question of the focus of the foundation's agricultural programme in Asia was still under debate. Bradfield met John D. Rockefeller III as well as Dean Rusk in order to question whether rice should be the exclusive focus of the work in Asia. Weaver presumably learned of these conversations and wrote to Rusk to explain his division's position on rice. Bradfield had told Weaver and Harrar, before departing for Asia in 1954, that focus on rice was too narrow. Weaver, as if anticipating the objection, said he and Harrar were 'about to start out to learn more about the status of this affair.... It looks very much as though we could come back with evidence that it would certainly be unwise to concentrate exclusively on rice. What the sensible balance will be, we can of course not say at the moment.'[57] In his reply to Weaver, dated 31 August 1955, Rusk wrote about a conversation he had with John D. Rockefeller III on the question of focus:

JDR 3rd told me of his talk with Bradfield, and I was planning to pass this along to you on Thursday. The Chairman's impression was that Bradfield probably felt somewhat restricted by the heavy emphasis upon rice in his terms of reference, and that Bradfield would welcome encouragement to take a broader look.

The chairman's own view, with appropriate deference to the judgement of our senior officials, is that a broader look is to be encouraged and that officers should not feel restricted by the weight of emphasis in the rice paper.

The fine print in the rice paper leaves the way open for a broader look. The underlying purpose of the interest in rice is to find what contribution we can make to the food situation in Asia. Although I suspect that rice is likely to play the predominant role in any such food program, it need not play an exclusive role.[58]

Moreover, Rusk reminded Weaver, the trustees of the RF now had in mind a heavy concentration on rice because of the strength of Weaver's and Harrar's advice. The trustees were also aware, he said, that 'the economic picture of production and consumption of rice had altered to a buyer's market' in Asia since the presentation of 'Research on Rice' in December 1954. Rusk also reminded Weaver that Bradfield and Chandler had already looked into this question in a more deliberate and careful way. The trustees would decide. Bradfield had declared a position which he held for the next twenty-five years, that only a holistic concern for rice's relation with all other crops was realistic. But the focus upon a single crop evidently proved attractive to the trustees and others because that was exactly the focus for research when IRRI was established in 1960.

Bradfield made his major report to the Rockefeller Foundation in July 1956. He assured them that there were greater opportunities for increasing food production in Asia than was commonly believed and that the rice trade and supply situation had much improved since the Rockefeller–Myers tour of 1953, particularly for Burma and Thailand.

> Rice improvement work, started since the War, is just beginning to bear fruit in many areas. Improved irrigation facilities, better supplies of fertilizer and chemicals needed for weed and pest control are also contributing factors. Continued research will be needed, however, to keep production up to levels required to feed their still rapidly growing populations.[59]

In support, Bradfield referred to the stimulation of rice research by the International Rice Commission and to the high standard of rice-breeding activities at the Central Rice Research Institute of India. Bradfield stated that more research was done in Asia on rice than on any other crop, but that 'the volume is not in keeping with the importance of the crop in any country, except possibly Japan'.[60] Agreeing that practically all of the earlier discussions concerning Rockefeller Foundation activities in the Far East were centred primarily on rice, he said, however, that in view of the opportunities for food increases in other crops like corn, soyabeans, and wheat, 'it does not seem advisable to confine the interest of the RF to rice. A cereal monoculture has always proven hazardous to both farmer and farm'.[61]

Bradfield's agronomic analysis of the main rice cultures in Asia is of particular importance in understanding the future work of IRRI: he said three types of rice cultivation had developed historically in relation to water: the rain-fed plains irrigation typical in much of India, the controlled irrigation typical of Java and Japan, and the upland un-irrigated rice culture typical in the Philippines.[62] In the first two rice cultures, improved cultivating practices, for instance the Japanese method then being promoted in India, could have a beneficial effect on raising yields. Arguing that 'new systems of cropping need to be worked out which are suited to the local environments, the needs of the farmer, the demands and rewards of the market, etc.', Bradfield pointed to the successful strategies of intercropping and double-cropping rice and corn, wheat, winter barley, potatoes, green manure, soyabeans, and clover.[63]

Where irrigation was not available, as in upland rice in the Philippines, yields fell off rapidly, he said if rice was grown for even two or three successive years on the same land:

> Many agronomists feel that upland rice should be replaced with other crops as soon as possible. Corn, for example, will produce twice as much food per acre as

upland rice in many places when grown under comparable conditions. From the evidence available it seems likely that the total area devoted to rice culture in the Far East will be gradually expanded but that the percentage of the crop grown under both the upland conditions and under rain-fed irrigation will gradually decrease. The percentage of grain under controlled irrigation will gradually expand as water resources are developed.[64]

It was presumably on this kind of analysis that IRRI was organized to focus on rice cultivation under controlled irrigation. It was also the only kind of cultivation which was familiar to American agricultural scientists. Bradfield was not alone in expecting the marginal cultivation of rice to decrease. Most rice cultivation in most Asian countries nevertheless remained under rain-fed or upland conditions. The subsistence peasant agriculture, so alien to the experience and values of American agricultural scientists, remained a persistent economic and political force.

Bradfield too did not recommend creating a single international centre but instead suggested concentrating on training and research at three regional centres with links to another eight institutions. These three regions would be grouped on cultural, linguistic, and political grounds, and the centres of concentration were to be the Indian Agricultural Research Institute at New Delhi, the University of the Philippines at Los Banos, and the University of Tokyo and Japan's National Institute of Agricultural Sciences. This was building on the best. The RF, he noted, was already co-operating with the US government in sending students from Thailand, Vietnam, Indonesia, as well as the Philippines to Los Banos; four or five people could be trained in these regional centres for the cost of training one person in the United States. All this research and training would require active co-operation from the Ford Foundation and the US government's International Co-operation Agency (predecessor of USAID), the FAO, and the Colombo Plan. 'Indications are that it will prove easy to avoid overlapping and easy to obtain closer co-operation between the Ford and Rockefeller Foundations'. Regarding the enormous agricultural programme of the ICA, Bradfield said that there was the greatest chance for competitive duplication between the ICA and the RF. While some ICA personnel were competent and a few contracts with US land-grant institutions had been productive, much of the work suffered, in his opinion, because of restrictive US government regulations, short-term projects, and incompetent American personnel. Bradfield stressed that there were three types of foreign experts needed in Asian agriculture: a few top-level advisers, advanced research scientists, and teachers 'free from some of the local traditions and prejudices'.[65] It was not, however, just expertise that was needed, he said, but efficient organization and greater motivation.

In addition to stressing the need to concentrate funding in the three regional centres of New Delhi, Los Banos, and Tokyo, Bradfield said there were six other cities where investments could profitably be made: Kaesetsart University, Bangkok; Lyallpur and possibly Dhaka, Pakistan; Bogor, Indonesia; the new Faculty of Agriculture developed by Texas A & M at the University of Ceylon; and the Central Rice Research Institute, Cuttack, India. His recommendations to concentrate on scientific research and training, and to set up counterpart relations between Americans and Asians in this task, was expressing the rationale for most of the RF's work. He mentioned, for example, the interest of the Indian Agricultural Research Institute in having an American co-dean for its graduate programme. 'By having a man from the USA familiar with this other scheme of organization [the integration of advanced training and research], he could be of great service in helping these institutions develop this more efficient type of organization'.

Having repeated that three regional focuses of work in Asia offered advantages not available in one centre, finally Bradfield posed the inevitable foundation question: what really was needed to develop the agricultural potentialities of the area?

There is little doubt but that adequate zeal could be generated in all of these countries, with the possible exception of Burma, in a short time by demonstrations within the country of what modern science can do toward improving their agriculture. Mexico was not very strongly motivated in 1941. The Rockefeller Foundation program there had had a strong motivating influence. Conditions in general appear to be at least equally propitious now for expanding the agricultural activities of the Rockefeller Foundation into the Far East.[66]

Bradfield used the situation in India as an example of the difficulty as well as the potential of generating 'adequate zeal'. India lacked scientists with advanced agricultural training. Even worse, there was the tendency to enshrine a hierarchy of command which inhibited younger scientists and teachers, leading to serious problems in motivation in taking up new work. The tendency of those receiving advanced foreign training to avoid practical research work was also noted. Such a situation was quite similiar to that in other parts of Asia, he said. The RF could thus act, as it had in Mexico, as a motivating influence in agriculture. This theme had become, and was to remain for years, one of the most important expressions of 'rice politics' in research planning: that 'getting agriculture moving' was not just the application of the principles of biology, but also lay in 'getting people moving' in Asia. That meant showing them how their problems should be defined and how the technical solutions should be organized, and thus greater efforts in communication and mobilization were needed in Asia than was imagined.

2.4 A Single Centre: The Process of Specification

From 1956 to 1958 the RF's support of research in Asia followed the patterns recommended by Weaver, Harrar, Bradfield, and Chandler. The policy of strengthening existing rice research institutions in Asia was continued and little else seems to have happened. Chandler later said about this period, 'I do not recall that the subject of an international rice research center was significant in our discussions until August 1958'.[67] Secretary Mapa's request in 1950 had still not resulted in any major investment by the RF in the Philippines, though the RF was active in a minor way in a number of small projects.

Then the subject of a single international centre for rice research was raised at a luncheon meeting at the Ford Foundation to which senior RF officers were invited. The occasion was ostensibly to discuss co-operation between the two foundations regarding grants to the College of Agriculture at Lyallpur, Pakistan. But conversation between the two vice-presidents, Harrar of the RF and Hill of the FF, turned to rice.

The following chronology has been prepared from various documents in the Rockefeller Archive Center, and corroborated and amplified by Robert F. Chandler, 'IRRI—The First Decade'.

1. August 1958, RF and FF luncheon, discussion of rice research.
2. October 1958, Harrar drafts 'International Rice Research Center'.
3. November 1958, Chandler meets with Philippines officials to examine project feasibility.
4. January 1959, Bradfield visits Los Banos: location decided upon.
5. June 1959, formal meetings in Manila between Secretary of Agriculture and Natural Resources and RF and FF.
6. September 1959, Chandler moves to Manila to establish IRRI; Secretary Rodriguez prepares memo on IRRI for Cabinet.
7. December 1959, Memorandum of Understanding signed; appointment of Jose Drilon, jun., as administrative assistant and of Luz and Associates as architects.
8. February 1960, Sterling Wortman moves to Manila as assistant director of IRRI.
9. April 1960, first meeting of IRRI's board of trustees, Manila.
10. February 1962, formal opening of IRRI at Los Banos by President Macapagal and John D. Rockefeller III.

One must ask why the Rockefeller and Ford Foundations were prepared to co-operate at this stage and why on the subject of rice research? They had quite different traditions and were attracted to each other for different reasons. An examination of the Ford Foundation's history in the 1950s first reveals major international investments and then turmoil and confusion. Started in 1936, the FF was reorganized in 1950 to extend its philanthropy beyond charitable support of Detroit hospitals. Under the presidency of Paul Hoffman, businessman and director of the US Marshall Plan in Europe, the FF made grants totalling $100 million for international development between 1951 and 1953, much of it in India. RF officials touring Asia in 1952–3 had called many of these FF projects 'second- and third-rate', with incompetent personnel and assumptions 'too optimistic for the Orient'.[68] Controversy between trustees and officials of the FF resulted in the shifting of the head office in 1954 from California back to New York for greater control, and then for two years the FF concentrated mainly on national projects. In 1955, for instance, $50 million was granted to US medical schools and colleges, while only small grants were given for international work, such as $600,000 to the Population Council, whose president was John D. Rockefeller III. After 1956, the FF appointed a new president, Henry Heald, who reportedly had 'no interest or experience in international matters' but was mainly concerned to sort out the conflict between trustees and officials.[69] Also appointed was the former Provost of Cornell University, Forrest Hill, as Heald's vice-president. He was particularly interested in international work, and eventually became chairman of IRRI's board of trustees. Now, 'as a result of staff proposals and the persuasive initiative of the Rockefeller Foundation and John D. Rockefeller III, the Ford Foundation began to give increasing attention to the interrelated problems of world population growth and chronic food shortages'.[70] The RF advisory committee on agriculture had pointed out as early as 1951 that when it was eventually necessary to co-operate with other foundations 'research in agriculture has a distinct tactical advantage when it competes for financial support'.

Following the 1958 luncheon, Harrar wrote that 'the FF might be willing to consider necessary capital investments and the RF might handle staffing and program because of its special experience'.[71] This was the key to the attraction between the foundations. The FF had three or four times more money than the RF. By 1968 the assets of the RF were still only one-quarter of the assets of the FF, that is $890 million, compared to the Ford Foundation's $3.66 billion. During the period of the Ford Foundation's reorganization (1957–8) its senior officers were clearly looking for an interesting, blue-chip, international philanthropic investment which could be supported without fear of controversy. It was reasonable to turn to the RF, the next largest US foundation and the one with the longest international experience. The financial power of the FF presumably changed the terms of the bet for a single centre which Weaver and Harrar had decided in 1954 was too risky.

The Cabinet of the Philippines government was told that the rice institute was being located in the Philippines in spite of the persistent request of other countries. What was so attractive about the Philippines? Harrar said that it was 'an important rice producing area but one where demand far outstrips supply'. 'Because of the similarity of climate,' added Chandler, 'results obtained in the Philippines would be applicable in such countries as Thailand, Burma, Malaya, Indo-China and Indonesia.'[72] In addition to these agricultural reasons there was another advantage, namely that there was an important school of agriculture at the University of the Philippines which had long been in contact with the United States.

Other major reasons why the Philippines was attractive was the already high level of US direct investment and the high rate of return. The foundations were also sensitive to the strategic and commercial importance of the US–Philippines relationship. This relationship was a profitable one for the United States at the time of the IRRI decision; between 1957 and 1960, for example, income from the Philippines to US investors was three times the value of US direct investments for the period.[73] In 1950, when Secretary Mapa wrote to John D. Rockefeller III, US direct investment was $149 million; by 1963 it had increased to $415 million. US investment in Philippines petroleum refining and distribution tripled from $50 million in 1950 to $150 million in 1963.[74] (The expectation of improved fertilizer supplies was an important ingredient in the focus of IRRI's research.) The 1959 Philippines Cabinet memorandum on the subject of IRRI said that 'with all the confusion that is now going on, this research foundation would also make Filipinos and Americans realize the significance of perpetuating the close affinity between our two countries'.[75]

Two alternative sites in the Philippines were contemplated for IRRI but were turned down before Los Banos was chosen. The Agricultural College of Central Luzon, wrote Chandler, had more land for research plots than the University of the Philippines at Los Banos (UPLB) and, even better, was located in the principal rice-growing area of the country. The Huks had been powerful in this region, but were not referred to after 1958 as a risk factor in the IRRI decision. Central Luzon college was later seen to emphasize training only and to be unprepared for research. On the other hand, the Bureau of Plant Industry had just offered to the FAO the Maligaya Rice Research Farm in another former Huk stronghold, Nueva Ejicia, but the FAO (presumably referring to its International Rice Commission) could not decide how to finance a UN rice research station. It is important to note that the International Rice Commission was also actively considering locating a rice research institute in Asia. Chandler was also warned by officials at the University of the Philippines that the Bureau of Plant Industry was 'strongly dominated by political expedience' and had produced little research relative to the funds available to it.[76] Thus the University of the

Philippines at Los Banos was in the RF's judgement 'the indicated location', partly because there were not really any attractive alternatives.

Los Banos had been host to Americans teaching agricultural science for fifty years prior to the establishment of IRRI. But following World War II there was a new generalized US commitment to restoring and modernizing the whole system of agricultural education in the Philippines.[77] The classical land-grant university model was used to change research and teaching at UPLB and to create a vigorous extension service. American professors worked in such disciplines as entomology, agronomy, and agricultural economics, and some scientists at IRRI established a continued close relationship with colleagues at UPLB, particularly with those having previous connections to Cornell University. For example, Nyle Brady, the director of IRRI in the 1970s, had been both a Cornell faculty member and a professor of agronomy at UPLB in 1953. It would simply not have been as convenient to establish IRRI without the earlier US involvement at UPLB, which had 'a progressive School of Agriculture' with 'the benefit of years of association with leading agricultural scientists from Cornell under an ICA college contract'.[78] Chandler soon reported that 'its operations were freer of political influence than is generally true of the Philippines'. In addition, UPLB officials told Chandler that 40 hectares of land would be made available immediately for the institute and an additional 40 hectares could be purchased if needed. Hill reported to the FF that D. L. Umali, Cornell graduate, professor of plant-breeding and later dean at UPLB, had been authorized to purchase the additional land for the FF and RF, in the name of the College of Agriculture.[79] UPLB had itself just received 1 million pesos for rice research from the government. These were all signs of the willingness of the university and the Philippines government to co-operate.

2.5 Specifying Research Priorities for IRRI: Yields Again

While assessing rice cultivation in the Philippines and negotiating IRRI's location with various government ministries and public bodies, the RF's agricultural scientists were beginning to specify the focus of future research in terms of their own current values. This focus emerged from their description of the problems which faced rice agriculture in underdeveloped countries in general and the Philippines in particular. Stating that 'rice is the single most important food crop grown today', Harrar argued in 1958 that 'it is grown mostly in areas thought of as being underdeveloped' with the result that 'methods are primitive and inefficient, production is low and prices relatively high'.[80] For those reasons, and because yields per acre and labour used were so much lower than the optimum in such countries, 'each year great economic losses occur as a result of agricultural malpractice including the use of non-improved varieties, improper soil and water management, inefficient hand labor and wasteful harvest and storage practices'.[81]

The Philippines was chosen not only because it had all the characteristics of other tropical rice-growing countries but because its inefficient methods of agriculture could not now produce enough to keep up with demand. Imports had been a regular feature of the Philippines food economy, but in 1958 a surplus had been declared as available for export. When Hill of the FF met with President Garcia of the Philippines, Hill reported that Garcia seemed to have 'a clear picture of the importance of rice in the economy'. Garcia talked about the possibility of exporting surplus rice to Japan, therefore 'anything that can be done to increase production and lower costs will be to their advantage'.[82] Chandler echoed Harrar's judgement about agricultural malpractice: 'I have been observing rice fields here in the Philippines and in no instance have I seen any evidence that top production is being achieved.... Cultural methods in the Philippines are poor and the use of fertilizers is extremely limited.'[83]

Given these assessments, what was IRRI's focus supposed to be? It should, in the most general way, solve 'this problem of rice production'. Calling for 'improvement of quality and quantity of rice available each year', Harrar added that the 'annual world yield of rice could readily be doubled if scientific and technical information now available could be universally applied'. Though it might seem utopian, he said, 'by extrapolation, the total yield might conceivably be doubled again as a result of fundamental research on those problems of rice production which are still only partially understood'. Any effort to 'produce information and materials ... with benefit to rice yields, would have important social significance' because of the 'enormous imbalance between demand and supply' of rice.[84] Note the belief in doubling yields through application of information 'now available'. At IRRI's beginning doubling yields would be the first level of attack on 'primitive and inefficient malpractices' in agriculture.

The second doubling of total yield was expected to result from new fundamental research focused on the rice plant. Harrar listed what he called 'a broad spectrum of important research problems' as follows:[85]

1. varietal improvement to develop well-adapted hybrids;
2. genetic and cytological study of mutation and outcrosses;
3. ecological and physiological study of the nutrition, growth, and reproduction of the rice plant;
4. the physics, chemistry, and microbiology of paddy soils;

5. the water temperature and mineral content of water effects on plant growth;
6. fertility problems of rice production;
7. host–parasite relationships and pest–pathogen control methods;
8. mechanization, handling, and storage;
9. economic studies relative to all phases of production and marketing.

This list, from experience in Mexico and from observations made since 1952 in Asia, constituted the ordering of priorities which was the focus of IRRI's work for the next twenty years. Only in the cases of mechanization, handling, storage, and economic studies was there even indirect reference to the social conditions under which the first doubling of rice yields could readily be achieved and these were not top priorities. The difficult problems of distribution are referred to once as marketing. This list followed logically, in Harrar's mind, from his initial characterization of the problem of low yields per acre. IRRI would conduct advanced scientific research, both basic and applied, which would further 'understanding of the rice plant as a living organism', a phrase repeated in the June 1958 memorandum 'Principal Points Agreed Upon'.[86] This very focus of work is evident in annual reports of rice research experiment stations in the southern US at the time, with the sole additional reference in the US cases to fertilizer and pesticide use and to multiplication of pure-line seed for rice growers' associations. Yield per acre was familiar to American agricultural scientists because it had become a legislative instrument included in the costs per acre, price support, and deficiency payment mechanisms of the USDA. Popular and scientific media of the period referred to 'yields per acre' as a test of the relative efficiency of new technological changes being pressed upon US farmers. Yields of rice and other crops had doubled in about fifty years. Now that the cause of these increases was better known, such a doubling could be elicited at will through direct intervention in other countries. The yield problem was standing in for the production problem which in turn had replaced the world food problem. IRRI's planners were interested in isolates, not patterns; in variables, not systems; in universality, not locality.

The concern for basic research was played down in discussions with the Philippines, and so reference to the rice plant as a living organism was dropped after June 1959. The definition of IRRI's focus presented by Rodriguez and Hernandez to the Philippines Cabinet was that it 'will be devoted primarily to the study of the rice plant with a view to increasing the production per unit area, both in quantity and quality, of this important staple in this country'.[87] The Memorandum of Understanding of December 1959 repeated the statement that 'yields are far below production potential'. Yield per unit of land was still perceived to be 'the isolable technical problem', the very point which Weaver, Mangelsdorf, and Harrar argued for in 1952.

As early as 1952 the RF staff had agreed that yield was an isolable problem of such major importance that a solution 'would find acceptance and application even under present circumstances'. These agricultural scientists knew that present circumstances were least amenable to direct change by an external agency like a foundation staffed by agricultural scientists. The RF and others had tried, but failed, to change present circumstances in China in the 1940s. They perceived most Asian governments they dealt with as weak and most agrarian circumstances as highly resistant to change, particularly by external agencies. They may have believed that even the efforts of the US government, along with its military and commercial powers, would or could do very little to change 'present circumstances'. But they still clearly believed that 'producing information and materials', in Harrar's words in 1958, would have important social consequences. The information and materials formed a rice-technology package which could be promoted and applied even under present circumstances, a package which would stimulate change in the agricultural economy, the agrarian conditions, and the nature of the country's economic relations as a whole. This package and change of circumstances would gradually displace the values which reinforced the primitive and inefficient agricultural methods. Even in this official correspondence, there were numerous references to the contests between the dynamic forces of science and the retarding influences of tradition and ignorance. New information and techniques, in the hands of people who wanted to use them, would one day win the contest.

What remained was to organize and govern IRRI to carry out these tasks, allowing for considerable refinement and specification of tasks by scientists. It was Dean Rusk's responsibility to raise questions about IRRI to which the RF's trustees would expect answers, about not only its organization but also its political relations. The questions raised in 1954 about 'indigenous developments rooted ... in the local problems' being more durable and reliable were now set aside. Rusk asks, 'are chances of genuine international cooperation enhanced by an institution known from the start as a Ford–Rockefeller enterprise as contrasted with the difficulty of interesting other countries in existing institutions in, say, India, Thailand or Japan?'[88] The decision to build a single centre contained the judgements that whatever was really meant by genuine co-operation little of it could be expected in Asia and that existing research institutions and the International Rice Commission of the FAO were unlikely to progress very far without the Ford–Rockefeller enterprise. The FAO's inability to make use of the Maligaya Rice Research Farm, even after a visit to Rome by the Philippines Secretary of

Agricultural and Natural Resources, was an example of existing barriers to certain kinds of co-operation. 'Would the physical facilities you have in mind', asked Rusk, 'be significantly superior to those you already know about?' The question of greatest significance to scientists in the Philippines was 'would the new institution take over from Los Banos such work on rice as is now going on there?'[89] The answer is that the facilities were to become superior, that IRRI eventually became one of the best equipped and funded agricultural science research facilities in the world. Its work eventually dwarfed the other work going on in the Philippines. IRRI later persuaded senior Philippines government officials that IRRI would do such a good job that building national research capabilities on rice need not be a high priority.

As for organization, Rusk had already advised Harrar that Chandler should be involved from the beginning because he would probably have to shepherd the IRRI project on behalf of the RF. Harrar had said that the staff leadership might have to come from the West, but Asian associates should be sought to enable the prompt formation of an international team working together in a project of at least hemispheric importance. Finally, Rusk asked, 'is the freedom and flexibility in staff appointments the crucial advantage [in starting from scratch]?' This was largely a rhetorical question because the FF had turned to the RF in part for the latter's pool of scientists and experience in staffing scientific projects of this size. The RF was a research-oriented foundation in a way in which the FF was not in those days. Hill told Philippines officials that the FF wanted the RF 'to assume responsibility for actual operation' of IRRI. That both the director and associate director of IRRI in the 1960s had previous experience with the RF shows how its initial links were much stronger with RF than with the FF.

On 9 December 1959, Rockefeller and Ford Foundation representatives, along with a representative of the Secretary of Agriculture of the Philippines, signed a Memorandum of Understanding to establish IRRI at Los Banos.[90] This memorandum, and IRRI's subsequent articles of incorporation, set out the framework within which scientists would be expected to work. The important point to be noted in the articles is that the institution would be incorporated for fifty years in order 'to conduct basic research on the rice plant, on all phases of rice production, management, distribution and utilization with a view to attaining nutritive and economic advantage or benefit for the people of Asia and other major rice-growing areas through improvement in quality and quantity of rice'. The Memorandum of Understanding of 1959 stated: 'In many areas, rice varieties are unimproved and cultivators make little use of modern techniques of production with the result that yields are far below production potentials'.[91]

The University of the Philippines at Los Banos would, on lease terms of 1 peso per year for twenty-five years, make certain lands available for IRRI, and IRRI effects and equipment would have an automatic tax- and duty-free status. A renewal of the lease for an additional twenty-five years was offered. IRRI salaries were paid in US dollars and were not to be taxed by the Philippines. If IRRI was to be disbanded in the future, then land and buildings would revert to the university. The 1959 memorandum stated that 'it is contemplated that staff members will be drawn principally from those countries in which rice is a major food crop'. Those younger scientists who received advanced training at IRRI would return to their respective countries to provide leadership in local and regional rice improvement programmes. The articles of incorporation refer neither to the desired composition nor to the return of trained staff to their countries. When IRRI began active research in 1962, there were scientists of seven nationalities working there; the first ten trustees appointed to IRRI's board in 1960 included four Americans, two Filippinos, and one person each from Japan, Taiwan, Thailand, and India. But IRRI's governance and the style of work and social life were to remain more American than anything else, a cross between the research division of a transnational corporation, a military base, and a diplomatic enclave. There was, in fact, nothing else quite like it for many years. Characteristic of the deep relations between the two countries, IRRI came to symbolize Philippines–American relations for both IRRI's critics and enthusiasts.

It may seem odd that these two foundations which did not draw their wealth directly from agriculture or Asia should establish IRRI and plan to contribute to the improvement of rice cultivation and thereby to the well-being and political stability of millions of Asian rice cultivators.[92] The RF and FF were not primarily interested in the land, labour, or rice of these cultivators but in the way their conditions, their production, and their behaviour impinged on American interests. If their cultivating conditions could be modernized and more productive, then their political behaviour might not pose a problem for the United States and its major institutions. Asian cultivating conditions and political behaviour were, it seems, perceived as a technical configuration which needed to be improved. Their commitment to increased production was buttressed mainly by four motives which coincided with the reasoning and interests of a number of Asian governments.

The foundations expected the new surplus would be largely appropriated by the coalition of classes and interest groups which supported vulnerable states in Asia. At the same time some of the new income would specifically benefit the controlling elements of the political status quo in rural Asia where the majority of the population lived. Some of the benefits would also trickle down, they hoped, so that access to food or to government welfare programmes would be extended because of the increased value of the appropriated surplus. And finally, if deficit countries could increase production and stop importing food, they would begin to participate differently in the world economy by importing manufactured and processed products. In these ways, the foundations believed, the tempo of economic advance could be revolutionized as Nelson Rockefeller envisaged in 1951.

Quite naturally it was left to the judgement of American agricultural scientists, believed to be specialists in these questions, to determine just how the existing technical configuration could be changed. The interests of the RF and FF were ultimately formed by the goals of investment and market expansion, military and strategic primacy, political stability, and the free flow of goods and information between the capitalist centre and the periphery. But agricultural scientists helped to direct their trustees' attention to the way in which 'the world food problem' could be solved through a programme of scientific research and improvement. This programme would address, on behalf of people dwelling at the centre, the conditions at the periphery which either limited the growth of the centre's systems or produced undesirable feedback effects on life at the centre.

In addition to their commitment to the widening of the national interests, what factors explain the depth of conviction among planners of IRRI and the scientists of the Green Revolution? In the 1950s these Americans were buoyed up with optimism: the defeat of Germany and Japan left them like the lords of the universe. Compared to the backward-looking European powers with shrinking empires, these confident Americans saw a future open to them, and saw a way to bring other countries out of their darkness into the light of this future. This kind of dualism (darkness and light) had wide popular appeal, and these planners were imbued with this dualism too. The best and the brightest were attracted to this future and to saving the world for a profoundly new kind of modernity. They had a high moral purpose in mind. They genuinely believed that hunger was morally offensive and unnecessary, and that it could be conquered in this way. Their moral and spiritual views were thus alloyed with a grasp of their interests, and the result was a powerful, if narrow, plan of action.

Moreover it was completely logical that this optimistic plan would succeed: why not? America had been transformed and the rest of the world could surely follow. Had they not solved the food problem in America? Was new American technology not the solution to agriculture everywhere? Their assumptions acquired the status of natural laws in their minds. When they looked at Asia, however, they saw few leaders and thinkers who understood this future. Little wonder that they created a single definitive centre, as Harrar and Weaver called it in 1954.

NOTES

1. The role of the Ford Foundation was important in IRRI after 1958. The Rockefeller Foundation Archives are opened after 20 years and the Ford Foundation Archives are open almost immediately. We have attempted to adjust for the inbalance in availability of documents from these archives.
2. Nelson Rockefeller, 'Widening the Boundaries of National Interest', *Foreign Affairs*, October 1951, 528.
3. Ibid. 529.
4. Ibid. 527.
5. Ibid. 528.
6. Ibid. 529.
7. Ibid. 538.
8. John Fairbank, 'The Problem of Revolutionary Asia,' *Foreign Affairs*, October 1951, 111.
9. Ibid. 108–9.
10. Ibid. 130.
11. P. C. Spender, 'Partnership with Asia', *Foreign Affairs*, January 1951, 206.
12. J. Soustelle, 'Indo-China and Korea: One Front', *Foreign Affairs*, October 1950, 64. See also numerous references to rice production, rice surplus, and politics in South-East Asia in the 'Pentagon Papers', published as US Department of Defense, *United States–Vietnam Relations, 1945–1967*, US Government Printing Office, 1971. See references to rice in statements by Presidents Truman, Eisenhower, Kennedy, and Johnson in ch. 7, 'Justifications of the War'.
13. Spender, 'partnership', p. 218.
14. Albert Ravenholt, 'The Philippines: Where Did We Fail?', *Foreign Affairs*, April 1951, 410.
15. Ibid. 410.
16. Economic Survey Mission to the Philippines, Report to the President of the United States, Washington, 9 October 1950, p. 55. Quoted in Al McCoy, 'Land Reform as Counter Revolution', *Bulletin of Concerned Asian Scholars*, Winter–Spring 1971.
17. David Horowitz and David Kolodny, 'The Foundations: Charity Begins at Home', in P. Roby (ed.), *The Poverty Establishment*, Englewood Cliffs, NJ: Prentice-Hall, 1974, p. 44. See also Peter Collier and David Horowitz, *Rockefeller: An American Dynasty*, New York: New American Library, 1977.
18. Lawrence Shoup and William Minter, *Imperial Brain Trust: The Council on Foreign Relations and United States Foreign Policy*, New York: Monthly Review Press, 1977, p. 106. See also Robert D. Schulzinger, *The Wise Men of Foreign Relations*, New York: Columbia University Press, 1984, ch. 5, 'The Experts' Cold War' and ch. 7, 'Southeast Asia, China and a Strained Alliance'. See also K. W. Thompson, *Cold War Theories*, Baton Rouge: Louisiana State University Press, 1981; Edward H. Berman, *The Influence of the Carnegie, Ford, and Rockefeller Foundations on American Foreign Policy: The Ideology of Philanthropy*, Albany, NY: State University of New York Press, 1983: 'The

developmental strategies championed by the Carnegie, Ford, and Rockefeller foundations were sincerely held and implemented with great seriousness of purpose.... [They] were neither politically neutral nor value-free, however, as foundation representatives frequently claimed. Rather, this rationalist–technocratic approach was rooted in certain assumptions from which it sprang' (p. 161).
19. Ibid., 63.
20. Warren I. Cohen, *Dean Rusk*, Totowa, NJ: Cooper Square Publishers, 1980, p.31.
21. Ibid. 72. Rusk wrote memos directing US strategic attention to 'reserve' Indo-China and South-East Asia from further communist 'encroachment'. See William C. Gibbons, *The US Government and the Vietnam War*, Part I 1945–1960 (see esp. ch. 6, 'The US Joins the War'), Princeton: Princeton University Press, 1986.
22. Cohen, *Rusk*, p. 83.
23. Waldemar Nielson, *The Big Foundations*, New York: Columbia University Press, 1972, p. 54. See also Harry Cleaver, 'The Origins of the Green Revolution', unpublished Ph.D. thesis in Political Science, Stanford University, 1974; Robert Carnove, *Philanthropy and Cultural Imperialism: The Foundations at Home and Abroad*, Boston: G. K. Hall, 1980.
24. For the history of the entire context (and numerous references to rice and to the Rockefeller Foundation), see Randall Stross, *The Stubborn Earth: American Agriculturalists on Chinese Soil 1898–1937*, Berkeley: University of California Press, 1986. Compare Gary Hess, 'American Agricultural Missionaries and Efforts at Economic Improvement in India', *Agricultural History*, January 1968.
25. Cynthia Hewitt de Alcantara, *Modernizing Mexican Agriculture: Socioeconomic Implications of Technological Change, 1940–1970*, Geneva: UNRISD, 1976. See also Jean-Marie Desroches, 'La Fondation Rockefeller dans la stratégie agricole américaine au Mexique 1939–1949' unpublished MA thesis, University of Montreal, 1982.
26. Cleaver, 'Origins of the Green Revolution'. It is clear that many of the questions of priorities and emphasis being addressed by the RF had their foundation in the long experience of US researchers who adapted to changes in agriculture and simultaneously promoted such changes, e.g. in the American South. See Charles Rosenberg, 'Rationalization and Reality in the Shaping of American Agricultural Research 1875–1914', *Social Studies of Science*, 7 (1977).
27. Nielson, *The Big Foundations*, pp. 31–55.
28. See correspondence between O. E. Lawrence and Warren Weaver of the RF in 1938–9 quoted in Gerald Swatez, *Social Organization of a University Laboratory*, Berkeley: University of California Space Sciences Laboratory, 1966, p. 185. For the history of the Rockefeller Differential Analyzer Project, see Larry Owens, 'Vannevar Bush and the Differential Analyzer: The Text and Context of an Early Computer', *Technology and Culture*, January 1986. pp. 63–95. Weaver was deeply involved in this development between 1935 and 1938.
29. Robert E. Kohler, 'The Management of Science: An Experience of Warren

Weaver and the Rockefeller Foundation Programme in Molecular Biology', *Minerva*, Autumn 1976, 280. See, particularly, 'Weaver's Investment Policies'. See also Donald Fisher, 'The Rockefeller Foundation and the Development of Scientific Medicine in Great Britain', *Minerva*, Spring 1978.
30. Advisory Committee to the Rockefeller Foundation, 'The World Food Problem, Agriculture and the Rockefeller Foundation', 21 June 1951 (Rockefeller Archive Center R.G. 1.2 Series 242D).
31. Cleaver, 'Origins of the Green Revolution', p. 322.
32. Kristie Macrakis, 'The Rockefeller Foundations and German Physics under National Socialism', *Minerva*, Spring 1989, 33–57. The RF memorandum cited by Macrakis (p. 33) is Rockefeller Foundation, 21 December 1934 (RF 900).
33. Population Council, *The Population Council: A Chronicle of the First Twenty-Five Years, 1952–1977*, New York: The Population Council, 1978, p. 9.
34. Secretary of Agriculture and Natural Resources, P. L. Mapa to John D. Rockefeller III, 11 September 1950 (Rockefeller Archive Center, R. G. 1.2 Series 242D; hereafter RAC. When these documents were provided they were unprocessed.)
35. C. Barnard to P. L. Mapa, 5 October 1980 (RAC).
36. Advisory Committee to RF for Agricultural Activities, 'The World Food Problem, Agriculture, and the Rockefeller Foundation', 21 June 1951 (RAC). The advisory committee comprised Stakman, Mangelsdorf, and Bradfield, who worked in the RF's project in Mexico, as well as Harrar and Weaver.
37. J. G. Harrar, Paul C. Mangelsdorf, and Warren Weaver, 'Notes on Indian Agriculture', 11 April 1952, pp. 25–6 (RAC).
38. Ibid. 14.
39. John K. King, 'Rice Politics', *Foreign Affairs*, April 1953, 454. King was a Carnegie Fellow and worked as research secretary to the CFR's project on 'United States Policy and Southeast Asia' (1954–5). He wrote *Southeast Asia in Perspective*, in 1956; he later had an intelligence career in the CIA, where he is reported to have opposed US policy in Vietnam.
40. King, 'Rice Politics', p. 458.
41. Ibid. 460.
42. Ibid. 447.
43. Ibid. 460.
44. J. George Harrar, memorandum to file, September 1953 (RAC).
45. Warren Weaver and J. George Harrar, 'Research on Rice', Appendix I of Minutes of meeting of the RF Board of Trustees, 30 November to 1 December 1954. Written 21 October 1954 (RAC).
46. Ibid. p. 2.
47. No evidence was offered for this sweeping generalizing; no reference to Japan was given.
48. Weaver and Harrar, 'Research on Rice', p. 4.
49. Ibid.
50. Ibid. 7. Reference to 'hybrid rice' is not to commercially available 'hybrid rice' (as in the case of corn) but to original crossing of different varieties (e.g. *indica–japonica* rices) for the purpose of revealing potential vigour.
51. J. George Harrar, 'Meeting Human Needs through Agriculture', in *Strategy*

Towards the Conquest of Hunger, New York: Rockefeller Foundation, 1971 (RAC).
52. Harrar and Weaver would have been aware, while writing the report, that the siege of Dienbienphu, begun in February 1954, had succeeded. French forces were on the way out in Indo-China as American operations were beginning.
53. Weaver and Harrar, 'Research on Rice', pp. 6–7.
54. Ibid. 8.
55. It is essential to understand the interplay between the RF's programme for wheat in Mexico (and criticisms of it), and the planned work for rice in Asia. The dominance of the RF in Mexico, and professional resistance to it, was paralleled by later experience in the Philippines. See Bruce Jennings, *Foundations of International Agricultural Research: Science and Politics in Mexican Agriculture*, Boulder, Colo.: Westview Press, 1988. See also D. H. Stapleton and E. Levold, 'The Diffusion of Agricultural Technology by Rockefeller Philanthropy, Especially by the Rockefeller Foundation in Mexico and Central America,' presented at conference on Science and Empires, UNESCO, Paris, April 1990.
56. Memorandum of meeting between W. Weaver, J. G. Harrar, R. Bradfield, and R. F. Chandler; 21 December 1954 (RAC).
57. Warren Weaver to Dean Rusk, August 1955 (RAC).
58. Dean Rusk to Warren Weaver, 31 August 1955 (RAC).
59. Richard Bradfield, 'Agriculture, and Agricultural Education and Research in the Far East' (A Report to the Rockefeller Foundation), New York, 1956, pp. v–vi (RAC).
60. Ibid., pp. vi–vii.
61. Ibid., p. viii.
62. Upland unirrigated rice is not typical of the Philippines, but Bradfield thought lack of irrigation was typical of upland rice.
63. Bradfield, 'Agriculture in the Far East', p. v.
64. Ibid. 125.
65. Ibid. 139.
66. Ibid. 148.
67. Robert Chandler, 'IRRI—The First Decade', in *Rice, Science and Man*, Los Banos: IRRI, 1972, p. 3.
68. Harrar, Mangelsdorf, and Weaver, 'Notes on Indian Agriculture', p. 26 (RAC).
69. Nielsen, *The Big Foundations*; see esp. 'Coming of Age in the Ford Foundation', pp. 89–90.
70. Ibid.
71. J. George Harrar to Dean Rusk, 6 October 1958 (RAC).
72. J. G. Harrar, 'International Rice Research Center', 5 October 1958; and Chandler to Harrar, 17 October 1958 (RAC).
73. Most information in the following paragraph is from George Taylor, 'The Challenge of Mutual Security', and Frank H. Golay, 'Economic Collaboration: The Role of American Investment', in The American Assembly–Columbia University (eds.), *The United States and the Philippines*, Englewood Cliffs, NJ: Prentice-Hall, 1960.

74. Esso, the Rockefeller oil company, built a large fertilizer complex which undercut two projects already announced by Philippines companies which terminated their projects after the Esso announcement.
75. 9 September 1966.
76. R. F. Chandler, diary entries, 20 November 1958 (RAC).
77. For a comprehensive analysis, see Peter W. Stanley, *Reappraising An Empire: New Perspectives on Philippine–American History*, Cambridge, Mass.: Harvard University Press, 1984.
78. Harrar, 'International Rice Research Center'.
79. F. F. Hill to Records Center, 2 July 1959 (FF Archives; hereafter FFA, 65–55). Umali became deputy minister of agriculture in the Marcos government in 1966, joined the FAO in 1971, and became assistant director-general of the FAO for Asia. For his views on how Green Revolution strategies favoured big farmers, see *Far Eastern Economic Review*, 29 February 1980, 54.
80. Harrar, 'International Rice Research Center'.
81. Ibid. 1.
82. F. F. Hill to Records Center, 2 July, 1959 (FFA 65–55).
83. Chandler to Harrar, 3 December 1958 (RAC).
84. Harrar to Chandler, 3 October 1958 and Harrar, 'International Rice Research Center' (RAC).
85. Harrar, 'International Rice Research Center', pp. 4–5.
86. Harrar to Rodriguez, 30 April 1959 (RAC).
87. Memorandum for the Cabinet, Government of the Philippines, 9 September 1959 (RAC).
88. Dean Rusk to J. G. Harrar, 9 October 1958 (RAC).
89. Rusk to Harrar, 9 October 1958 (RAC).
90. 'Principal Points Agreed upon between the Representatives of the Rockefeller and Ford Foundations and the Secretary of Agriculture and Natural Resources', 4–14 June 1959 (RAC).
91. Ibid.
92. Rockefeller companies do have operations in Asian agriculture. Subsidiaries of IBEC (owned by the Rockefellers) merged with Charoen Pokphand in 1970, which is a vertically integrated food business with about 70 operations. This merger brought IBEC in direct contact, e.g., with peasant chicken farmers. *South*, September 1982, 59–60. For an analysis of the expansion of CP which followed this merger, see *Far Eastern Economic Review*, 3 March 1988.

3
IRRI's Mission: To Develop Technology and Test Methodology

> the annual world yield of rice could readily be doubled if scientific and technical information now available could be universally applied ... by extrapolation, the total yield might conceivably be doubled again as a result of fundamental research on those problems of rice production which are still only partially understood.
>
> George Harrar, Rockefeller Foundation, 1958.

IN 1973 two new and dramatic forces began to affect IRRI. The world food economy was turned upside down by the OPEC oil price increase, and rice prices more than doubled on the world market. Although rice politics again became important, and demand for higher yielding varieties increased, it was clear that IRRI's expected big jumps in production had not been realized or had not been sustained. At the same time the effects of the new Consultative Group on International Agricultural Research (CGIAR), created in 1971, appeared in the budgeting process at IRRI: programmes and projects would have to be justified to a larger number of donors, some of whom were only shopping for small segments of 'research investment'. Eventually the CGIAR imposed an unfamiliar process of long-range planning on IRRI scientists.[1]

Into this situation came as the new IRRI director Nyle Brady, whose grasp of organization was different from Robert Chandler's—he was more a managerial scientist, in tune with the larger scale IRRI. All of these factors laid the ground for traumatic change at IRRI, for challenges to earlier definitions of its mission, and for attempts to preserve elements of its original interests. In fact, the debate about IRRI's mission in the late 1970s repeated the choices presented to IRRI's planners in the 1950s, particularly in the questionable decision to build a single definitive research centre. This would define problems globally and work on all aspects to provide 'final solutions'. No other independent research centre would be needed.

The story of IRRI's evolution has been, in part, the exploration of the definition of agriculture offered by the RF in 1951; this was the first principle in the strategy toward the conquest of hunger, namely that

IRRI's Mission: Technology and Methodology 63

'agriculture is nothing more than the application of biology and other natural sciences to the art of growing food'. IRRI's mission was to develop technologies to increase rice production and to sell these technologies in rice-cultivating regions. The development and selling functions necessitated a third function, to test these technologies. IRRI's director-general described IRRI's mission thus in 1974: 'Basically our scientists feel, and our Board agrees, that we must first develop a strong-core base before over-extending ourselves into extension and production oriented activities. In other words, they feel that before we start trying to sell our products, we must first develop and test them.'[2] Brady's statement is couched in terms heard commonly at IRRI in discussions about IRRI's 'mission'. The term 'mission' was carefully chosen: IRRI was conceived and built in an era of mission-oriented projects by two foundations which had studied and understood the theory and practice of mission-oriented projects; they expected to apply science and technology to conquer hunger in Asia, just as NASA had mobilized science and technology to launch satellites and reach the moon, as military bases were to guard American interests, and as research divisions of corporations had to modify the automobile and petrochemical industry in such a manner that their sales would increase.

IRRI's mission is interpreted here from documents, annual reports, books, and interviews with about thirty persons working at IRRI, from the director-general to research assistants, and a number of scientific observers at IRRI. The focus, then, will be on IRRI's view of itself, its own preferred interpretation of its mission. This is not to say that IRRI's interpretation of its mission has never shifted. Indeed, what the mission should be has been contested within IRRI as well as criticized from without; the various professions and disciplines supposed to co-operate in IRRI's mission naturally have had different views, given their interests and traditions. But these differences have not interfered with the predominant understanding of IRRI's basic mission. IRRI's values referred to here are the individual and institutional ideas which guided the selection of criteria on which choices were made and justified. The evidence selected illuminates IRRI's view of itself and was chosen to achieve a critical assessment of IRRI's evolution. As the mission evolved, blindspots of neglect and oversight were revealed about the complexities of agriculture in Asia unaccounted for in the 1951 definition of agriculture quoted above. The analysis of planning for IRRI shows that complexity was contemplated but eventually disregarded. We intend to show how IRRI's critical scientific responsibility in regard to these blindspots was limited by its own organization and values. Whether such limitations were justified in the long run and whether they could be overcome are the key questions which the reader must bear in mind. They certainly led to a crisis in the mid-1970s from which IRRI did not recover during the decade.

The structure of IRRI's mission should not be taken for granted. Brady's remark in 1974 reminds us that developing the products, new rice technologies, was considered the most fundamental work and indeed occupied most of the budget. But it is disingenuous to suggest that IRRI would wait before selling its products. In fact selling IRRI technologies in Asia and promoting the idea of research as a solution to the problem of tropical rice began almost immediately after IRRI opened. This selling took much time and attention, and its vicissitudes caused greater and greater emphasis on testing the technologies. Vigorous testing methodologies were sought which could strengthen the credibility of IRRI's claims for the technologies in the eyes of Asian governments, researchers, and farmers. The structure of this mission had much to do with IRRI's conception of science and its ultimate application. In this conception the only source of new technology is to be found in basic research: application of new technology inexorably 'draws down' the existing ideas of basic research, and this new technology is just as inexorably applied to predefined problems. A vacuum is created. The flow of basic research into new technologies must therefore be continuously replenished, and that has been IRRI's first priority.

3.1 New Plant Types and the First Variety

In 1950 the government of the Philippines wrote to the RF expressing its desire for new higher yielding rice seeds; the first priority for IRRI's scientists was to develop those seeds. Believing it unlikely that production increases would come about from bringing new land under cultivation because there was little new land, and convinced that a big jump in yield per hectare was close at hand, scientists began to search for something that became, in IRRI folklore, akin to magic beans. There were essentially two concepts running side by side: one, that a rice variety with nearly universal application would surely raise production, and the other, that a rice variety which performed well under optimal conditions would surely succeed best. Its first annual report, announcing IRRI's priority to search for new rice seeds, was addressed to the first concept:

It is entirely possible that within five years a variety of rice will be available that will yield almost any time and anywhere in the torrid zone. Such a variety would, of course, be insensitive to variations in length of day, would have resistance to lodging and to more common diseases and hopefully to certain insects, would have a dormancy period of a few weeks and would respond to increases in soil fertility levels.[3]

IRRI's Mission: Technology and Methodology 65

Given the idea of new varieties as the leading factor in the great leap forward, it was natural that the plant breeders' objectives and requirements would enjoy precedence over IRRI's other work. To the extent that new varieties should be designed to respond to 'improved cultivating practices' like chemical fertilizers, the study of agronomy in rice cultivation was considered important, but everything else was to be subordinated to a kind of genetic engineering. In order to bring IRRI up to date on the state of the art of rice genetics, a major conference was held in 1963, with fifty-nine scientists invited, seven from the USA, and the others from the Philippines, Taiwan, Japan, Thailand, and Sri Lanka. This 'summit conference' of geneticists and breeders did not, even in the recorded discussion, raise questions about adaptability or yield, but rather focused on interspecific crosses, chromosome morphology, the interpretation of the sterility of the *japonica–indica* hybrids crossed in the 1950s, linkage maps, and the like. When 'the areas of research meriting strengthened or new research' were summarized at the end of the conference, the first issue was 'extensive genetic analysis of economic traits, especially of physiological and quantitative characters of importance to plant breeders'.[4]

The search for a breakthrough characterized the announcement in IRRI's 1964 conference by Americans Peter Jennings and Henry Beachell of the need for modifying existing plant types. These two IRRI plant breeders, one of whom, Beachell, had worked as trainer–adviser in India in 1955, observed that the *indica* varieties then in use 'offer little hope for substantial yield increases through the uses of nitrogen fertilizer or other applied technology'.[5] They proposed a new short-duration and short-statured variety, 'based on the assumption that there will be an increased availability of nitrogen fertilizer at a cost that the farmer can afford'.[6] They also proposed to modify cultural practices such as transplanting methods, the rate and time of applying fertilizers, weed control, and mechanization. They would develop a variety with harvesting characteristics consistent with the assumption that 'mechanized methods would continue to spread'.[7] It would have milling characteristics (or consumption characteristics) consistent with the assumption that 'the bulk of the rice in South-East Asia today is probably consumed without any consistent demand for a specific grain size and shape'.[8] They assumed irrigation would be available, but did not discuss it at length.

The new plant type 'should mature in the shortest possible time' [from seed-bed to maturity in about 100 days]. When questioned, Beachell said, 'it is possible that varieties [like Belle Patna, a 100-day variety he developed in Texas] might yield satisfactorily in the tropics'.[9] The most significant assumption repeated consistently by Jennings and Beachell at this conference was that there were 'widely adapted varieties', or 'non-sensitive varieties of early maturity, with resistance to disease, which should have a wide range of adaptation with respect to latitude'. This was stated, for example, in answer to politely critical reproaches by Valasco, of the Philippines, and Weeratna, of Sri Lanka, that such assumptions of wide adaptability were contrary to the experience of most tropical rice breeders.[10] Jennings and Beachell must have been near the point where specific IRRI varieties had been developed that embodied these characteristics; at least IRRI allowed preliminary distribution a year later, in 1965, of the variety later named 'IR 8'. Although they assumed that this variety would perform well in many environments, they did not then seem to have a clear appreciation of the limits which chemical and water requirements would place on the actual adoption of the variety by farmers. In other words, their assumptions were too narrow for the complexity of cultivation conditions in Asia.

Jennings and Beachell, who directed IRRI's breeding work, were building their experiments to modify plant type upon earlier research done on fertilizer response in Japan, dwarfing in Taiwan, and hybridizing *indica* and *japonica* in India in the 1940s and 1950s. Previous efforts to alter the rice plant's architecture had apparently resulted in a research consensus which IRRI incorporated, particularly Tsunoda's thesis that yielding ability and response to nitrogen fertilizer were related to plant type.[11] Scientists in the Philippines, particularly those at the nearby University of the Philippines at Los Banos, warned IRRI in the early 1960s that the high costs and the ecological consequences of chemical requirements for IRRI's varieties, as well as the dramatic shortening of plant stature which would make them susceptible to flooding, were going to limit their utilization by Asian farmers. According to one study, Philippines scientists

> believed that the plant had to be changed but not to the extent that IRRI desired... [they] did not dispute IRRI's goal of increasing rice production. They disputed the extent of that goal, the attainment of which would result, they argued, in changes in farming too abrupt for most cultivators. One of them indicated that he told the Institute that its program would benefit no more than one-third of Asia's rice producers.[12]

Such doubts, and IRRI's determination to produce an almost universally applicable variety, were soon to come to a head in the debate about IR 8.

There was great pressure on IRRI to produce results; yet by 1965 they still had not officially named and released a variety from their promising lines. Chandler formed a seed committee to name a variety and recommend it to the Philippines Seed Board, the agency which must approve all varieties for commercial use in the Philippines. The most promising line in 1965 was IR 8-288-3, but the IRRI committee decided to test it further, signing a memorandum of agreement in 1965 with the Bureau of Plant Industry to conduct tests in farmers' fields. However, prior to and during these official manœuvres, IRRI had been distributing 2,000 2-kilo bags of IR 8 seeds directly to Philippines farmers, as well as testing it in India, Pakistan, Thailand, and Malaysia. In 1966 IRRI decided to name and recommend IR 8 to the Seed Board: scientists on the board's rice improvement working group had misgivings about IR 8's grain quality and taste, the qualities which had to be approved by the board in addition to yield performance. On one hand IR 8 was incompletely tested; on the other hand farmers and seed producers were already growing it. A compromise evolved: instead of the Seed Board classifying it as certified seed, it was listed as 'good seed' until two years later when it was, after more testing, certified.[13]

Institute scientists acknowledged at the time that IR 8 was very susceptible to bacterial leaf blight and rice blast fungus and that because the grain was prone to break and crack, recovery rates of full grains (head rice) were relatively low.[14] But these drawbacks and external criticisms notwithstanding, IRRI released IR 8 because of reports of extraordinary yield performance in a number of locations in Asia. IRRI's *Long Range Planning Committee Report* (*LRPCR*) in a retrospective analysis stated that:

Even before IR 8 was named, its limitations and drawbacks were known. It had poor grain quality, lacked resistance to common rice diseases, and could be damaged by many common rice insects. Its high yield potential, however, created among producers, extension workers, and governments an immediate demand for its release. Despite its drawbacks, it was released for farmers' use.[15]

The fourth draft of *LRPCR* included farmers in the list of those demanding release of IR 8; from the evidence available it seems that only farmers near the research station at Los Banos knew enough about IRRI's work to demand anything. We may therefore conclude that IR 8 was released primarily to satisfy the purposes of governments, foundations, their extension officials, and a few seed producers in the Philippines. Many observers, including IRRI scientists and rice researchers in South Asia and the United States, said that IR 8 was released long before it should have been, but that there was strong pressure within IRRI and the two supporting foundations for the release of a product.[16]

IRRI's scientists had worked for years in the expectation that varieties would be released and were well aware that HYV wheat seeds from CIMMYT in Mexico (created by a reorganization in 1963) were being popularized and promoted in Pakistan by Borlaug and others. IRRI officials, such as Chandler and Wortman (director and deputy-director), travelled in Asia before IR 8's release and knew how the Ford and

Rockefeller Foundations were committed, in India and Pakistan at least, to fulfil promises to do something about food shortages of the 1964–5 season. The foundations were also heavily involved in national planning, agricultural development advising, and in specific infrastructural change, such as the huge Intensive Agricultural Development Program (IADP), underwritten in India in 1961 by the Ford Foundation, and the All India Crop Rice Improvement Program (AICRIP), underwritten in India from 1964 by the Rockefeller Foundation and USAID. Agricultural development officials of Asian governments, who were seldom agriculturalists or scientists themselves, had been repeatedly told to expect big results from the new technologies which IRRI was developing. The evidence from East Pakistan discussed below attests to this. These people construed their problem in such a way that such new technologies looked like perfect solutions, and IR 8 was quickly promoted in 1965–6 to satisfy them. IR 8 also had dramatic effect among foundation trustees who were being asked to continue to finance not only IRRI but also the new agricultural research institutes in Mexico and in Nigeria, institutes which would cost millions of dollars.

IRRI's achievements were disseminated in the late 1960s among agricultural scientists by journal publications and conference presentations (for example, Beachell attended meetings of both the Rice Technical Working Group in Arkansas, and the UN's International Rice Commission in Louisiana). IRRI was also visited by politicians and their accompanying news people (such as U Thant of the United Nations and Spiro Agnew, US Vice-President). At the same time as some IRRI staff were expressing caution about the capacity for sustained yield of the new rice varieties, a swell of official enthusiasm, inspired but not controlled by IRRI, was moving in Asia. In 1969 two members of IRRI's board of trustees reported on the 'dedicated efforts' of plant breeders: N. Parthasarathy of India and S. Dasananda of Thailand hailed the 'phenomenal breakthrough' and 'phenomenal achievements' at IRRI.[17] IRRI received the prestigious Magsaysay Award for International Understanding in 1969. On the front page of the *New York Times*, along with a photo of Chandler standing in an experimental plot, it was explained how IRRI varieties were going to change rural conditions in the Philippines and probably elsewhere.[18] Amid the bad news about the war in Vietnam, here was good news indeed.

Foundation officials and IRRI scientists were starting to talk about the consequences of the higher yielding varieties:

Let's call it the second generation of development problems associated with rice—rice marketing, rice technology and economics.... My question is to what extent does IRRI plan to cope with these questions.... Increasingly we seem to be calling upon the LSU group in this area. Should we somehow combine LSU and IRRI, or will the private sector become a dominant force?[19]

In meetings to plan a 1969 conference to decide the direction IRRI should take in the 1970s, F. F. Hill, who was by now chairman of IRRI's board of trustees, expressed his interest in a third generation of problems going beyond food surpluses.[20] Indeed, following 'the IRRI outlook conference' in late 1969, the same issue of the *New York Times* (following the lead of *Foreign Affairs*) warned that imminent overproduction of rice in Asia was seen by both American and Asian experts as leading to a glut of rice and loss of an important export market.[21] Most second-generation problems did not long endure in the 1970s: the vast surpluses that could not be processed, milled, or properly stored appeared only temporarily in one or two places; the collapse of rice prices, which already were kept deliberately low by governments, did not occur; and other factors, such as the war in Vietnam, kept the rice trade brisk for those who already had a share of the market. By 1973, partly because of rising oil prices and poor Asian harvests, the price of rice on the world market had nearly doubled over the previous year.

The LRPC at IRRI took quite a different view of these questions ten years later: the 'production impact' of IR 8 was perceived as 'modest', a realization that, it said, initiated the 'second phase' of IRRI's growth:

The initial success of IR 8 brought challenges and problems. The miracle rice was not well suited to all environments, and equally high yielding varieties suited to poor environments were in demand. In fact, the total rice production impact from the products of IRRI's first decade were modest. Normal resistance to new ideas, fear of risks, discouraging development policies, and institutional barriers often prevented farmers from using the new rices. The inability of the new technologies to fit the demands of the many different environments in which rice is grown was also a major factor.[22]

Despite these shortcomings, IR 8 had yielded well in trials over the first few seasons, and thus satisfied the immediate need of government agencies which promoted it to show that they were 'doing something'. IR 8 was succeeded in 1969 by IR 20, which had wider resistance to disease and insects and then by IR 22 and IR 24, which both yielded well and had better grain quality than IR 8 but had little disease resistance. There is some ambiguity in this interpretation by the LRPC of the history of these breeding objectives. Given the profile of the new plant type in 1964, and the belief in widely adaptable varieties, why was the range of improved characteristics of these varieties so limited? They were particularly suited for irrigated environments only, and then required precise application of water and chemicals. Perhaps 'widely adaptable' referred only to latitude, longitude, or some other general characteristic. But it is ironic, to put it mildly, to note 'the inability ... to fit' of these varieties when they were developed on the premiss that they would have wide adaptability and that they would not need to be tailored to fit different environments.

3.2 Beyond IR 8

As a second decade of research approached, IRRI and its underwriters tried to assess the recent past and to plan their next work. Judging from interview recollections and current documents, there appear to have been three predominant moods at that stage: a hope that the major rice production problems might be over; a concern that IRRI was unprepared to focus on post-production questions (processing, marketing, etc.); and thus the feeling that this work might be done more effectively by the private sector. Nevertheless a doubt remained that the production problems really were over. At this stage, in 1969, IR 8 and IR 6 were the only cultivated IRRI varieties and were being adopted in many countries by ambitious farmers who were obtaining more than double their usual yields. And IR 20 and IR 24 were in their 'release' stage. Although there really did not appear to be any reason to doubt that production problems were over, some scientists were saying that there would be severe constraints on farmers who might want to adopt new varieties, including the unpredictable risks associated with them, the high cost of their requirements, and their intolerance of marginal conditions. A FF official summarized the 1969 planning conference as follows: because of IRRI's superb location, he said, its findings are primarily applicable to well-endowed production areas which 'represent only 10 to 20 percent of the world's tropical rice acreage'.[23] Moreover, 'the production package for the favored situations still requires further improvement through research and testing'. Again, Chandler was asked, is the improved plant type 'near-universal' in its adaptability and applicability? Is it just production practices which have 'to be tailored to specific environments?' Finally, asked the foundation, could the 1969 conference report define 'what portion of the rice producing area is as yet inadequately served by more advanced varietal and production practice technology?'[24] In revising the planning conference report, Chandler and Barker were again asked:

If one's primary objective is to increase output and lower unit cost, then he would focus his production oriented research on the good soils in regions where rice was well adapted. If in addition, he has as an objective the raising of incomes, living standards, social and economic involvement of the rural disadvantaged, then presumably he focuses heavily on the problem situations.[25]

The conference recommended increasing the productivity of suboptimal and marginal lands and studying the ecological, social, and economic environments in which rice was grown. While IRRI was to avoid direct involvement in debate about the agricultural and economic policies of nations, it should be co-operating more with the research and development scientists of rice-growing countries in Asia. The report also hints at the necessity to direct IRRI's attention away from its own experimental plots and laboratories.[26] But most significantly, there was a measure of acknowledgement that new varieties from IRRI would not be likely to yield almost any time and anywhere in the torrid zone.

Whether one doubted the universality of the new varieties or not, it was clear that if a dramatic increase in production took place, IRRI was not studying how to deal with its consequences. Having asked 'what important rice production problems remain to be solved?', scientists then analysed the work of IRRI's small farm machinery programme, water management programme, and the agricultural economics group. Possible conflict of interest with private millers, processors, and commodity traders who already profited from inequalities in distribution of land, of income, of storage and milling facilities, and of market information, were debated. The political risk to US institutions like LSU, which might join IRRI to study these second-generation problems, was discussed: the long-standing and very real antagonism of sectors of the US rice industry towards offering technical assistance to their competitors in Asia was raised again in this debate.[27] The 1969 conference report stated that IRRI should concentrate on building post-harvest machinery for Asian rice farmers, but not to design or construct large-scale US-style milling and processing technology.

The increased emphasis upon more ecological zones for rice cultivation, noted by the LRPC, did not disturb IRRI's mission to develop new rice technology. In fact it made the existing line of work more urgent, in the perception of those engaged in the rice-breeding and testing programmes. A steady stream of new varieties destined for different cultivation zones was desirable, and various scientific departments at IRRI began to compete to develop, name, and release new varieties. Satisfaction with the new varieties was also expressed in IRRI's administrative system, and within the Ford Foundation the occasion of the release of IR 20 and IR 22 prompted the remark: 'here's a PR opportunity in the US as well as abroad'.[28] And Chandler recorded early in 1970 that the response to IR 22 was so strong that IRRI could not fulfil all the requests for seed, for example, from India and Nicaragua.[29] There was a confident and liberal policy towards allowing other international agricultural research institutes, such as CIAT in Colombia, to use rice genetic breeding material from IRRI. CIAT, where Jennings now worked, could use IRRI genetic material independently of IRRI, simply giving credit to IRRI in the first announcement. 'Even though they have IRRI blood in them,' wrote Chandler, 'go ahead and do what you want to about it.'[30] There is no doubt that IRRI was known by its varieties and that for other people to 'do what they want' was potentially provocative.

For complex reasons, discussed in the following section, IRRI's own research organization was reformed in 1973–4 and the actual naming and release of new rice varieties ceased abruptly and traumatically in 1975, except in the Philippines. For fifteen years developing of new varieties had been IRRI's first orientation, but ten years after the first tests of IR 8, with fifteen IRRI varieties and forty-one experimental lines released through national research programmes in Asia, pressure was brought, apparently from other countries, through IRRI's trustees, to force the decision, unpopular at IRRI, to stop naming new rice varieties as IRRI varieties. While the largest share of the budget continued to be allocated for breeding and other genetic activities, the glamour attached to the creation of new rice plants was largely gone. From within IRRI itself, there were warnings that the scientific concentration had been far too narrow.[31]

In the judgement of the LRPC, this breeding activity, in which eighteen original land races and varieties of rice from the Philippines, Taiwan, India, Indonesia, USA, and China served as the gene pool for the fifteen released IRRI varieties, 'minimized the genetic diversity of the world's rice crop'. Hargrove, Coffman, and Cabanilla went even further, stating in 1979 that this reduction of genetic diversity, first noted by IRRI between 1972 and 1976, was of sufficient relevance 'to demand a prompt broadening of maternal genetic base of modern rices'. Their work caused quite a stir at IRRI because it revealed some of the uncritical decisions taken all along the way by busy scientists. This reduction of genetic diversity, scientists countered, could probably be found in any breeding programme for other crops in countries like the USA or Japan, and was 'not necessarily cause for alarm'. However, citing the analogy of the US southern corn leaf blight disaster, Hargrove and colleagues recommended encouraging strong and autonomous local rice-breeding programmes in Asia and identifying and using alternative sources of dwarfism.[32] Finally, they cautioned, 'rice scientists should at least realize that modern efforts in rice improvement and associated technology clearly have the potential to carry the world into unintended and unforeseen problems'.[33]

The debate about unintended and unforeseen problems did not take IRRI by surprise, but it did reveal considerable differences of values and priorities. After all, IRRI's founders in the late 1950s, having contemplated the weakness of local solutions to local problems, decided to build a single definitive centre. In the late 1960s, IRRI had to decide to breed some new varieties for marginal conditions in addition to those fitted for highly controlled, optimal cultivating environments. The differences of opinion, twenty years after IRRI's founding, were about whether or not, according to one view, IRRI should continue to concentrate on the highest and quickest possible yield increases and let the social and political chips fall where they might. The other view held that by dispersing researchers to study various ecological zones, questions of equitable distribution, and non-adopting farmers in suboptimal conditions, IRRI could succeed in leading the way for the national research programmes. It appears from our interviews that those people holding the latter view believed that the advantages of concentration had already been quickly achieved and continuing this course simply would not have significant effects on total production increases. People with the former view believed that the dispersion was in fact a dissipation of resources which, although it could eventually have some practical effects, might destroy IRRI as they knew it.

3.3 Shifts in IRRI's Organization

When Nyle Brady began work as IRRI's second director in 1973, he was charged with the immediate task of clarifying IRRI's objectives in time for the 1974 budget request. F. F. Hill, who was chairman of IRRI's board and vice-president of the FF, and who had been a strong supporter of Brady's appointment, wrote that 'IRRI's image is somewhat blurred; not disastrously so but still blurred.'[34] If Brady would produce a clear statement of IRRI's missions and the programmes proposed to carry it out, Hill was confident that any financial difficulty could be overcome. He stressed his and the Ford Foundation's view of the mission: to work hard on increasing yields on irrigated land and to pay attention to upland unirrigated land, the key to yield increases being 'improved production technology'. Both the IRRI staff and those who were financing IRRI's work, said Hill, were not as sure of the mission as they should have been; extending the institute staff's attention to unfavourable rice environments (such as upland, unirrigated, rain-fed) was important but not as important as convincing financiers that the original mission was still sound.

To understand this mission further one must examine IRRI's organization before 1973. Starting as a small research team of twelve, scientists were all chosen by Chandler and deputy-director Wortman. Although by the mid-1960s there were scientists of seven nationalities on the staff, each one had been trained (that is received his doctorate) in a US university. Most of the first Americans at IRRI recognized that they did not know much about rice or about Asia, but they did have a model of modern scientific agricultural research, something that looked like an American experimental station.[35] Thus the classic cluster of agricultural sciences found at research stations in the United States was transferred to IRRI—plant-breeding, agronomy, pathology, soil chemistry, and entomology—with later additions of agricultural economics (1963), communication (1963), agricultural engineering (1966), and water management (1968). All of these new additions needed to prove their value to the classic cluster, in terms of IRRI's mission to develop new technologies, just as they were having to do in the US. This need limited the role of these satellite groups, in the opinion of one long-time observer of IRRI:

The first head of the department of water management told me near the end of his first couple of years that he still felt it very necessary to establish his credentials in the pecking order of discplines at IRRI to show that studying water was a 'legitimate' subject. This is how limitations were placed from within our division and from other divisions on what water studies would do, on its objectives.[36]

Scientists in the classic cluster of disciplines had their own hierarchy, which placed breeding and genetic manipulation on top. Within the cluster plant breeders had the greatest prestige, but could not work without the knowledge of experts in pathology, entomology, and so on. Only the cropping systems division challenged the supremacy of the founding research groups in the classic cluster, and that challenge was central to the crisis in which IRRI found itself by 1980.

Chandler's review in 1972 of IRRI's first decade reiterated the priority of varietal breeding and its attendant disciplines.[37] Various departments evidently understood that varieties came first at IRRI because during the 1960s they also made their own varietal crosses, coded EE for Entomology and P for Pathology. It is not known how pathologists or entomologists produced new plant varieties, but IR 20 was the result of work by entomologists, with very good resistance to tungro virus and performing well in the Philippines. One IRRI scientist said, 'there were divisional jealousies about IR 20, and some breeders tried to stop its release as an IRRI variety'.[38] A compromise was reached in which IR 22 and IR 24 developed by the plant breeders were released simultaneously with IR 20 developed by the entomologists. 'Both IR 22 and IR 24 yielded well for a time but were failures in the long run. This taught the breeders a lesson,' said one entomologist.[39] Hierarchy and rivalry were inherent in IRRI's organization during the period at the end of the 1960s after it evolved away from a small, tightly knit research group controlled by Chandler.

When Brady arrived in 1973 he found three almost separate varietal breeding programmes. He set about to fulfil Hill's instruction to define once again the mission and how it should be achieved. The consensus he created on reorganization is outlined in his 1974 letter to USAID quoted above, that a strong base of research should be the priority: this alone would lead to increases in rice production. Criticism that IRRI neglected transfer of its results or that it was not fully oriented to support production on farms would have to be met, Bradly believed, by extending more research to more cultivating environments. The separate varietal breeding programmes were soon amalgamated in 1974 in the genetic evaluation and utilization programme (GEU), creating the largest co-ordinated research effort at IRRI: thirteen interdisciplinary research groups with five scientists each, guided by the GEU operations committee. The founding groups in the classic cluster preserved their front-line status despite the trauma of amalgamation: in the budgets from 1974 to 1977 they were consistently allocated more than 50 per cent of the total core research budget. These groups were first to present their work to the FAO's Quinquennial Review Team in 1975, and their work is reported before all others in IRRI's annual reports and research highlights. In addition to receiving half the research budget, they continued to perform most of their work on-station and to employ the greatest number of assistants through the 1970s. They continued to dominate the methodological outlook into the 1980s.

The commitment for a single centre for rice research made in 1958 resulted, twenty years later, in one of the largest agricultural research institutions in the world. A picture of IRRI prominently displayed on the walls is an aerial photo capturing the many fine buildings and the rice fields which surround them, beginning as 80 hectares and increased to 334 hectares by 1980. From its initial explosive opening, IRRI grew gradually in budget and size: in 1962 there were eighteen senior scientists; their number slowly grew to thirty-six by 1976. Twenty more were added in the next four years, along with almost 300 support staff. By 1980 there were fifty-eight senior scientists with 1,543 staff (1,230 men, 313 women). Although some of this increase in size was directed to expanding IRRI's work outside the Philippines and outside Los Banos (like the cropping system research), most of it went to maintaining the experimental farm, the genetics collection, the research plots and laboratories, and the training, as well as the social infrastructure necessary to keep this large staff functioning.

During the 1970s there was a rapid increase in total budget, core funding plus outreach:[40]

    1974 $ 5.0 million
    1975 $ 8.5 million
    1976 $ 9.1 million
    1977 $11.2 million
    1978 $13.8 million

In 1980 the Director-General was requesting a budget for the coming year of over $19 million, a 22 per cent increase over the previous year. The great increase in IRRI's size was made possible when the two foundations brought many other donors into the picture through the creation, in 1972, of the CGIAR. In fact both the RF and FF were intimately involved in the manœuvres which led up to the founding of the Consultative Group on International Agricultural Research (CGIAR) under the aegis of the World Bank. Some new donors gave their support to specific programmes and had specific scientists in mind, others to machines and to buildings, and a few to the core budget in general. Despite statements that more effort should go to transfer to other countries, the major share of the funds was spent at Los Banos.

Along with the large size and number of donors came a certain vulnerability to inflation and changes in donor interests and capacities. For example in 1980 the Director-General reported that the 25 per cent inflation rate in the Philippines had created severe difficulties at IRRI. Scientists could not be appointed to vacant positions,[41] and of salaries those for labourers had to be increased 44 per cent and for technical/secretarial staff 33 per cent over 1979.[42] Owing to IRRI being reclassified as a 'business concern', electricity costs rose 90 per cent. In addition, six months into 1980 petrol (gasoline) prices rose 40 per cent. The difficulty filling scientific vacancies was not the only problem created by IRRI's large size and vulnerable budget: despite the increases in salaries, the Director-General also admitted that 'many of the most competent of our young local staff are attracted to other national and international organizations'.[43] Although a number of these difficulties would also affect a smaller research institute, IRRI's great vulnerability lay in its very large size and the concentration of most of its resources in one spot. This budgetary vulnerability had inevitable consequences upon research strategies because it was difficult to insulate scientists from interdepartmental competition for scarce resources, and difficult to persuade other researchers in Asia that increasingly costly methodologies at IRRI were easily replicable elsewhere.

With the growth in size and budgetary competition a key variable affecting a scientist was the structure of relations between the administration, the scientists, and their staff. Essentially the relationship was between foreigner scientists and Filipino staff, because only a few senior scientists and administrators were from the Philippines. In many ways the differences between the two groups were structurally reinforced. For example, most scientists and all Americans in the first decade at least were offered 'foundation contracts' in which they were appointed to one of the two foundations and seconded to work at IRRI. All others were contracted to IRRI itself. This distinction declined in the 1970s when no new foundation contracts were offered. Payments of salaries and benefits to scientists and some other foreigners were made in US dollars at New York, through the Agricultural Institute Suboffice of the International Institute of Education.[44] Scientists transferred sufficient funds to Los Banos for household expenses, and the opportunities for saving, even after tax, were very favourable in comparison with those of the staff, who were paid in pesos at Los Banos. Another example of structural distinctions is the very wide wage differential at IRRI, whereby scientists earned up to $4,000 per month, top technicians up to $600 per month—more than the salaries of the professors on the adjacent campus of the University of Philippines—yet other technicians in the same laboratories at IRRI earned as little as $60. Such a range is not unusual in Asia, and rewards and incentives have been used at IRRI in order to boost productivity and enhance differences within the staff. Salaries at IRRI were known to be as high as any in the Philippines, and throughout this period there were filing cabinets full of applications from people waiting for jobs at IRRI. Nevertheless the structural differences between scientists and those staff who worked on their experiments, differences expressed not only in terms of salaries, the provision of subsidized housing and cars with drivers for scientists, but also of the educational opportunities for their children, retirement benefits, and travel, etc., were having a significant effect.[45]

In addition to these structural differences, there were varying models of team research: in some cases the senior researchers seldom went to the field, and in others they were there regularly. A number of technicians, research assistants, and research students were given the leading role in designing experiments and monitoring the results in order to plan the next step, while others, at IRRI for an equal time, were given little such autonomy. Explanations offered by scientists for the wide variation in the style of operation include culture ('American scientists tend to like to work closer with their teams than most Asian scientists who have another conception of authority'), travel ('some people want to remain in control at Los Banos, others are travelling in other countries so their workers must be independent'), generation ('as men get older they'd like things to get easier'), and achievement ('some of the research teams are well known to be superior to others').

Given the preponderance of American scientists at IRRI during this period, scientists–staff relations were sometimes mirrors of the complexities of Philippines relations with the outside world, and particularly with the US. Short-term visitors and long-term observers all noticed a very wide variety of attitudes in the rural and urban public toward IRRI, ranging from very enthusiastic to very critical. This complex of desire and wariness about IRRI, with its echo in the relation between scientists and their staff, is complicated by two other factors. The first was the close proximity of IRRI to key figures in the government of President Marcos, such as Arturo Tanco, Minister of Agriculture, who took a very prominent role in IRRI's board. IRRI inevitably had to rely upon the good offices of the Marcos government which became increasingly unpopular following the imposition of martial law in 1972. The other is that responsibility for rice research which had previously been done by Philippines institutions was transferred, both implicitly and explicitly, to IRRI. Research capability in the Philippines, therefore, was widely believed to be underdeveloped, because of an established intellectual dependence upon IRRI scientists.

3.4 Methodological Problems at IRRI

Agricultural scientists the world over have been essentially experimental and practical, preferring to practise what can be called a 'plot–lab methodology' on experimental farms, where cultivating conditions are carefully controlled, plants are grown in essentially horticultural conditions, tests and other operations are carried out in nearby laboratories, nurseries, or greenhouses. There is an interdependent relation between the field plot and the laboratory. Arrangements at IRRI did not differ from this pattern in any significant way.

From the scientists' point of view, questions about methodology are at the heart of their vocational and professional discourse, especially debates about the most efficient and most effective means of operationalizing and testing intellectual questions. The classic features of scientific activity in agriculture are careful recording of all phenomena (both expected and unexpected), the thorough use of controls, and the tenacious repetition of short-run phenomena. Debates about methods establish the bounds of the professional community and magnify the stresses within it, perhaps causing those bounds to shift. Scientists at IRRI, in addition to discussion of working priorities, have endlessly debated the following methodological question: which techniques might result in proof of greater increases in yield per hectare?

The two prolonged methodological debates at IRRI were about how yields could be increased and how research should be organized to carry this out. While not everything at IRRI was subordinated to these issues, many other questions of methods were in fact wrapped up in these two big questions. Given IRRI's commitment to maximizing yields per hectare, and the widespread belief in the superiority of biological engineering for developing new varieties over agronomic and cultivation factors, there was room for considerable methodological manœuvring on how yield increases could be achieved and demonstrated and how research should be organized.

The dominant form of experimental development and testing at IRRI was the plot–lab methodology, and most of the resources were therefore concentrated in the experimental farm at Los Banos and the adjacent laboratories and nurseries. But there had long been calls for 'off-station' research in farmers' fields. These calls became stronger with the growth

IRRI's Mission: Technology and Methodology 79

of the cropping systems division, which presented a new paradigm for research at IRRI. The methodological turbulence at IRRI was caused by most scientists' commitment to a single centre for rice research. This commitment was the condition under which all planning of research was done. Some of the inertial effect of prevailing methodologies can be illustrated by the slow shifts of IRRI's organizational practices, as well as the resistance to the rise of cropping systems research. Though there is no intention here to imply there was methodological anarchy, there was an internal debate with each side seeking external allies, none of which is reflected on the surface of IRRI's official publications.

Among those six activities which the Long Range Planning Committee said gave IRRI a comparative advantage over other research centres were mission-oriented basic research and the development and testing of research methodologies for rice research. Although the first annual report in 1962 had stated that 'the overall research program of the Institute is oriented toward both basic and applied research', by 1979 an institutional modesty seemed called for about earlier claims, in order to situate IRRI tactfully somewhere between the scientists of the US and Japan and those of the developing world. This modesty was a response to suggestions abroad that perhaps IRRI was not doing basic research: 'The type of mission-oriented basic research appropriate for IRRI might not be considered basic by western scientific standards.... It does not ... push back frontiers ... it builds a foundation for increased rice farm productivity, but its direct use in the production process is not immediately evident.'[46] Scientists were actually caught between the desire to work on glamorous basic problems for which disciplinary rewards and prizes might be awarded, and the demand to justify experiments whose direct use in rice production was not obvious to, say, a financial officer of the Asian Development Bank. Mission-oriented basic research covered the ground neatly, and the ambiguity about frontiers and foundations in the preceding quotation shows it. But why was such basic research necessary? This is because, in the view of IRRI scientists (in 1979) 'ultimately all the technology that can be developed for a problem with the current state of knowledge will develop', but the current state of knowledge was depleted, or in IRRI's words 'the relative lack of world research on rice means there is much less directly transferable technology and much less basic knowledge available'.[47] Given these two perceptions, there had emerged a crisis of depletion in the link between new technologies and rice research. From this point of view IRRI considered that to develop correct new methodologies was of strategic importance.

This somewhat mechanical view of knowledge, as a box in which there was now little directly transferable technology, contrasts sharply with the view of some IRRI planners in the 1950s who believed that there were already technologies which, if properly transferred, would solve the world food problem. In that earlier view, enough was known to scientists; it only had to be adapted and applied. The bigger problems were social, political, and institutional, beyond the gates of the experimental farm. Such bigger problems, such as land tenure and water supply, which surrounded the cultivation of rice were not things which their scientific methodologies seemed to address, nor were scientists by training (and disposition in many cases) prepared to face them. Those who did face these social problems seldom found them tractable or that there was much institutional reward or sympathy for this kind of work. The biological frontiers seemed both more attractive and more tractable, but for this work new technologies would be required.

There was an additional reason why research methodology was strategically important to IRRI, particularly after 1975 when IRRI chose to cease naming IRRI varieties. Methods themselves were like products which would be developed, tested, and sold. This issue is analysed in Chapter 4 below. IRRI's self-appointed position between the basic researchers in elite institutions in America and Japan, and agricultural researchers of, say, India or China, or Indonesia and Burma, conferred upon its methodologies a special relevance. And because one aspect of IRRI's mandate had been the training of new researchers, scientists believed that they were engaged in the transfer of professional techniques, a gradual upgrading in the transfer of professional techniques, and a gradual improvement of research standards in many rice-growing countries. An earlier draft of the LRPC's report stated that 'there is a tendency for over-worked, under-trained Third World scientists to replicate research methodologies used elsewhere. Thus "model methodologies" became extremely important'.[48] This reference to Third World scientists was omitted from the final report, but interviews at IRRI and in South Asia show that the impact of new methodological models developed at IRRI was in many scientists' minds. A number of South Asian scientists said they believed IRRI could establish methodological standards (such as spacing between test rows) because it had both the skills and resources to do so and that others would be obliged to follow. While there was nothing conspiratorial about this, they asked questions about those people or research centres who chose not to follow IRRI's methods. Would such people and centres be included in various multiple-site trials co-ordinated by IRRI? Would they be eligible for international research funding? There appeared to be a number of reasons for overworked Third World scientists to replicate research methodologies developed at IRRI. Some national scientists in Asia believed in the superiority of IRRI methods because some of them produced excellent results, because training was received at IRRI in their use, because their colleagues or peers elsewhere did so, and because they received recognition for doing so. In IRRI's first twenty years it communicated its methodology throughout the rice research community of Asia through conferences, visits, training, and multiple-site testing. Since such methods were not without their deficiencies and inappropriate or irrelevant assumptions, scientists elsewhere were exposed to a taste of the debates occurring at IRRI about methodology. What positions were held in these debates?

Clearly one school of thought believed that achieving maximum yields on the experimental farm would demonstrate the potential of a variety and thus have good effects at the farm level—confidence in the potential of a variety itself was, they argued, a key part of the drive to increase yields and thus production. Maximizing the potential yield was something useful scientists could achieve at Los Banos. A second school believed that although yield potentials were important, showing what a variety was capable of, there were numerous agronomic factors which contributed to yield increases. Most of these factors could be tested on the experimental farm, including techniques to control crop losses. These two schools agreed that yield was the key problem, to be studied best on the experimental farm and not elsewhere, and took a 'technology first' view of agricultural change. In both schools of thought it appears that knowledge of existing rice cultivation systems was not a prerequisite, and if acquired was done on an avocational and not professional basis. Criticism of the idea of a yield horizon that could be pushed almost to infinity by genetic manipulation and tested on the experimental farm under unique conditions appears to have been seen as threatening. For example, one visiting scientist described his experience in 1978 when he challenged this basic assumption:

The whole premiss here is genetic improvement, but for long periods of history increases in production have been due to agronomic improvement, not genetic. I did a graph plot using IRRI's data from Maximum Yield Trials, and it showed an eventual plateau and decline in yield. Brady was annoyed when I showed it to him. He did not like it at all. But I think we have to question yields and their causes, and most specially in IRRI's own trials.[49]

Another, third, view of methodology was that yields on farmers' fields could best be increased through training, motivation, and demonstrations among people who actually cultivated rice on their own fields. This approach included cultivators' visits to the plots and labs at IRRI, as well as IRRI staff's work on their own farms. In fact many thousands of farmers have visited IRRI. This methodological school advocated maximum co-operation between IRRI and national production programmes like Masagana 99 in the Philippines. Although to increase yields on the research station was interesting, they said, its effect was difficult to measure. IRRI's objective, they argued, should be to show how these practices and yields could be achieved on farmers' fields.

And finally there was a fourth school of thought, centred on but not exclusively attached to the cropping systems work. Its view was that yield increases are to be understood from the vantage point of the cultivators in terms of their ecology and economy within their own cropping system. Overall increases in production or in income may not require higher yields per hectare in every case, but could be achieved through new crop rotations, crop substitutions, altered cultivation practices, and the reduction of losses. The emphasis here is on the cumulative effect of small improvements; to these people maximum yield trials on experimental farms do not really constitute much more than a demonstration that the research plots have unique conditions which cannot be replicated elsewhere.

This unusual character of the experimental farm is found in most other countries. But the gap between conditions at IRRI and on Asian rice farmers' fields is many, many times larger than, say, the gap between conditions of rice researchers' fields at Beaumont, Texas, and those of nearby farmers. The gap was pointed out by Wolf Ladejinsky, the foremost American expert on Asian agriculture, on his first visit to IRRI in 1963. Again, in 1978, a visiting scientist working at IRRI explained the consequences of this gap in the following manner:

We are really developing model methodologies here at IRRI because we cannot do much else. The tough work is on farmers' fields, and this is well-known, but our present methodologies are only useful on the experimental farm, or on other research farms.

The trouble at IRRI is that everything here is a model method. If IRRI sprays pesticide they spray like crazy with firehoses. If IRRI does weeding, they remove everything. One result is you get lots of different bugs in lots of stages and conditions. Nothing here is representative of farmers' fields.[50]

This gap was evident over the entire first twenty years of IRRI's history. Clearly one element of this debate about methodology is the status of the rice farmer in relation to scientists. In one view the farmer was partly hidden, passive, and thought to be ignorant. In another view the farmer was more active, knowledgeable, and progressive.

The other major methodological debate at IRRI was partly a corollary of the first one, particularly in the 1970s, and concerned the predominance of IRRI's experimental farm and associated establishment, and their massive cost. At one pole there were people committed primarily to a centralized research organization employing the plot–lab methodology. They believed such a centre to be absolutely necessary to carry out mission-oriented basic research, to attract the money necessary for big science, and to attract good researchers in the future. At the other pole people were committed to a dispersed organization studying rice in different agronomic and cultural environments. They believed IRRI should be more of a service organization to strengthen national rice research programmes in Asia and elsewhere, and would attract the necessary funds and scientists as a result of the direct benefits which could be demonstrated on the farmers' fields.

During the 1960s, when IRRI was smaller, there appear to have been few challenges to the centralized model. It had, after all, been chosen by the two foundations which footed most of the bill. In the 1970s, with an increase in size and participation of a number of donors through the CGIAR, the contest between the centralized model and the dispersed model took a more definite form among scientists because of the rise of the cropping systems methodology and research programme. Throughout the entire period, however, the establishment at Los Banos in fact dominated the budgets and planning of IRRI scientists. In part the debate generated a view of IRRI that the whole undertaking should be made to pay for itself, perhaps through the sale of rice seed, perhaps in terms of the value of increased rice production. Subsequently this view seemed impractical, and (imperceptibly perhaps) another view of IRRI gained ground as a locus of control in a world-wide research system, a locus which is good in itself, regardless of cost. Not just the experimental farm, the plots, and laboratories, but also the huge genetic collections which have to be continually planted, classified, and sorted, among IRRI's many other functions, have made it a world centre for rice, as intended.

The problem of model methodologies, the organization of the experimental farm, and the relations between staff and international scientists are part of IRRI's central methodological tension—whether to study what occurs on real rice farms and pursue experiments in those unpredictable conditions (known as the weaker model) or to continue to widen the base of knowledge of rice under more controlled research station conditions (known as the stronger model). In the former, weaker model there is dispersed activity in less controlled environments with little visibility except to rice farmers, who have not been IRRI's main clients. Under the latter, stronger model there is scientific visibility and control, and a major factory has been built up at Los Banos which produces technologies and tests methodologies. This difference recalls the original cautions of Weaver and Harrar about 'a single definitive centre' in 1954. A major investment in the single centre obviously had attendant stresses. But this is the model of research in most developed countries, including the United States, from which IRRI scientists were drawn first. Approval and understanding from those quarters were important for IRRI. Furthermore, it was considered easier to develop new technologies, particularly to conduct biological engineering, in a single centre using the controlled plot and lab methods. Note that IRRI and its scientists have been shifting slowly from this preoccupation over the years, just as they have slowly heeded Bradfield's concern expressed in 1956 about 'the hazards of a cereal monoculture'. This central tension and slow shift in interpretation of IRRI's mission can best be seen in the evolution of the Cropping Systems Division at IRRI, a division which stated in 1978 the iconoclastic challenge that it is an illusion that agricultural production can be served by massive centrally planned research.

Cropping Systems as a Problem for IRRI's Methodology

People preoccupied with the establishment at Los Banos seemed to have perceived research on cropping systems as allied with other forces which, in combination, could lead to the dilution of the central collective effort. Research off the station, research on farmers' fields, research with national production programmes, and IRRI's outreach projects in other countries were some of these other forces. Their perception was not entirely mistaken. The earlier preoccupation within the establishment could hardly be sustained in light of the pressures upon IRRI to extend its work to other locations and to other research problems. The inadequacy of building on the best revealed itself slowly. The fear of dilution was rooted in a simplistic notion of how concentrated scientific research would solve the world food problem.

In 1973, shortly after his appointment as director-general, Nyle Brady helped write 'A Proposal for Broadening the Mission of IRRI'. The proposal announced that 'IRRI's research is beginning to reveal several of the reasons why inter-cropping is popular among small farmers who have only limited resources'.[51] It acknowledged the gap between what could be obtained under ideal conditions on the experimental farm and the yields obtained by the farmers and urged that IRRI's research should be expanded to include the study of this gap and to promote study of intercropping at the same time. During this expansion, 'while the physical and biological aspects will receive priority in the initial stages of the program, they will be considered in a framework of probable economic and social feasibility based on farm demonstrations as well as experiment station tests'.[52] This was expansion, not reorientation, said Brady, Athwal, and Hill, trying to avoid confrontation. There were a number of important self-limiting commitments in their proposal, designed in part to allay fears in the CGIAR that IRRI's new Cropping Systems Division would compete with the recently created ICRISAT and other international centres and in IRRI that the new division would get out of hand:

The extent of expansion should not be such as to have the multiple cropping program dominate the institute. IRRI's main reason for being is to do research on rice. We would not want the tail to wag the dog.... IRRI will not broaden its cropping program to approximate the activities of a general purpose experiment station.... IRRI will not set up breeding programs on upland crops grown in rotation with rice, but instead will evaluate and test upland crop varieties produced elsewhere.... Studies of the impingement of institutional arrangements and of off-farm macro-economic forces such as national political and economic policies on crop systems will not be undertaken by IRRI.[53]

In a proposal to USAID one year later, IRRI made a more formal commitment that the Cropping Systems Division 'should not exceed 25 percent of the total IRRI operational budget'.[54] Written by R. R. Harwood, an agronomist, in consultation with scientists De Datta, Gomez, Herdt, and Pathak of IRRI, the proposal acknowledged the leadership of Taiwan in cropping systems research, followed by Thailand and Sri Lanka. Brady's covering letter for this proposal stated that the cropping systems work would begin in the Philippines and be modelled on the Masagana 99 Movement, then be moved to Indonesia (where IRRI already worked supported by USAID funds), and eventually to Bangladesh.[55] Finally, there were other limiting commitments made to USAID: 'it would seem wise to delay the initiation of extensive applied research and extension efforts until a stronger research base has been established. Developmental and fundamental research on rice should not be diluted by overly rapid growth of the cropping systems work.'[56] This limit was established by IRRI scientists fighting for the supremacy of the core cluster of disciplines which drew their strength from the commitment to the experimental farm and associated laboratories at Los Banos. If there had to be a shift in commitment to the problems of the less fortunate rice farmers, then scientists would prefer to apply their traditional methods of conducting most of the tests on the experimental farms. Cropping systems appeared to them as a dilution of their high standards of control, and the methodological debate became a vehicle for discussing the general direction of IRRI's methodology. The debate also provided an opportunity for those who were concerned with the separate issues of work off the Los Banos experimental farm, work on upland and non-irrigated rice, and outreach work with the national rice research programmes in the rest of Asia.[57]

The disagreement on the reasons for the rapid growth of the Cropping Systems Division at IRRI coincides roughly with an insider's view and an outsider's view; these views were presented most forcefully by Randolph Barker, then of IRRI, and David Hopper, then President of the International Development Research Center (IDRC) in Ottawa, but they characterized more general positions.[58] The insider's view is that IRRI scientists knew the difficulties which surrounded the use of the IRRI rice varieties and were beginning to assess the political implications of the concentration on technologies for irrigated areas only. IRRI was thus looking for donors who would finance the cropping systems programme because it was broad-based and promised significant increases in production. The proposal received good response in Canada, and negotiations were at an advanced stage, including a search for Canadian scientists to participate. This idea of assisting Asian countries in food production, thereby reducing their need for food imports, came at a time of very low wheat prices. Thus political representatives of Western Canadian wheat growers caused the delay. IRRI was put in the position of having to argue its merits all over again and to persuade Canadian officials to press the proposal.

The outsider's view is that IRRI officials were reluctant to allocate existing funds to the multiple cropping programme because it was unconventional and presented a paradigm for research which was more complex than the one rooted in the experimental farm and the laboratories. Only when the CGIAR took up the proposal seriously, and agencies like IDRC pushed it and were prepared to pay for major expansion, including the cost of the network of researchers in various countries on which it was based, were IRRI scientists and officials prepared to make room for the new group. The insiders state that the outsider's view 'typically exaggerates' the innovative role of the CGIAR investors and negates the role of the front-line scientists. The outsiders state that the insider's view does not account for the rigidity and self-interest of the dominant research groups. Both viewpoints concede, however, that the Cropping Systems Division's expansion was a very significant departure in IRRI's methodology and that it was acceptable only because it held the promise of big increases in production.

The rest of the story has two parts: the first shows how the unconventional views of one scientist in the 1960s, Richard Bradfield, illuminate the prevailing thought of researchers at IRRI and the conditions under which a new methodology was introduced there; the second part shows the steps taken by others to contain the implications of this departure from the prevailing thinking.

In his first report from Asia to the RF in 1956, Bradfield wrote:

New systems of cropping need to be worked out which are suited to the local environments, the need of the farmers, the demands and rewards of the market, etc.... it does not seem advisable to confine the interest of the RF to rice. A cereal monoculture has always proven hazardous to both farmer and farm. A diversified system of farming built around rice as the monsoon crop seems to be indicated. To develop suitable systems for the different areas will require a great deal of research and education.[59]

By 1964, Bradfield was working regularly at IRRI and offered 'Some Unconventional Views about Rice Culture in S.E. Asia' at an IRRI seminar, in which he compared rice production in the Philippines and California. In his opinion the mode of rice production throughout South-East Asia had to be changed. He asserted that all operations involved in growing all crops will be minimized and mechanized where possible. This can be done under the conditions prevailing in the larger extensive rice growing areas of the US and Australia. To do it on the small intensive farms (1–3 to 4–6 hectares) of S.E. Asia with their monsoon climates will require much further research and modifications in age-old practices. But it is an indispensable step in attaining our major objectives.[60]

Bradfield then argued that small farmers must save time from rice cultivation for other crops and that IRRI should help them save time because IRRI should not see their labour as cheap and inexhaustible. 'Don't be indifferent,' he told IRRI scientists, 'don't perpetuate the entrepreneur or landlord who is able to exploit a large force of underpaid labourers.'[61] Three years later, Bradfield again argued that small farmers should save effort and time and, since the land is costly, should not allow an inch to lie fallow. This is the ethic of continuous cultivation which the IRRI Rice Garden came to represent. In addition, he argued for judging productivity by taking into account the total harvest from all crops, not the yield of any one crop, and even for reducing the yield of one to gain in the yield of another. Just when IRRI's public relations fortunes were soaring with the success of IR 8, Harrar, president of the RF, recalled telling Bradfield, 'we have to do one thing at a time'. This enthusiasm of his seems to be a reason why Bradfield's proposals encountered resistance.[62]

However, Bradfield regularly argued for fuller adoption of the cropping systems idea, using his position to speak to as many interests as possible. It appears he made his strongest requests about 1970. He certainly rose at the meeting called in 1970, attended by a large number of upland cropping specialists to plan ICRISAT, and made a strong plea for study of rice-based upland cropping systems. This appeal was followed immediately by a letter to the next chairman of IRRI's Board of Trustees, Clarence Gray, then of the Rockefeller Foundation, in which Bradfield proposed that a new international institute (then to be called the Institute for Upland Crops) should be located at IRRI. Bradfield's own research work was all on the experimental farm at Los Banos. Bradfield's letter to Gray stressed that if the new institute, which eventually became ICRISAT in India, was not located at IRRI, 'independent systems for rice and upland crops will be developed instead of integrated systems, and [there will be] unnecessary and undesirable duplication of efforts at both institutes'.[63] Bradfield reiterated the judgement that 'many students of the problem feel that the land now planted to upland rice should be planted to other crops such as corn or sorghum .... The food production potential of this huge area will not be attained until the present system is drastically modified.' Bradfield faced the question of resistance to this proposal within IRRI. He was open about the reservations ... that some have regarding the influence that this merger would have upon IRRI. Will it 'dilute and compromise' the work on rice? This question is important. No one, I am sure, wants to take any step which would impair the work of IRRI on rice in any way. In my opinion, it would probably have just the opposite effect. This will be influenced by the course which IRRI decides to take in the next decade.[64]

Bradfield seems to have already accepted the judgement that the idea was unconventional. Though he had important questions for Los Banos, he had not been able to build a sizeable research group to study multiple cropping. At the time, the small Multiple Cropping Division consisted of Bradfield, one Canadian visiting associate agricultural economist who arrived in 1970, one assistant agronomist, and their research assistants. A response from the RF to an earlier proposal noted the lack of personnel; Cummings wrote to Chandler in 1968 that 'the initial outline prepared by Dr. Bradfield had not provided sufficiently for the understudies ... who might be given the opportunity to work with Dr. Bradfield and who might find the talent for carrying such studies forward into the future after Dr. Bradfield is ready to hand them over to others'.[65]

After 1970, however, the momentum for the expansion of the multiple cropping programme grew as pressures within IRRI to respond to increasing dissatisfaction with the limitations of the Green Revolution. Bradfield, before he retired, argued at the 1970 IRRI Research Program Review that the number of hectares of irrigated rice in Asia was only half that of the non-irrigated rain-fed rice areas. While he still expected that irrigation would increase in scale, there were many areas in which new irrigation would be costly or impossible to provide. Bradfield now asked:

Can we afford to pay so little attention to the problems of 2/3 of the less fortunate, more impoverished rice farmers of south Asia who farm 100m. hectares of the total 150m. hectares of rice grown in the area? It seems to me this vast opportunity might pay larger dividends to more people than adding a few more quintals per hectare to the production of the irrigated areas than can be obtained by the use of methods already developed.[66]

These criticisms of IRRI's original mission were supplemented by the conviction of one founder-member of CGIAR in 1971, the IDRC of Canada, that IRRI's single-crop orientation and plot–lab control methodology aimed at irrigated rice was completely inadequate for its mission.

The best point from which to observe the implications of the new Cropping Systems Division begins at the end of 1975. IRRI's board then agreed to stop the naming of varieties, and the Technical Advisory Committee of the CGIAR arrived to evaluate IRRI's progress. When the TAC Quinquennial Review Team reported on the Cropping Systems

Division in early 1976, the division had already reached IRRI's imposed limit of 25 per cent of the total budget, according to TAC.[67] The division had described itself to TAC as being in 'the observational–descriptive phase', but TAC's Review Team cautioned that its 'ambitious coverage may spread the efforts and skills of the seven senior staff too thin'. If most of its work was site- and location-specific, asked TAC, how can it be 'generalized to the development of principles'? 'Their role is not merely to analyze traditional practice, but to challenge, expand and change it ... the work of an international institute must be capable of extrapolation and impact beyond local test sites. This is the standard against which the cropping systems program should be assessed over the next five years.'[68] These difficulties were not overlooked by IRRI scientists wary of the rapid growth of this division. The Review Team said that 'given the evangelical fervour of the cropping systems concept and network', there was some danger that it could become isolated from other research scientists at IRRI. Other scientists saw that the systems concept and the network to promote it were indeed well received in Asia and that its budget had increased to meet new costs. Its enthusiasm was disliked at IRRI and it appears that TAC's team did not coin the term 'evangelical' for it; some other IRRI scientists did. In the view of many other scientists, to whom the concept of agricultural ecosystems was alien, the budgetary limitations on the division were greatly appreciated. After all, ecosystems had formed no part of the training of agricultural scientists in US land-grant universities during the 1950s and 1960s, except perhaps in the occasional instance of agronomy. It was apparently difficult for some IRRI scientists and some TAC team members to reconcile the systematic description of cropping methods in many different locations with the development of generalizable principles, though it is clearly a general principle that research must address the specificity of diverse cultivating environments and the probability that a new balance of crops might be more productive.[69] This question of a new balance of crops is the question which the cropping systems division posed, after a careful study of the existing balance.

The Cropping System Division's methodology was critical not only of biological scientists' routine practices, but also of economists studying the constraints upon the new rice varieties and upon yields. One scientist spoke of a continuous clash between the two groups because cropping systems research builds the study of constraints into the design of the work whereas the economists studied constraints after the fact. An illustration of the clash lies in Harwood's explanation in the letter to foundation officials of the difference between them:

The constraints group is working solely within established cropping patterns, identifying the reasons why yields are not higher. Our program is concerned entirely with the identification of areas where productivity can be increased by modifying the pattern.

... In other countries for the first time this year our two programs have often competed for manpower resources, however, and problems are arising. There will have to be closer coordination of effort from Los Banos.[70]

Economists like Ruttan were questioning whether the new methodology could ever allow scientists and farmers to make realistic optimum cropping predictions, given the larger number of variables involved. Harwood assured them it would eventually be possible.

Criticism by the TAC Review Team did not deter the Cropping Systems Division from stressing the principle of diversity and specificity. At the 1978 Annual Research Program Review, the presentation by the division's team leader, Zandstra, stated that when technologies do not account for and fit this diversity developing them wastes resources and increases the risk of negative effects. He said that the division's programme thus had to have its own scientists at many research sites operated by national research programmes. The belief in the value of the extensive research network of the division had increased, he said, and 'IRRI's present level of support to the cropping systems network is not adequate.'[71] Zandstra later reported that the reaction of other scientists had been defensive because many of them had a rather narrow focus.[72] He had been warned by colleagues that the language of a draft of the division's report should be toned down and had done so, but still found resistance to its goals. 'Cropping Systems represents a new direction at IRRI,' Zandstra said, 'but I'm not sure we've turned the corner yet.' The key point is that the division's position in the Program Review had specifically criticized the controlled methodology:

> The illusion that agricultural production can be served by massive centrally planned research does a disservice to the farmer.... The success of IRRI's cropping systems program depends on the acceptability of the research methodology it generates and on the success with which this methodology is employed in the national programs. The notion that the cropping systems methodology can entirely be generated at IRRI is false.[73]

Following this presentation, Zandstra reported being told privately by IRRI's director-general that if he continued to talk like that he could 'lose his audience at IRRI'.[74]

Cropping systems is an old problem at IRRI which others previously had been able to ignore. Building on the unconventional views of a senior scientist, Bradfield, a group of young North American agronomists trained to think of the interaction of ecological and social systems—a new kind of agronomist—was thrust into the middle of older discipline-oriented scientists, both Americans and Asians, who had built their careers around IRRI, its laboratories, and its experimental farm. Some of these older scientists perceived the cropping systems' advocates as evangelical; their own work was committed to rice pests, rice soils, or to similar specific factors and variables which they separated from cultivating systems in order to study them under ideal conditions. Despite limitations placed on its growth and internal opposition, the Cropping Systems Division soon reached its budget ceiling and argued that its budget was insufficient because the real task lay beyond Los Banos with national research programmes and with farmers in their fields.

Yet the internal 1979 Long Range Planning Committee report reflected the continuing resistance to the division's growth, for it recommended that cropping systems should now seek a more complete integration of its research with other divisions to reduce the danger of its isolation from other divisions. Also the division was told it had to 'increase its role as problem-specifier for programs generating technology for rice [genetics as well as crop management-related]'.[75] The committee repeated the importance of the division's work on IRRI-managed sites.

Even within the division inadequate and 'unscientific' information were causing problems. When assessing the whole decade of the Cropping System Division's work, in which he was directly involved, Harwood concluded that:

We are presently going about agricultural development with an extremely limited range of technology options as compared to those which we could be employing. We attempt changes without really understanding what the impact of those changes will be. We are puzzled by the widespread lack of acceptance of our development efforts and by our occasional drastic failures. Obviously we cannot cover the great spectrum of diversity and complexity of agriculture with the approaches we have used up to now. We are beginning to head in the right direction, but the change has been painful and has come extremely slowly. I for one am not pleased with our rate of progress, and especially with the sluggishness of the scientific community in coming to grips with the problems of agricultural structure.[76]

This resistance or sluggishness was a direct expression of a methodological commitment. On the one hand, other scientists stated that conditions in farmers' fields were uncontrolled and results there were thus of little scientific value. On the other, there was an implicit belief, as one long-time observer of and former scientist at IRRI said, in the existence of a universal peasant society and a single type of traditional agriculture. 'This means that IRRI doesn't care about variation. To the extent they take evidence at all, they see the Philippines as the best evidence for their belief in this universal peasant culture. They use labels like "tradition" and "custom" but variation within these gross categories plays no role in their dominant methodology.'[77] Perhaps this belief in a universal peasant cultivator, along with the resistance to the complexity of farmers' fields, were simply aspects of the desire to work only on manipulable variables in a controlled plot and lab methodology.

The Brown Plant Hopper as a Problem for IRRI's Methodology

The extraordinary proliferation of the brown plant hopper (BPH) as a major rice pest in different countries in Asia in the 1970s presented IRRI with a special opportunity and a special dilemma. There is some evidence to suggest that as an institution IRRI misunderstood the BPH phenomenon, and it is useful to explore the possible explanations for this misunderstanding—that IRRI's disciplinary and departmental structure restricted the responsibility to interpret the phenomenon to a small number of scientists, that the lack of an ecosystem concept in the main disciplines rendered the BPH phenomenon unintelligible, and that major political and economic forces surrounding IRRI deflected researchers' attention away from a deeper interpretation of evidence which IRRI already had about the BPH problem.[78]

The first evidence for this misunderstanding is contained in a chart published in the 1971 *Annual Report*, showing that where diazanon is heavily sprayed the result is very large number of BPHs. In this case 2 kg. were sprayed on rice, 79 and 118 days after transplanting, and the result was up to 1,000 insects per hill (vs. 4 per hill in control plots). The heavy dosage was applied under IRRI's policy of maximum protection for high yield plots so as to give the genes for yield the opportunity for full expression. (This could be called the doctrine of maximum protection.) Although the spraying in fields was done to suppress the stem borer, it was already well known that heavy spraying of diazanon caused 'hopper burn', that is it stimulated the infestation of BPH.

The complexity of the BPH attacks on rice was being misunderstood because what was actually being suppressed were the BPH's natural enemies—many families of insects, the most important being spiders. In recent years IRRI scientists have been studying spiders in the rice field at Los Banos: 51 spider species were found in 1984, 120 species in 1986, and 263 species in 1989. It seems that research on spiders was conducted from 1962 in the Philippines (by entomologists), and IRRI was aware of it. Not until 1977, however, was this physically arduous kind of research begun at IRRI and it focused on insect predators as natural enemies of the BPH. By then major commitments had been made to suppress the BPH chemically and to breed BPH-type resistance into new varieties of rice—both of which commitments appear (retrospectively) to have been unnecessary. The discovery of the rich and complex relationship between insect natural enemies and rice pests was beginning to unfold at IRRI only in the 1980s.

Why was such a significant risk to HYV rice not understood at IRRI?

Why was the relationship of BPH and diazanon spraying contained in the 1971 *Annual Report* missed? Why were the first outbreaks of BPH in 1972 not studied by ecologists from IRRI? Why was it not seen that when the agro-ecological situation is disturbed the result is wild fluctuations in the pest–prey relationship? These interesting questions are difficult to answer decisively, but they are important questions nevertheless.

Because of the disciplinary specialization at IRRI, only one department and one senior scientist had the responsibility to "see" the relationship between BPH and diazanon and spiders. At the time pests were defined as a nuisance which could be controlled through heavy spraying. The first objective, after all, was increased yield per hectare, and preoccupation with high yield obscured the possible insight that there was a completely different path open to them, a path in which natural predators would play the crucial role.

In 1973 and 1974 the BPH infestations spread rapidly and IRRI's need to understand this spread was increasing every month. By 1975 IRRI was beginning to insert BPH-resistance into its breeding programme, thus severely limiting the utility of its previously developed varieties. During these very years IRRI was being presented with the expectation that it should be studying its work in the context of ecosystems, but it yielded very little to those expectations. Ecosystem thinking did not prevail among the entomologists, the only ones who were "assigned" at IRRI to think about pests. Almost everyone else was thinking about yields.

One consequence of this approach was that IRRI pursued the concept of a BPH biotype: it developed the concept of a series of BPH biotypes (number 2, number 3, etc.) which did not correspond to BPH in the field. These "reference varieties" were bred artificially so as to test resistance to each one in the yield plots. The ultimate futility of this approach was shown in Indonesia, where even IRRI's very successful IR 36 variety was eventually abandoned.

Finally, it should be remembered that the simplistic attachment to spraying more and more diazanon was part of a very large and expensive programme of action in the countryside surrounding Los Banos: the government programme Masagana 99 provided a lot of money for spraying and suppressing the BPH along with other rice pests. The role played by IRRI in Masagana 99 was to provide the logic and rationale for the action programme. Given the multi-million dollar investment in Masagana, feedback to IRRI from Masagana did not encourage IRRI to look outside its own accepted paradigm.

This may be a case revealing a difference between technology as solution and science. Studying the pest–predator relation in this case would have been a good application of science, in which science would then understand technology (pesticide) as the problem, not the solution.

3.5 IRRI's Early Efforts to Gauge the Constraints and Consequences of New Rice Technology for Large vs. Small Farmers

In this section we examine one aspect of IRRI's methodological problems, namely the attempt to evaluate the social impact of HYV rice, including the possible *differential* constraints and consequences of the new rice technology on small and large farmers in Asia. Versions of these issues were addressed year after year at IRRI; they were published in most *Annual Reports* between 1970 and 1976 and show that those treatments of the question are less than successful. Moreover these treatments cause us to ask whether IRRI was making a significant contribution to understanding the relationships between farm size and HYV rice. We have no conclusive answer to this question but the evidence which follows suggests a hypothesis about IRRI's institutional stance toward the wider social and economic effects of the HYVs. Our hypothesis is that IRRI was unable or unwilling to make a serious study of the effects of HYVs on farm size and intentionally misled itself (and perhaps others) about the extent of its knowledge of this subject, roughly between 1970 and 1976.[79]

It is not easy to make a clear conceptual distinction between issues which relate to the *constraints* on the adoption of modern technology which face small and large farmers, on the one hand, and the *consequences* of the adoption and spread of that technology for farmers with different size holdings, on the other hand. For example, a lack of adequate credit for small farmers to finance required inputs may be a constraint on their adopting modern technology, but the unavailability of funds may at the same time be a consequence of the increased overall demand for agricultural credit precipitated by new cultivation methods. Thus, in keeping with the *Reports* themselves, we made no effort in this section to keep discussions of constraints and consequences separate.

Written by the Agricultural Economics Department at IRRI, these sections of the *Annual Reports* (*ARs*) show that people at IRRI were expected to study the widening implications of introducing technological change in cultivating environments. This became more pressing in 1969–70 when the possible impact of the Green Revolution (including surplus production in some countries) was taken up by IRRI, including the way to lower costs and increase earnings of small farms. But these expectations were only just finding tangible expression in the 1970s.

In the 1970 *Annual Report*, IRRI addressed the farm size issue more or less directly. In connection with a study conducted in 1969 on changes in rice technology in Java, the Agricultural Economics Department said: 'It is important to establish whether farm size itself has an impact on input use and on farm income per hectare. That is, does the smaller farm have a relative disadvantage for obtaining modern inputs and increasing yield per hectare?' (p. 186). Notice that although the second sentence in the quotation is purported to pose, in the form of a detailed question, the issues raised in the first sentence, actually the question is far narrower than the original statement. For the question focuses on the ability of small farmers to *increase yields*, while the original issue was the relation of farm size to farm *income per hectare*. If the only problem here were infelicitious wording, then the difficulty would be innocuous. However, we believe that much more is at stake. This is but one small example where a broad question, such as income per hectare, was 'reduced' to some related but narrower question, such as yield per hectare. Our criticism is not that some 'factor' of the original problem was chosen as the subject of a study; our criticism is that IRRI too often failed to keep issues clear and proceeded as though there are no differences between narrower and wider issues.

For purposes of the study, the farms were grouped into three size categories: small (less than 0.35 ha), medium (0.35 to 0.7 ha), and large (greater than 0.7 ha). Table 3.1 gives the results of the study for the two villages involved.

The entire text relating to the interpretation of Table 3.1 reads:

For Porong (Table 10), the returns above variable costs per hectare of the small farms are significantly different from those of the medium and large farms (t = 2.45 and 2.79, respectively). The difference appears to be due to differences in expenses for hired labor and in yields. Small farms do not seem to be at a disadvantage in procuring cash inputs other than labor, however. In fact, for all categories except insecticides, small farms had larger inputs per hectare than large farms.

For Prembun, the returns above variable costs per hectare for small farms are significantly different from those for medium farms (t = 2.15) apparently because of yields which are also significantly different. Here again, however, small farms seem to have no disadvantage in procuring inputs. For all cash inputs, use per hectare is higher on small farms than on medium or large ones. (1970 *AR*, p. 186)

A number of questions come readily to mind regarding the usefulness of this study. For example:

1. If the purpose of the study is to determine the effect of farm size on income per hectare, why were the costs of machinery, animal labour, credit charges, and so on, not included? These costs are likely to vary among farms and may well be significantly related to farm size.

2. It is worth noting that relatively few farmers surveyed were actually planting new varieties (only 11 of 60 in Prembun and 2 of 60 in Porong in 1968). Why is no information given on whether small or large farmers tended to, or were able to, adopt modern varieties, methods, and machines? It seems to us peculiar to study the problem of the

TABLE 3.1. *Relationship of farm size to labour, capital, and return above variable costs. Porong East Java, and Prembun, Central Java, 1968 wet season*

Size of farm*	Mean size (ha)	Farms (no.)	Grain yield (t/ha)	Capital inputs (Rs/ha)			Total	Hired Labour (man-days)	Return** above variable costs	
				Seeds	Fertilizer	Insecticides			Rs/farm	Rs/ha
PORONG										
Small	0.2	35	2.9	1,810	2,550	140	4,500	210	8,364	41,820
Medium	0.45	16	3.2	1,430	3,590	350	6,360	140	23,580	52,400
Large	1.16	9	3.5	1,120	2,430	160	3,720	120	65,076	56,100
PREMBUN										
Small	0.17	25	1.4	1,000	2,580	200	3,780	100	2,368	13,930
Medium	0.49	21	1.6	990	2,290	80	3,360	132	9,369	19,120
Large	1.37	14	1.4	960	1,540	10	2,510	170	15,974	11,660

Notes: * Small: < 0.35; Medium: 0.35–0.70 ha; Large: > 0.70 ha.
** Price of rough rice is Rs 18/kg. US $1 = Rs 378.
Source: After table 10, IRRI 1970 *Annual Report*, p. 186.

availability of modern inputs in an area where only 13 of 120 farms (about 11 per cent) were using modern varieties.

3. Why were yields for small farmers lower and hired labour costs higher? Of the factors considered these two seem to account for most of the difference, but no effort is made to understand their nature.

All in all, this study seems to be of little help. It certainly did not begin to address the general question of the significance of farm size *vis à vis* the new rice technology. Nor did it support IRRI's apparent view, noted below, that increased efforts to develop small farm machinery was an appropriate strategy.

It is interesting to note that in the same 1970 *Report*, a study comparing four farm types in Albay, Bicol (Philippines) is discussed. In this study, 'return above variable costs' was also used but here expenses included not only fertilizer, insecticides, and hired labour costs but also the value of shares for harvesting and threshing. No explanation was offered as to why different factors were included in the two studies. (Were shares of the crop not used as payment in Java?) Also interesting is the following excerpt from the Philippines study: 'For two types of farm (lowland rice and lowland rice in rotation), functions were also fit [sic] according to farm size. Farms with one hectare or less were classified as small farms, while those with more than one hectare were classified as large farms' (1970 *AR*, p. 184).

Yet the tables given make no use of the small/large subdivision and no further mention of large and small farms is made in the text. Thus, the large/small distinction was introduced and then completely ignored. (It is noted earlier that lowland rice farms in Bicol which rotated rice with other crops tended to be smaller than farms of other types, but probably because of better management these smaller farms were able to obtain higher farm incomes.) Thus neither of the 1970 studies which made references to farm size have shed much light on the general problems.

In the following year, 1971, the Agricultural Economics Department's chapter in the *Report* contained the result of a survey started in 1970 of 513 farmers in Gapan, Nueva Ecija, Philippines. Part of the analysis of that survey appeared in a subsection entitled 'Farm Size and Tenure' and which begins this way: 'In many parts of Asia, farm size has an important influence on the rate of adoption of the new technology. It has been observed that large farms with a stronger financial base are the first to adopt, and some writers argue that large farms are gaining an economic advantage over small farms' (1971 *AR*, p. 95). The conclusion of the study regarding the effects of farm size was this:

Over the years there has been a consistent difference in yield according to farm size in Gapan. Small irrigated farms have had higher yields than large ones. This is true even though there is no difference in rate of adoption of high yielding varieties or rate of nitrogen input among farm sizes. One hypothesis which we have not yet fully tested is that the small farms tend to be located on the more fertile soils. The only other factor besides yield that seems to be influenced by the size of the farm is the rate of mechanization. (1971 AR, pp. 95–6)

(We have not been able to locate anywhere in subsequent *Reports* where the hypothesis about the fertility of soil on large vs. small farms is tested.) In spite of the apparent correlation between small farm size and high yield, the study compared the five highest and five lowest yielding farms and found that: 'Surprisingly, there was no significant difference in nitrogen input and farm size between the top and bottom farm group ...' (1971 AR, p. 97).

Now the first thing which must be noted here is that a comparison of yields per hectare on large and small farms is not enough to address the question raised at the beginning of the section about whether or not large farms are gaining an economic advantage. The important thing, certainly, is not yield but returns over costs. As was clear in the 1970 study we discussed above, the cost of labour and the value of shares paid in return for labour can be crucial and these factors were left out of account in the study. Furthermore, there is the issue of mechanization, which is one way in which a large farm can 'outperform' a small farm economically speaking (as North American agricultural development shows), even where the large farm has lower yield per hectare.[80]

This issue is actually discussed briefly in the 1970 *Report*, which notes that even though mechanization does not provide as much economic advantage to rice culture as it does to other crops (notably sugar-cane), size of farm does influence the rate of tractor adoption. Unfortunately, however, no data are presented which help to assess the magnitude of the advantage large farms attain by mechanization.

Finally, in 1971, a section of the Agricultural Economics chapter was devoted to a discussion of 'Government programs and mechanization'. There it is noted that the

Central Bank of the Philippines has negotiated two loans with the International Bank for Reconstruction and Development.... While higher incomes generated by the use of high yielding varieties have probably permitted more small farmers to purchase hand tractors or rent tractor services, it seems clear that the large increase in tractor purchases after 1965 could not have occurred without the two Central Bank loan programs for mechanization. (1971 AR, pp. 87–8)

But what is one to conclude from this? Was it also true that higher incomes generated by the use of high yielding varieties probably permitted more large farmers to purchase four-wheel tractors and that with or without the loans a trend toward mechanization would have occurred? What were the relative advantages to small and large farms of mechanization?

In spite of the fact that IRRI seems to have shown that small farmers

IRRI's Mission: Technology and Methodology

were not at any general or serious disadvantage in HYV cultivation, the 1972 *Report* indicated a continued interest in the development of small, low-powered machines for small farmers. In 1972 the Agricultural Economics Department took another look at the two loan programmes discussed in the previous year (see above), this time with reference to the impact of the programmes on patterns of mechanization in the Philippines. It was noted that these loans went mostly to large farmers, but again no attempt is made to analyse the impact on the economic viability of large vs. small farms (1972 *AR*, p. 83).

The 1972 *Report* also describes a study conducted by the Agricultural Economics Department in three villages in Thailand in 1971. Unfortunately the discussion of farm size is highly confusing and very brief. In full it reads:

To test the impact of farm size on yield, we examined two relationships for each village: (1) yield of modern varieties in relation to the proportion of the farm area planted to modern varieties, and (2) yield of modern varieties in relation to the actual area planted to modern varieties.

Only in village A was the relationship between yield and area planted to modern varieties statistically significant. Yield varied widely among farms, however. In villages A and B, yields over 4 t/ha were achieved only on farms that planted less than 40 per cent of their area to modern varieties and only on farms planting less than 2.5 hectares to modern varieties. In village C, yields in excess of 3 t/ha were achieved only on farms planting less than 10 per cent of their area to modern varieties. Thus, there appears to be an important relationship between area planted and yield. (1972 *AR*, pp. 53-4)

This passage is difficult to follow. In the first paragraph a distinction was made between the percentage of area planted to modern varieties and the absolute area planted to modern varieties. However, the second paragraph begins and ends with reference to 'area planted', which is ambiguous with respect to the distinction between percentage and absolute area planted. Thus the first and last sentences of the second paragraph seem contradictory and it is not clear which of the other bits of information in the paragraph support these two claims. This study addressed only part of the general issue of the relationships between HYVs and farm size. The issue about farm size and yields was only part of the question since ultimately the concern was with the economic viability of small-scale farming in the light of improved rice varieties and technology. The study in Thailand might seem to suggest that small farmers were somehow gaining an advantage over large farmers. However, that conclusion is not warranted. For one thing, no data are presented which enable a comparison of the effects of the adoption of new varieties on large vs. small farms. Secondly, as we have said, the real question concerns not simply yields but returns above costs.

The 1973 *Report* for the first time contained a section entitled 'Research Highlights', and in it we find the following paragraph:

> The scarcity and high price of fertilizers and other agricultural chemicals add to the mounting problems of all farmers, particularly those with low incomes. With credit uncertain and costly, the small farmer cannot compete for the scarce supplies of chemicals and other inputs needed to capitalize on the productive potential of new rice varieties. (1973 AR, p. xiii)

No reference is made here to any research conducted by IRRI or by other investigators, and no mention is made of the differential effects of the scarcity of inputs or of credit or other problems of the small farmer in the Agricultural Economics chapter. There is no entry in the index of the 1973 *Report* to suggest that work was done on farm size that year.[81] It does seem reasonable to suppose that small farmers would be the ones who were hit the hardest by increases in production costs, credit shortages, etc. Yet, on the whole, IRRI reported very little systematic research in these areas. In spite of the virtual absence of any relevant research on the importance of farm size (apparently, anyway), the 'Research Highlights' section claims that 'the problems of the small rice farmer continue to receive primary attention of the IRRI scientists' (1973 AR, p. xiii).

In connection with a study of 'farmers' reporting bias', conducted in Gapan, Philippines in 1972–3, it was discovered that the farmers studied tended to underestimate their yields, on the average, by some 45 per cent! On small farms, farmers underestimated their total production and overestimated the area of their farms, while on large farms, farmers underestimated both total production and area. There was no significant difference between the underestimation of yield per hectare for large and small farmers, however, since large farmers underestimated total production to a greater extent than small farmers. What is worrisome in this is that 'in most studies yield estimates are obtained by asking farmers their total rice production and their rice area' (1973 AR, p. 196).[82]

In the 1974 *Annual Report* we see the Agricultural Economics Department's results of a study undertaken in 1971 on the changes in Asian rice farming related to the new technology. The study involved over 30 investigators from 6 Asian countries, gathering information from 36 rice-growing villages in 14 study areas. Among other things the study attempted to assess the relative benefits from improved technology to large and small farmers. The study noted:

> There is a close association between absolute farm size and the adoption of labor-saving technology.... Rotary and hand weeding are more commonly practised on small farms, while herbicides were used only on sample farms larger than 1 ha. Tractors and threshers are more common on the larger farms.

By contrast, yield increasing technology—fertilizers, insecticides, and modern varieties—were [sic] widely used on all farms, irrespective of size. (1974 *AR*, p. 272)

Unfortunately, nothing was said about the importance of the difference in the use of labour-saving technology, even in economic terms. Did the larger farms gain a significant economic advantage this way or not?

Besides absolute farm size the study examined relative farm size.[83] The agricultural economists 'hypothesize that *relative* farm size within villages would be a significant factor in determining access to, and benefits from, the new rice technology only where there was a comparatively inequitable distribution in size of farm operating units' (1974 *AR*, p. 272). To measure the degree of equity in size of holdings, a Gini coefficient was calculated for each village. The coefficient approaches 1 for villages where the bulk of land is operated by a few large units and the rest by many small units. The coefficient approaches 0 for villages where all holdings are of nearly equal size. It was observed that 'there appeared to be no relationship between the degree of equity in the size of distribution of operating units and the average size of farms in a village' (1974 *AR*, p. 273).

It is difficult to present a summary of the information given in the rest of the subsection called 'Farm size and the new technology'. The problem is not that there is too much information—quite the contrary. What there is is very sparse and confusing. The text of the section takes less than one full page, and one promising paragraph, which seems to relate to the hypothesis quoted above, reads as follows:

The impact of relative farm size was compared among villages by classifying farms as large or small based on the median of farm size within each village. Villages were grouped according to the level of the Gini Coefficient. Within these village groups, farm samples were pooled making a series of two-way tables between farm size and other factors related to adoption, input use and benefits from the new technology. (1974 *AR*, p. 273)

When one looks for these two-way tables one finds instead of a grouping of villages by the level of Gini coefficient, rather different groupings. For a study of adoption rates for new technology, villages were divided into four groups, some groupings based on geography and others on Gini coefficient (see Figure 3.1). The results of this study of rates of adoption are stated as follows:

The relationship between farm size and adoption is most striking in Pedapulleru where the new varieties were planted on more than 90 percent of the large farms, but less than half of the small farms. Small farmers in Group II lagged behind the large farmers at the start, but caught up rapidly in the fourth and fifth year. Clearly, the size of the Gini Coefficient is not the only factor that influences the degree of lag in small farms. (1974 *AR*, p. 274)

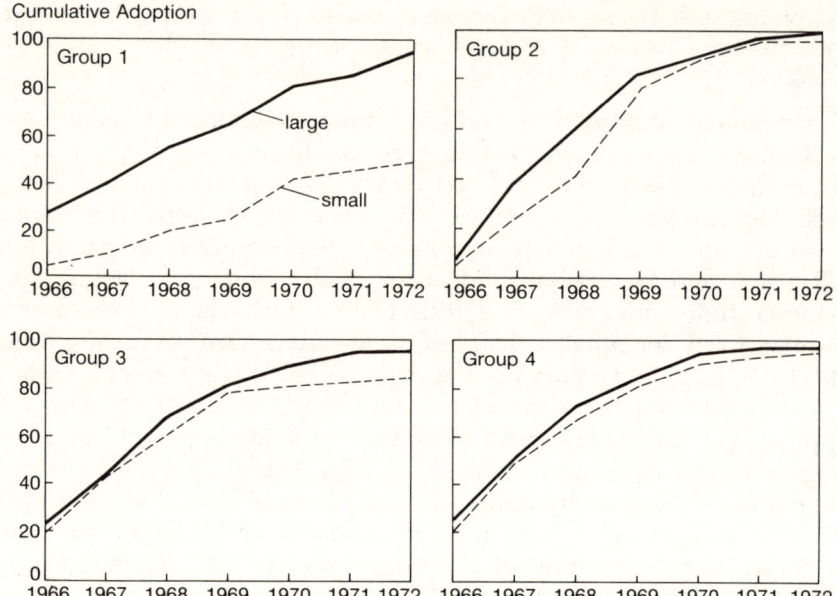

FIG 3.1 *Factors in adoption of modern rice varieties. Gini coefficients, farm sizes, and years of adoption of modern varieties.*

Source: After table 6, IRRI 1974 *Annual Report*, p. 273.

No attempt was made in the text to explore what other factors influence the degree of lag and nothing was said which enables one to see what significance (economically or otherwise) this time lag had. Did early adopters have an advantage or disadvantage or neither? Just what was the impact of farm size on rate of adoption? Again we find IRRI's tendency to leave virtually unanswered some important questions directly related to the purposes of a study.

The next topic dealt with in this section of the 1974 *Report* was whether small or large farmers were affected differentially as regards profit from rice production or level of living. Nothing is stated, but one must assume that the same grouping was followed in this part of the study as in the previous one, even though there are only 2 of 10 villages included in Group II, 5 of 6 in Group III, and 7 of 10 in Group IV.[84] The results of the study are given in IRRI's table 10 reproduced as our Table 3.2. The textual analysis reads:

Most of those who reported a higher level of living attributed the change to higher rice profits. The percentage of farmers who reported an increase in profits from rice tends to be higher than the percentage who reported a higher level of living (Table 10).

In groups I and II a significantly greater number of larger farms, compared

TABLE 3.2. *Farmers benefiting from modern varieties, by selected village groupings, 1971–1972*

Village group*	Villages (no.)	Gini coefficient	Farmers reporting benefits (%)**	Farm size of those reporting increases		Chi square significance level
				Small (%)	Large (%)	
PROFIT FROM RICE						
I	1	0.56	34	26	74	0.001
II	2	0.38	75	43	57	0.05
III	5	0.38	52	49	51	n.s.
IV	7	0.24	52	51	49	n.s.
LEVEL OF LIVING						
I	1	0.56	28	23	77	0.001
II	2	0.38	48	35	65	0.01
III	5	0.38	38	44	56	0.10
IV	7	0.24	45	50	50	n.s.

Notes: * I = Pedapulleru in India. II = Kandarpur, Barain in India. III = Marcos, Tab-ang, Sunayawan in the Philippines and Odahu, Nganjat in Indonesia. IV = Malumba, Canipa, Bulucaon, Beynt Nuwebe in the Philippines, Rai Rot, Nong Sarat in Thailand, and Meranti in Malaysia.
** Increases in profits and higher level of living, principally as a result of adopting modern rice varieties.
Source: After table 10, IRRI 1974 *Annual Report,* p. 274.

with small farms, reported an increase in profits from rice and in level of living. Farm size was not associated with increased benefits in Group III and IV. (1974 AR, p. 274)

Given the fact that IRRI indicated an explicit awareness of the concern many observers had expressed about relative benefits to large and small farmers at least as far back as the 1971 *Report,* it is surprising to find so little effort expended in understanding this fundamental issue in the study.

In addition to the questions about the methodology (for example the grouping of villages) which we have mentioned already, there is one crucial respect in which IRRI's treatment is suspect. Large and small farms were distinguished using the *median* farm size for each village. We assume that this means that half the farms were classified as large and half as small. But surely, at least in those cases where the Gini coefficient is relatively large, this will distort the results by including many farms one would wish to call 'small' in the category of large farms. This would result in an underestimation of any detrimental effects on small farmers.

Concerning substantive issues it is interesting that Table 3.2 suggests that larger farmers are more able to convert higher rice profits into a higher level of living. In all four groups a lower percentage of small farmers reported an increased level of living than reported higher rice profits. This suggests that concurrent with, and possibly consequent upon, the introduction of modern rice technology was a tendency for large farmers to fare better than small ones. Of course, this conclusion is only suggested by the data in our Table 3.2 above, but it seems odd that there is no mention of that possibility.

For the first time, a chapter entitled 'Consequences of New Technology' appeared in the 1975 *Annual Report*. It begins with a report on a study initiated in 1975 'to assess the impact of new rice technology on income distribution' (1975 *AR*, p. 324). The first part of the study dealt with the effects of improvements in technology on the incomes of large and small farmers in the Philippines.

The method followed was fairly straightforward. A mathematical model was presented and then applied, by the specification of parameters calculated from empirical data, to the Philippines. For purposes of examining various cases a small farm was taken to be one for which the ratio of marketable surplus to total production is 0.2 and a large farm is one where this ratio is 0.8: it was assumed that the absolute amount of home consumption per person was the same for large and small farms and that the increased production due to technological advances resulted in price fluctuations undisturbed by government intervention. Price elasticities of total demand and supply of rice were assumed constant for the relevant range of analysis. Cases were considered where the technology affected small and large farmers differentially (i.e. where there is a different shift in the respective supply curves). Specification of the parameters for the Philippines gave the results shown in Table 3.3.

On the basis of these results IRRI claimed: 'Our analysis showed that improvements in technology tended to equalize incomes among farmers in the Philippines' (1975 *AR*, p. 324). Taken by itself that conclusion is misleading since it suggests that technological improvements have in fact resulted in an equalizing trend among large and small farmers in the Philippines. The study does not contain any (direct) empirical data on income trends, however.

It is quite clear that the question of the tendency of technological improvement to produce income shifts is much more complex than this study suggested. Some technological improvements may be available only to large farms, and others only to small farms, and others to both. Changing input requirements may put stress on the availability of credit, and small farmers may find themselves unable to purchase needed inputs. Increased production capabilities may result in increased land values which enable large but not small farmers to increase their

TABLE 3.3. *Estimates of the differential impacts of technical progress in rice production (10% shift in supply function) on small and large farms in the Philippines*

	Parameters specified			Change (%) in		
	K = 10	n = 0.3	p = 0.4	Cash rev.	Prod. cost	Cash Income
1						
Small	0.2	0.4	10	14	−2.9	4.3
Large	0.8	0.4	10	−7.1	−2.9	−4.2
2						
Small	0.2	0.3	10	2.9	−2.0	4.9
Large	0.8	0.5	10	−8.6	−3.8	−4.8
3						
Small	0.2	0.4	7	−1.4	−3.7	2.3
Large	0.8	0.4	14	−3.1	−1.7	−1.4
4						
Small	0.2	0.3	7	−0.1	−2.7	1.6
Large	0.8	0.5	14	−4.6	−2.5	−2.1

Source: After table 3, IRRI 1975 *Annual Report*, p. 328.

holdings. However, in spite of these and other simplifications, the study does suggest that, other things being equal, small farm size *itself* does not necessarily work against a farmer. It does not show that improved rice technology had, in the Philippines or anywhere else, produced an equalizing trend or even any pressure towards equalization of incomes.

IRRI itself, in another publication, seems amply aware of the complexitites of the farm size issue.[85] Yet this awareness found no expression in the Annual Reports. As we have seen, IRRI turned time and time again to the farm size issue only to report some minor study or analysis of marginal relevance. How does one reconcile these meagre results with IRRI's repeated expressions of concern about small farmers and with statements, such as the ones we have been quoting, which seem calculated to make it appear to readers of the *Reports* as if IRRI has made serious attempts to investigate the effects of implementing its technological discoveries and the constraints on their further adoption? Was farm size a real issue, as many observers seem to think? Did IRRI really care about such questions? Examining IRRI's budget, as we do elsewhere, allows us to see how little was allocated to study of the consequences of new technology. (See Table 3.4.)

Finally, in 1976, IRRI reported the result of a pilot project in Barrio Turbuan, Philippines, for a study of the impact of new rice technology

TABLE 3.4. *IRRI allocation of core budget research funds* (% of core resources)

	1974	1975	1976	1977
Genetic evaluation and utilization	38	39	40	40
Rice-based cropping system	13	16	21	21
Control and management of rice pests	14	11	10	10
Soil and crop management	18	12	9	9
Machinery and development testing	4	11	8	8
Constraints on rice production	3	2	3	4
Environment and its influence	6	5	4	4
Consequences of new technology	0	0	2	2

Source: IRRI, *Long Range Planning Committee Report*, 1979.

on the 'various facets of rural economy, such as employment and income distribution' (1976 AR, p. 302). The data for four large and four small farmers—farmers who cultivate more than 2 ha of rice were considered 'large'—surveyed in the project included detailed information on expenditures, income, savings, fixed assets, and so on as of 1 June 1975 and 31 May 1976. This promises to be a very worthwhile and useful study but clearly the information for so few farms over the period of a single year is of very limited value. There were other indications of IRRI's willingness to expand research into consequences; for example IRRI organized the Conference on the Economic Consequences of the New Rice Technology held in late 1976. That conference had among its objectives 'to develop a plan for research on the socio-economic consequences of the new rice technology in the next 10 years' (1976 AR, p. 81).[86] But until then there is no evidence that IRRI studied the differential impact of the HYVs on villagers in terms of their gender, and clearly this should have been a focus, given the special roles of women and men in rice economics. There is also little evidence of attention to the relation of HYV rice and different cultural environments.

In conclusion we return to the issue with which we began: did IRRI make a significant contribution to understanding the relationships between farm size and HYVs? And we return to the hypothesis about the institutional stance of IRRI towards the wider social and economic effects of HYVs: that IRRI was unable or unwilling to make a serious study of the effects of HYVs on farm size and intentionally or unintentionally misled itself and perhaps others about its knowledge in this area. We think this hypothesis is supported both by the evidence from the *Annual Reports* discussed above and from our observations made later at IRRI. When one of the authors (Levy) visited IRRI in 1978, a senior administrator and scientist remarked that technological progress was at least as beneficial to small as to large farmers. When asked for references which supported this conclusion, the official showed a slide which was part of an overview presentation in the Director's office. Since the slide was also called 'Differential impact of technological progress in rice production on small and large farms in the Philippines—1976', it is obvious that this slide was drawn from Table 3.3. That is, the slide contained the last column of the table. There are, however, two important differences: the slide omitted the term 'Estimates' from the caption and replaced the term 'case' with 'Case Study'. Thus the impression created by the slide was that it represented field data. In contrast, it is clear from the table and the accompanying text that the figures are generated by a mathematical model and thus the results are only indirectly connected with field data.

Of course taken in isolation this observation is trivial. But seen against the background of the discussions in the *Annual Reports* it provides significant support for our hypothesis that IRRI was unwilling or unable to make a serious study of the effects of HYVs on farm size, and intentionally or unintentionally misled itself and others about the depth of its knowledge in this area.

3.6 Traumatic Change and Continuity in Approach

Certain continuities show how steady the interpretation of IRRI's mission was after Brady took over as director. Despite discontinuities, forced amalgamation, and a traumatic change in rice naming policy, the stability of the classic cluster, and the hierarchy of disciplines appear to have been largely conserved, both contributing to the inertial effect of prevailing methodologies. The amalgamation meant that budget allocation was no longer directed to 'divisions', but to 'problem areas'. An attempt was eventually made, as we will now show, to rate these problem areas in terms of their pay-off in increased rice production. The CGIAR forced these considerations into the budget process. But the commanding role of genetic manipulation, of biological engineering, continued undiminished, despite the board's 1975 decision to stop naming varieties. As IRRI's associate director said in 1978: 'Yes, perhaps there is a hierarchy in rice research. If there is, breeders are at the top, because varieties come first. Of course, other scientists don't like it if breeders think they can solve everything, the way some thought in the past.'[87]

The traumatic change in policy was that ten years after the first tests of IR 8, great pressure was brought upon IRRI to cease naming and distributing its own varieties. While aimed at both IRRI's board and scientists, this pressure was resented most by the scientists in the classic cluster because their reputations, and their own estimates of self-worth, had been earned by actual release of new varieties as much as by the publication of scientific papers and other intellectual achievements. According to IRRI's values, plant-breeding was an agricultural researcher's highest achievement because biological engineering, like the social engineering called for by the boards of the RF and FF, could actually solve the world food problem. After all, a plant breeder (Borlaug) had already received a Nobel Prize for Peace for this reason in 1970. Some IRRI scientists engrossed in varietal development were angry and depressed when the board agreed in 1975 to cease naming and distributing varieties.

This major change in policy was announced two weeks before the arrival of the FAO's quinquennial review team chaired by plant breeder M. S. Swaminathan of India. (Swaminathan succeeded Brady as director-general of IRRI in 1982.) A survey of Asian rice breeders had shown a fairly widespread feeling that IRRI should not monopolize the glamour of new varieties. They were chagrined that farmers who were asking for new varieties were simply asking for 'IRRI rice'. IRRI varieties often contained a specific national germ plasm, and other breeders who created rice varieties apparently experienced professional jealousies over IRRI's success.[88] The results of the survey were not then publicly acknowledged; instead IRRI explained the change in terms of the 'marked expansion in national rice improvement programs'. In any case, IRRI had been sending its breeding materials to Asian research programmes for years. This fact strengthens the view that the pressure resulted from nationalist sensibilities, sensibilities further heightened by the impression that IRRI scientists might already consider their work to be the 'foundation of wisdom' about rice.

The Quinquennial Review's report reiterated the importance of IRRI's primary mission, recommended further reorganization, and strongly endorsed the new policy to release and distribute all of its material through the national research programme.[89] The search for the new technologies continued under these conditions. The pattern of budget allocation from 1974 to 1977 shows that problem areas were ranked in their importance in terms of new technologies (see Table 3.4) and that the GEU programme, along with the study of pest control management (which is entomology, part of the classic cluster) received about 50 per cent in each year. Budget provisions for 1978 to 1980 show that the pattern was not expected to change.[90]

While the continuities between IRRI's organization in the 1960s and 1970s can be established (the increase in scale notwithstanding), there was another new factor emerging. This was the requirement to plan far ahead and to stick to priorities when IRRI was competing for funds with a dozen other international agricultural research centres. In 1972 Chandler explained IRRI's success in terms of the freedom of action (that is, finances) offered by the two foundations, the *esprit de corps* among scientists, and the political skills of his executive assistant (Drillon) in dealing with the Philippines Congress and government officials.[91] The LRPC agreed with this view and to it added the advantage of the simplicity of IRRI's organization. As one senior scientist said in 1978: 'financial decisions were easy to make in the 1960s because we just gathered twenty people together in a room, and it was done'. The scientists' view of a 'golden age' of camaraderie and money contrast sharply with their comments in the 1970s about cumbersome increase in size, routinization of work, and loss of personal contact with other researchers. To this must be added the unpleasant requirement to sort out and decide on research priorities on a medium-term basis. This requirement stemmed from the Consultative Group on International Agricultural Research (CGIAR), which arranged funding for all international centres like IRRI, and was also stimulated among scientists by their experience with the FAO's quinquennial review in 1975.

The Long Range Planning Committee at first found it difficult to reach agreement on criteria by which to judge future allocations. Consequently, in 1977 the committee asked senior scientists to score all research problem areas and indicate the most appropriate weighting system which could be used to aggregate these scores and thus rank research problems. In an early draft of the report in January 1978, the committee said: 'It proved impossible to achieve a consensus on either the interpretation of the criteria or on the scores to be assigned to each problem area. It also proved difficult to get scientists to score research problems with which they were not directly associated.'[92] This can be interpreted through the possibility that the favourite research problems of the dominant departments and programmes were being challenged by others, including those with a strong new paradigm like cropping systems or agricultural economics. This difficulty was presumably related to the proposed rule that research budgets should be allocated on the basis of the rice productivity increases which could be attributed to specific research problem areas. This rule would raise 'problems of estimation', said the committee, because attributing increased production in the case of the solution of a single problem where no other unsolved problems exist is quite different from the case where other unsolved problems do exist. What is more, the committee said: 'Almost by definition, the more basic the research, the less is known about its potential productivity effect and its payoff dimension. Thus, the allocation to more basic research areas must be based on criteria other than those used in the productivity approach.'[93] Oddly enough, the final report had a happy ending. There is no reference to earlier difficulties:

> There was a high degree of agreement between the ranking of budget allocation and the ranking of those outside (that particular) research problem area.... From this experience it is difficult for the IRRI senior staff to conclude that there is a need for drastic allocation [sic; probably should read 'reallocation'] of funds among problems. On the contrary, the allocation seems to fit the perceived rankings of expected research benefits rather well.[94]

What also fit rather well was the consistency between the thinking about increases in yield, on which IRRI was founded, and the ranking and weighting of research problem areas: the top listed and weighted criterion for ranking research, 'increased yields per crop of rice', was followed by 'increased number of rice crops or associated upland rice crops per year'. This thinking was consistent with the requirement that research should be allocated to problem areas which, if solved, would lead to greatest increases in rice productivity. The committee's difficulties illustrate the long-standing tension between criteria for choice based on the supposed economic productivity effects of research and criteria based on what scientists believed would meet with the approval of other scientists and research institutions.

What has been described above is how the conceptions of science's role in agriculture reinforced IRRI's organization to produce new rice varieties. All other work was subordinated, in concept and budget, to the production of new varieties and new technologies. The operative criteria remained consistently those which would increase yield per acre, while attention shifted from exclusive focus on most favourable conditions to include less favourable cultivation environments. IRRI's organization, and the assumptions on which it rested, were not uncontested. There *were* different visions of how its mission could be fulfilled, visions which arose from different conceptions of the relation of the source of new technologies and the method by which they must be tested. But above all there was little grasp of how research strategies could address the evident local complexity of rice cultivation: the question of local situations was often not raised at all. This blindspot caused problems, as we shall later show.

NOTES

1. The CGIAR's formation is discussed in detail in Chapter 4.
2. N. C. Brady to Joel Bernstein, Deputy Administrator, USAID, Washington, DC, 21 June 1974: this was a covering letter to a request for support of IRRI's cropping systems programme.

3. IRRI, *Annual Report 1961–62*, Los Banos, 1962, p. 12.
4. IRRI, *Rice Genetics and Cytogenetics*, Amsterdam: Elsevier, 1964. Editors announced that 102 persons attended, but only 59 were named.
5. Henry Beachell and Peter Jennings, 'Need for Modification of Plant Type', in IRRI, *The Mineral Nutrition of the Rice Plant*, Los Banos, 1964, p. 29.
6. Ibid. 30.
7. Ibid. 32.
8. Ibid.
9. Ibid. 33.
10. Ibid. 34.
11. D. S. Athwal, 'Semidwarf Rice and Wheat in Global Food Needs', *Quarterly Review of Biology*, 46, 1 (March 1971).
12. Edmund Oasa, 'The International Rice Research Institute and the Green Revolution: A Case Study on the Politics of Agricultural Research', unpublished dissertation, University of Hawaii, Hondulu, 1981, pp. 198–9. Oasa's thesis is based on two years' presence in Los Banos and continuing interviews with IRRI staff.
13. Ibid. 190–3.
14. IRRI, *Annual Report 1966*, Los Banos, 1967, p. 66.
15. IRRI, *Long Range Planning Committee Report*, Los Banos, 1979, p. 3. Hereafter LRPCR. See also previous drafts of this report. This committee is quoted extensively here because, in the words of Director-General Brady, it 'successfully involved the entire senior staff in the determination of future goals and priorities'.
16. Interviews: IRRI, March 1978; Beamount, Texas, July 1977; Stuttgart, Arkansas, June 1977; Cuttack, India, January 1978.
17. Report of the Programme Committee of the Board of Trustees, IRRI, 3 February 1969, FFA PA 65–55.
18. '"Miracle Rice" Improves Life of a Filipino Family,' *New York Times*, 6 April 1970. This followed Clifford Wharton, 'The Green Revolution: Cornucopia or Pandora's Box?', *Foreign Affairs*, April 1969, pp. 464–76. Wharton was vice-president of the Agricultural Development Council. For a view from within the US Department of Agriculture, see Lester Brown, 'The Agricultural Revolution in Asia', *Foreign Affairs*, July 1968.
19. L. S. Hardin to F. F. Hill, 24 October 1968, FFA PA 65–55.
20. D. E. Pfanner, to file, 15 January 1969, FFA PA 69–516.
21. 'Rice Boom in Asia Raises Doubts', *New York Times*, 6 April 1970. For a deeper view, see Wolf Ladejinsky, 'The Ironies of India's Green Revolution', *Foreign Affairs*, July 1970.
22. LRPCR, p. 3.
23. L. S. Hardin to R. Chandler, 9 October 1969, FFA PA 69–516.
24. Ibid.
25. L. S. Hardin to R. Barker, 17 November 1969, FFA PA 69–516.
26. IRRI, *Rice Research and Training in the 1970s*, Los Banos, 1969. See also Oasa's analysis of this conference, 'IRRI and Green Revolution', pp. 267–86.
27. Based on interviews in June–July 1977, at Louisiana State University, Texas A & M University; at US rice research stations in Crowley, Beaumont, and Stuttgart, in February 1978; at Rice Processing Engineering Centre, Kharag-

pur; at Central Food Technology Research Institute, Mysore, and the city of Madras. See also the following foundation documents: L. S. Hardin to R. F. Chandler *et al*. ('Analysis of IRRI's Role in the 1970s'), 11 March 1969, FFA PA 69–516; J. Bresnan to E. S. Staples, 14 May 1969, FFA PA 65–55; D. C. Finfrock to G. F. Gant, 4 August 1969, FFA PA 69–516; H. G. Schaller to G. F. Gant, 18 August, 1969, FFA PA 69–516; 'Later Generation Agricultural Development Problems', 20 October 1969, FFA PA 69–516; 'Notes from the Rice Processing Meeting,' 23 March 1970, FFA PA 65–55; F. J. Miller to E. Staples, 30 March 1970, FFA PA 65–55.
28. L. S. Hardin to F. F. Hill, 18 December 1969, FFA PA 65–55.
29. R. F. Chandler, quoted in F. F. Hill to D. Bell *et al.*, 18 June 1970, FFA PA 65–55.
30. R. F. Chandler to U. J. Grant, 28 November 1969, FFA PA 65–55.
31. For an American perspective, see 'Inadequate Collection, Maintenance, and Storage of Germplasm Could Lead to Reduced Genetic Diversity' (ch. 2), in US General Accounting Office, *Report to the Secretary of Agriculture: Better Collection and Maintenance Procedures Needed to Help Protect Agriculture's Germplasm Resources*, Washington, DC, 4 December 1981.
32. Many of IRRI's varieties contained the dwarfing gene borrowed from Deo Gee Woo Gen, a rice from Taiwan. This enabled IRRI varieties to have short stature and stiff straw.
33. Thomas Hargrove, W. R. Coffman, and V. L. Cabanilla, *Genetic Interrelations of Improved Rice Varieties in Asia*, Los Banos: IRRI, January 1979, IRRI Research Paper 23, p. 12. For a very critical, sometimes inaccurate view of IRRI's relation with plant-breeding in India, see Claude Alvares, 'The Great Gene Robbery'; also interview with R. H. Richharia, *Illustrated Weekly of India*, 23 March–5 April 1986. Such stories suggest that the nationalist sensitivities in Asia described by RF officials in the 1950s were still quite strong in the 1980s. See also Thomas Hargrove, 'The Diffusion of Genetic Materials and the Objectives for Their Use among Rice Breeding Programs in India', *Field Crops Research*, 1 (1978), 197–213. Hargrove *et al*. were describing an undoubtedly important type of loss of genetic diversity, but a potentially more important loss has been caused by the promotion and spread of the HYV rices which has pushed traditional varieties out of the fields of a number of countries. This living varietal diversity was being conserved largely at the community level. It appears that IRRI believed it could compensate for this loss. But can it?
34. F. F. Hill to N. C. Brady, 22 May 1973, FFA PA 67–237.
35. During the search for Chandler's replacement in 1972, the International Vice-President of the FF and chairman of IRRI's board wrote to the Indian government's minister responsible for science and technology, C. Subramaniam, saying that two Indian researchers had been recommended for the position of IRRI director, and requesting a formal nomination and confidential evaluation. One of those mentioned in the letter, M. S. Swaminathan, did become IRRI's director in the 1980s. Although we do not know whether Subramaniam did nominate one or both of the recommended people, this shows that non-American directors were being contemplated by IRRI's board at the time. Brady was considered Hill's favourite candidate,

but the potential strain between IRRI and India was developing at this time, so the foundation's gesture to Subramaniam had many diplomatic implications. F. F. Hill to C. Subramaniam, 1 December 1972, FFA; PA 67–237.
36. Discussion, Banff, Alberta, February 1979.
37. Robert Chandler, 'IRRI—The First Decade', in *Rice, Science and Man*, Los Banos: IRRI, 1972, p. 5.
38. Interview, Los Banos, April 1978.
39. Interview, Los Banos, March 1978 and E. Oasa, personal communication, August 1979.
40. Sources: TAC, *Quinquennial Review*, 1976, p. 86; CGIAR, *Cumulative Reports*, 1977; IRRI, *Research Highlights*, 1979, pp. 112–14.
41. 'Informal Summary of Proceedings: International Centres Week and Consultative Group Meeting', 27–31 October 1980, Manila, Philippines.
42. In order of size of staff, the establishment at Los Banos was dominated by Buildings and Grounds and the Experimental Farms: these two departments' responsibilities included the roads, drainage, irrigation, greenhouses, laboratories, and field staff in addition to guards, housing for scientists, drivers for vehicles, and so on. Evidence from the annual reports reveals how IRRI faced the difficulties in protecting its rice crop, a problem which plagues all tropical rice farmers to some degree, because the experimental farm produces rice for seed—for example 144 tons in 1977. To maintain the grounds on which departments or programmes run their research plots required a large staff simply to keep pests away from the rice. By 1970 IRRI had to build an electric fence in order to kill 4,500 rats, and six years later rats were so numerous that poison, trap-gates, and new fences were required. In 1974 50 per cent of wage costs on the fields were spent in frightening away birds, and again in the following year, 43 per cent—the highest portion of those annual budgets. In addition, in 1976 an outbreak of virus dramatically increased the use of pesticide on the experimental farm. These are the unanticipated costs of a large rice-growing centre, illustrating what can happen with continuous production of a single crop on a very large research establishment.
43. 'Informal Summary of Proceedings: International Centres Week and Consultative Group Meeting', 1980, p. 159.
44. Comptroller General of the United States, *Report to the Congress: Participation in International Agricultural Research* (ID-77-55), Washington, DC, January 1978, p. 20.
45. A strike by staff against IRRI management in 1970 appears to have been a watershed for working relations there. One consequence of the strike, besides salary increases, was, according to long-term observers, the creation of distrust between staff of various ranks with respect to joint action over changes in IRRI. For one account of the strike, its relation to the devaluation of the peso, the demands made, and IRRI's response, see Robert Chandler, *An Adventure in Applied Science: A History of the International Rice Research Institute*, Los Banos: IRRI, 1982, pp. 99–100.
46. *LRPCR*, p. 32.
47. Ibid.
48. Draft IV, Long Range Planning Committee Report (January 1978), p. 51.

49. Discussion, Los Banos, April 1978.
50. Discussion, Los Banos, April 1978.
51. N. C. Brady and D. S. Athwal, with assistance of F. F. Hill, 'A Proposal for Broadening the Mission of IRRI', Los Banos, 1 July 1973.
52. Ibid. 24.
53. Ibid. 9, 22, 23.
54. IRRI, 'The Cropping Systems Program in Relation to the Direction and Goals of IRRI', June 1974, p. 9.
55. Nyle Brady to Joel Bernstein, 21 June 1974.
56. IRRI 'Cropping Systems Program' pp. 7, 9.
57. See the pioneering study of Gelia T. Castillo, *All in a Grain of Rice: A Review of Philippine Studies on the Social and Economic Implications of the New Rice Technology*, Los Banos: Southeast Asia Regional Center for Graduate Study and Research in Agriculture, 1975.
58. Barker, an agricultural economist at IRRI for ten years, offered these views at a session of the Conference of the Canadian Council of Southeast Asian Studies, University of British Columbia (UBC), Vancouver, November 1979. Hopper, an agricultural economist, and then vice-president for South Asia at the World Bank, offered these views in discussion with Barrie Morrison in Ottawa in 1977.
59. Richard Bradfield, *Agriculture and Agricultural Education and Research in the Far East*, New York: Rockefeller Foundation, 1956, pp. vii–viii.
60. Richard Bradfield, 'Some Unconventional Views about Rice Culture in S.E. Asia', IRRI Seminar, 16 April 1964, p 7 (text deposited in IRRI Library).
61. Ibid. 15.
62. Quoted in Oasa, 'IRRI and the Green Revolution', p. 318.
63. Richard Bradfield to Clarence Gray, 2 October 1970.
64. Ibid.
65. Ralph Cummings to Robert Chandler, 16 April 1978.
66. IRRI, '1970 Research Program Review—Maximum Cropping Project', 9 February 1971, pp. 10, 13.
67. TAC, *Quinquennial Review*, 1976, p. 29.
68. Ibid. 30.
69. On the distinction between site-specific and environment-specific research, see R. R. Harwood, 'Centralized Research and the Complexity of Social Agriculture', in R. S. Anderson *et al.* (eds.), *Science, Politics and the Agricultural Revolution in Asia*, Boulder, Colo.: Westview Press, 1982.
70. R. R. Harwood to N. Collins, copy V. Ruttan, 21 January 1976, FFA PA 65–55.
71. IRRI, 'Cropping Systems Program, 1978 Internal Program Review', 25 January 1978, p. 3.
72. Discussion, Los Banos, March 1978.
73. '1978 Internal Program Review', pp. 1, 3.
74. Discussion, Los Banos, March 1978.
75. IRRI, *LRPCR*, p. 57. For evidence that an interdisciplinary approach on farmers' fields was tried in other IRRI divisions, see the 1980 'Internal Program Review of Agricultural Economics', especially the focus on the consequences of rice technology: the small farmers' organization project in

Nueva Ecija which evaluated integrated pest management (1 February 1980, IRRI).
76. Harwood, 'Centralized Research'.
77. Discussion, Banff, Alberta, February 1979.
78. Possible explanations based on interviews in Asia; on UN Food and Agriculture Organization, *Integrated Pest Management in Rice in Indonesia*, Jakarta, 1988; and Wolfgang Linser, 'Sophistication of Rice Cultivation in Indonesia', paper at Conference on Alternate Sciences, University of British Columbia, October 1989. See also Donald L. Plucknett *et al.*, *Gene Banks and the World's Food*, 1987, esp. ch. 9, 'A Case Study in Rice Germplasm: IR 36'. For a study of these issues in North America, see John Perkins, *Insects, Experts and the Insecticide Crisis*, New York: Plenum Press, 1982.
79. Before proceeding to the details, we should forestall a possible methodological misgiving. The *Annual Reports* do not contain scholarly articles but summaries of such material. Might it not be the case that many of the shortcomings that we note have arisen in the process of summarizing in a few pages a great deal of detailed work? We have made considerable effort to consult the primary documents and as noted we have sometimes found that the faults are remedied therein. However, even if the above question were answered affirmatively, this would only occasion other questions: What purposes were these reports intended to serve? For what audience were they prepared? Why go to such lengths to produce such elaborate volumes if they do not accurately reflect the work of the Institute? If the volumes were not meant to be judged as technical literature—in spite of their use of complex statistical tables, diagrams, and technical language—then wasn't the complexity of presentation in them an obfuscating misuse of science and technology for which the authors bear responsiblity?
80. It seems that the information for the study was generated by the research team associated with a large study undertaken in 1971–3. This study was funded largely by IDRC of Canada and its findings are reported in *Changes in Rice Farming in Selected Areas of Asia*, Los Banos: IRRI, 1974 (hereafter '*Changes*'). The Thailand section of *Changes* does not analyse the impact of farm size on yield, however (*Changes*, pp. 243–63).
81. A check of the *Annual Report* indices for the years 1971–6 revealed only two references (both in 1974) to work done on the effects of farm size. This at least raises the question whether IRRI was much concerned with the question of farm size during those years.
82. The 1975 *Annual Report* (pp. 265–6) investigated a new technique for estimating grain yield over large areas. The method involves harvesting selected panicles rather than complete crops.
83. A clear reason for focusing on *relative* size is given by Gelia T. Castillo in her contribution to *Changes* called 'Diversity in Unity: the Social Component of Changes in Rice Farming in Asian Villages'. Farm size *per se* has little meaning. It acquires significance only when viewed within the context of the community, the productivity of the land use, population pressures, the tenure system, and the social and economic values attached to land ownership. Half a hectare of land in Java may be "equivalent" to 2 hectares in Thailand. In the Philippines 2 hectares of land in Nueva Ecija does not

mean the same thing as 2 hectares in Cotabato. "Large" and "small" are relative terms conditioned by the above-mentioned factors' (p. 352).
84. A look at *Changes*, from which these data seem to have been taken, reveals that with one exception (Meranti, Malaysia) the villages were a subset of those in the groups of the adoption study. Only those villages in which between 20 per cent and 80 per cent of the farmers reported a higher profit from rice or a higher level of living were included and that probably accounts for the difference in the number of villages in the groups.
85. See *Changes*. For example, in Gelia T. Castillo's contribution, 'Diversity in Unity', she writes: 'An analysis of changes in rice farming which narrowly concentrates on rice production alone does not really tell the story because of complementary, supplementary, and inseparable elements such as non-rice agricultural enterprises and non-farm employment' (p. 351). Dr Castillo was not on IRRI's staff. She was a Professor at the University of the Philippines, Los Banos.
86. See also *The Consequences of Small Rice Farm Mechanization on Production, Incomes, and Rural Employment in Selected Countries of Asia: A Project Proposal*, Los Banos: IRRI, 1978.
87. Interview, Los Banos, April 1978.
88. Oasa, 'IRRI', p. 380—based on 1978 interview with Thomas Hargrove, who conducted the survey for IRRI's Office of Information Services. See, T. Hargrove, *Scientific Communication among Rice Researcher* IRRI Research Paper 13, 1978. There is no evidence in the few available documents of this pressure. The Minutes of the Executive Committee of Trustees, 9, 10 December 1974 and the Report of the Program Committee, 29 January–6 February 1975 do not mention cessation of IRRI 'brand-name' varieties. The confidential draft of a report to the International Board for Plant Genetics by A. B. Joshi (India) and A. H. Bunting (UK), 5–11 June 1975, and the written comments by four IRRI scientists on the Joshi–Bunting visit do not mention it. Nevertheless the board of trustees changed the policy late in 1975. The absence of this issue in documents confirms the utility of studying oral history as well as archival history. Of course, there may be important documents we have not seen. There is a strong possibility that the variety-naming issue stood in for resistance in some countries (possibly India was among them) to the influence of IRRI advisers, and resistance to IRRI's influence on the allocation of international rice research funds.
89. See 'IRRI Statement on the Release and Naming of New Varieties', Annex III, in TAC, *Quinquennial Review*, Rome: FAO, 1976.
90. IRRI, proposed budgets, 1978 and 1979, FFA PA 77–320 and *LRPCR*, 1979.
91. Robert Chandler, *'IRRI—The First Decade'*, p. 5.
92. Draft IV, LRPCR (January 1978), p. 51.
93. *LRPCR*, p. 46.
94. *LRPCR*, pp. 48–9.

4
Communicating IRRI's Mission

> ... before we start trying to sell our products, we must first develop and test them.
>
> Nyle Brady, 1974

TECHNOLOGIES and methodologies were IRRI's main products; thus distributing them through the programmes called 'outreach', and through the training courses at Los Banos, the co-operative research networks, and the more general dissemination of information was essential to its mission from the beginning. Customers for these products were government officials in Asian ministries and agencies, agro-industrial corporations, political representatives of organized agrarian interests, international institutions which underwrote costs of agricultural development and research projects, and the visual and print media. The message had to be broadcast to many different publics. As early as 1963, when it opened its Office of Communication, IRRI was called a world information and communication centre for rice genetics. Increasingly it was realized by IRRI that the passive dissemination of genetic material would not serve its mission. Thus the LRPC stated in 1979 that in various Asian countries improved rice technology was not reaching the farmer because of communication obstacles and called for more effective and active communication strategies.

It was disingenuous of Director-General Brady of IRRI to say, in 1974, that IRRI's product must be developed and tested *before* it was sold. Though in this case he referred to the cropping systems programme, in general this classical separation of functions was visible only to IRRI's scientists. The evidence shows an active selling process began when IRRI opened, and from the point of view of IRRI's clients its mission always combined these three functions: development, testing, and distribution, simultaneously. One reason for IRRI's location in the Philippines was to facilitate continuous communication and promotion activities in Asia. IRRI's existence presumed the promotion of its products. Since most aspects of agricultural extension work were held to be of lower status than research by scientists and technologists, there were many at Los Banos with little enthusiasm for this part of IRRI's mission, and they wanted it done by other specialists. Nevertheless, outreach activities, co-operative research programmes, on-site training, and communication and research networks were all essential in IRRI's mission, whether or not they were interpreted as mere dissemination; while seen as something of a waste of a scientist's time, they were essential to his livelihood.

In some ways this selling activity resembled patterns of selling IBM computers and Lockheed fighter aircraft: lobbying, travelling, meetings, spreading enthusiasm, and cultivating relationships. Because clients were not agricultural scientists alone, the messages were oriented to diverse audiences: international research underwriters, Asian government agencies, or cultivators. Thus claims for the advantages of IRRI's products were shaped to the appropriate audience, but always spoke to much wider interests. This approach arose from the clear and consistent perception among IRRI's senior staff and government that its mission could not be achieved by command and instruction. There had to be a reshaping of consciousness in which IRRI's specific procedures and programmes would become perceived as natural and reasonable. There had to be a transfer of IRRI's cultural norms as well as a transfer of technology. Because new values, such as the advantages of scientific specialization, were being transferred along with the technology, there was a concerted encouragement to Asians to rethink the role of science and technology in agricultural development and to raise systematically the status of its practitioners. At the same time both governments and cultivators were struggling to raise their own relative status and influence within the Asian food production and distribution systems.

It is a striking feature of commercialized consumer societies that it is easier to stimulate belief in consumption and expectation of its benefits than actually to satisfy the felt needs from which the belief is supposed to spring. In this mission to promote the spread of modern commercial agriculture, IRRI wanted to have consumer–cultivators believe in the consumption of new technologies and to expect wide-ranging benefits from them. It seems to have been easier to communicate the expectations of new agricultural science and technology than to instil confidence in research results. It has definitely been easier to communicate an expectation of the technical fix for the problems of governments than to satisfy these expectations. Preoccupation with 'the product' has shifted attention which should otherwise have been focused on the product's underlying relations and assumptions. IRRI focused attention on the product and rode forward on the waves of expectation it generated in Asia. This is what was conveyed by the once popular phrase 'the revolution of rising expectations'. These waves of expectation met a complicated fate. Some people got something, but it often turned out to be unlike their dream of the future. Others found their expectations continued to outreach their satisfaction. The majority of farmers got little or were even harmed by the fulfilment of the others. In every country with which IRRI had working relations, governments continued to feel the pressure of these expectations. Because their food sufficiency problems proved extremely difficult to deal with, and because these problems had severe political consequences, governments continued to look to IRRI for assistance.

4.1 IRRI's Concept of Research

After lobbying with various government officials in the Philippines, the first task for IRRI's builders was to communicate their mission to the senior rice researchers in Asia. One of the ways open in 1960 was through the International Rice Commission (IRC). In 1958 the IRC had recommended an international research institute, but its concept did not find financial support in the United Nations. It is not clear from interviews and documents that IRRI fulfilled the expectations of IRC members. Asian researchers on the IRC questioned an American-directed institute staffed by scientists unfamiliar with tropical and Asian rice cultivations.[1] It was Dean P. C. Ma of the National Taiwan University and an early IRRI Trustee who introduced IRRI to the IRC meeting in 1960, hoping to lay the groundwork for co-operation between it and the IRC.[2] It became part of IRRI's interpretation of its mission to state that it built on a degree of co-operation which had begun before 1960, even though various RF consultants suggested during the 1950s there was little willingness among Asian governments to co-operate with one another. But there were even plans for an alternative IRC-centred institute. In fact the interest of Asian countries in co-operating with one another and with international agencies had been repeatedly demonstrated by their positive response to projects of the FAO and other agencies, many of which predate IRRI.[3]

The IRC was, however, composed of both rice-exporting and rice-deficit nations, and its official recommendations inevitably reflected their differences. Two examples are the issues of legal government control upon chemicals used in agriculture, and national pursuit of rice self-sufficiency; the acceptance of either recommendation by IRRI would have limited its mission. In 1961, at the IRC's Seventh Session in Saigon, it was agreed that 'the FAO should provide appropriate assistance to Member Countries of the tropical region for the framing of suitable legislative measures to prevent and control the indiscriminate use of dangerous agro-chemicals'.[4] Agrochemicals were fundamental to IRRI's approach to increased rice production, and legislation to control their use would affect the assessment of IRRI-bred seed varieties as well as vital American commercial interests. With respect to self-sufficiency, at a meeting of the Subcommittee on the Economic Aspects of Rice in Rangoon in 1962, the IRC noted 'the risk of instability arising from the conflicting policy goals of exporting and importing countries', and recommended ways of reducing this risk, including crop-diversification by rice exporting countries and 'less rigid policies of self-sufficiency by importing countries'.[5] It was IRRI's objective to assist countries to achieve food self-sufficiency within a few years; this kind of recommendation, if pressed on IRRI's board of trustees, would have undercut the value of IRRI's products to client governments (most of whom were members of the IRC). However, both these recommendations were only of symbolic importance. The IRC could not press for either of them, a sign of its political weakness.

There was disagreement within the IRC as to how best to relate to IRRI. Members sought to avoid diminishing national sovereignty and eroding the UN's political forum, but meanwhile sought to take advantage of the scientific opportunities. The IRC in 1961 finally revealed its incapacity to decide and simply approved this terse resolution: 'That the possible relationship [between IRRI and the IRC] be expressed in general terms so that the Director-General of FAO and the Board of Trustees of IRRI would be free to give the matter further study'.[6] IRRI's conferences were used to invite rice researchers to Los Banos: some were members of the IRC, but IRRI pressured various governments to allow scientists who were less senior but more active in research than IRC representatives to attend these conferences. In this way IRRI built up a broader network of research contacts than would have been possible simply through the IRC. As Associate Director McClung later said, 'the IRC ... provided a helping hand in establishing IRRI as an international agency. From the beginning IRRI scientists were accepted and had full opportunity to make their contribution'.[7]

In the end it was the IRC members which had to seek the opportunity to make their contributions to IRRI's work. The IRC became a spent force within ten years and by 1976 was moved, reduced to a few cardboard packing boxes, to Rome where it was supervised by an FAO official with many other duties. The only sign of its rehabilitation was the 'possibility' that the IRC might act as a mechanism to co-ordinate aspects of the Trilateral Commission's 1978 proposal to double investment in rice irrigation in South and South-East Asia.

Besides contacts with Asian researchers and officials through the IRC, the two senior officers of IRRI, Chandler and Wortman, conducted a series of visits in 1961 and 1962 to most rice-producing countries in Asia. They thus met rice researchers and saw their institutions, to some of which the Ford and Rockefeller Foundations had already made grants. They also met the senior ministry officials concerned with research, extension, and food supply. Through conferences at IRRI and these continuing visits, IRRI's mission was presented to a wide variety of Asian audiences. In South Asia, Chandler and Wortman met with the highest political officials (such as ministers of agriculture, heads of government) through the contacts established for them by the resident directors of the two foundations' programmes. Their definition of IRRI's mission was repeated in the first *Annual Report*: 'The administrators of the Institute are confident that these advantages (of facilities and staff) will produce new knowledge which eventually will be reflected in greater productivity on the farm, and in the well-being of the people of the rice-growing countries of the world'.[8]

The conception of science and its applications which underlay the organization of IRRI's work has had different interpretations through the years, but the general assumption was that new science and technology would lead inevitably to increases in aggregate rice production. Knowledge from the lab and plot would diffuse out to the progressive cultivator who would overcome conservative traditions to adopt the new technology. While governments might press reforms and impose development programmes, scientists could only release rice varieties and thus leave their utilization entirely to the vagaries of the agrarian systems. Much more certain for the agricultural scientists was that every new discovery moved knowledge progressively and inexorably forward from an existing base. Technological development would need to rely on the solid accumulated capital of earlier discoveries. There would be little need for feedback loops from the cultivators to reveal the consequences of the new discovery, because its important properties (that is, important to future discoveries) could be evaluated by the same criteria, and in the same field plots and labs, in which the new technology was itself developed. Such a philosophy is broadly consistent with the conception of the relation between science and technology expressed by Weaver and Harrar in the RF's 1954 'Research on Rice' and with that expressed in 1979 by the LRPC:

The importance of mission-oriented basic research in IRRI's program will increase as technologies that capitalize on existing knowledge are discovered. This will occur because as research is focused on a problem, the easy solutions and sometimes the most productive technologies are discovered first. As research continues, ultimately all the technology that can be developed for a problem with the current state of knowledge will develop. Only if new knowledge becomes available will it be possible to make further advances.[9]

It was this idea about the sole source of new knowledge and the singular role of researchers which IRRI sought to communicate to other Asian scientists from the very beginning.

4.2 An Office of Communication

Through these conferences, visits, and consultations IRRI officials realized that the government-sponsored delivery systems for new agricultural technology in Asia had poor relations with the rice cultivators. There was already appreciation in the foundations that traditional American notions of extension were not taking root, even after years of promotion in India and the Philippines. In the Philippines, the RF's Council for Economic and Cultural Affairs had already provided an American adviser to establish the Central Luzon Training Center for Community Development in 1954 and had funded a UPLB to study 'factors inhibiting the acceptance of improved agriculture and rural living' during the 1950s.[10] The need for a vigorous approach to selling IRRI's technologies, not their 'mere dissemination', was realized at the time of IRRI's official opening in 1962, and an Office of Communication was established in 1963 in response to advice that an active dissemination approach would not succeed. For example, in December 1962, Wolf Ladejinsky visited Los Banos and reported to the FF that:

It is difficult for me to see that, in the conditions as they are now in the countryside, the great majority of Filipino farmers are either terribly interested or materially in a position to put to use the results of the IRRI experiments. Mere dissemination of the findings for ultimate acceptance in a manner in which they reach, say, an American farmer, will probably not do for the average Filipino farmer. The acceptance is more likely to come through an effective multi-pronged attack against his disabilities.[11]

While supportive of IRRI's mission, Ladejinsky pointed to the failure of official community development efforts, which had been backed heavily by the US government but which could not be counted on to deliver IRRI's seed and cultivating recommendations to the farmers. He urged IRRI's founders to pressure the Philippines government to face the crisis within agriculture. He acknowledged there was strong local criticism of the style in which IRRI had so far developed, including its exclusive nature, but expected this would disappear when IRRI began to deliver on its promises. Within IRRI, his advice to go beyond 'mere dissemination' was the advice on which the Office of Communication was founded and which it was giving to other IRRI scientists. But specialists in agricultural extension and communication were, in the early 1960s, and still are, considered lower-status than agricultural scientists in the US land-grant system, the system from which most people at IRRI drew their assumptions about status. Therefore scientists seem to have kept these communication activities at arm's length, when possible.

The first person to be appointed to the Office of Communication was Frank Byrnes. Fifteen years later he recalled that when he arrived, IRRI's publications and communication effort was 'obsessed with the facilities and the money, and the whole focus was on IRRI. I tried immediately to get them to focus on the problems of rice cultivators'.[12] The 1963 *Annual Report* listed IRRI's clients as 'scientists, educators, rice farmers and the public'. While IRRI could reach many of these groups directly in the Philippines, it said, 'the problem becomes more complex outside the Philippines and necessitates development of other avenues of communication'.[13] Part of the problem was addressed by the arrival later that year of William Golden, an extension specialist from California's Department of Agriculture, instructed by Wortman to train production-oriented rice specialists who would be sent to train and work with extension personnel, first in the Philippines, but eventually in other parts of Asia. It was also in 1963 that the first special budget for IRRI's 'outreach' work (a phrase coined by F. F. Hill) was $326,000. By the end of the 1960s this sum would be more than a million dollars annually. This outreach budget was separate from what came to be called 'the core budget' of IRRI, most of which was spent in the laboratories and the experimental plots at Los Banos.

Obviously IRRI needed to develop an effective communication programme if it was to sell the new technologies. This seemed to call for scientific-sounding terminology. So in 1964, the Office of Communication said that it had been possible 'to begin demonstrating the practical differences between a behavioral science approach to information dissemination and training, and the more traditional concepts associated with most publication, public relations, information and extension programs'.[14] IRRI organized field days and invited thousands of visitors. Meanwhile it developed its production-training courses, even offering to train as 'farm technicians' inmates from four prison farms on Palawan. Communication was begun with industrial firms involved in agriculture, such as milling and equipment companies. But the need to rely on intermediaries was quite apparent: 'for many reasons, the Institute cannot communicate directly with the millions of rice producers ... nor does it consider such direct communication desirable or necessary'. Some of these intermediaries were obviously the mass media, with which, said the Office, 'two problems constantly arise'. One problem was to avoid extravagant claims for research, and the other was to avoid becoming involved in the internal affairs of the Philippines.[15] These two problems culminated with the media promotion of 'miracle rice' in the last years of the decade. IRRI did indeed develop other avenues of communication outside the Philippines and posted its own established 'liaison scientists' in Pakistan in 1965 and in India and Sri Lanka in 1967. A training course for 'production-specialists' on IR 8, the first outside Los Banos, was held in 1966 in East Pakistan.

The Office of Communication continually stressed that IRRI's scientists should concern themselves with selling their technologies, in addition to the simple establishment of research results: 'The Institute's research on the rice plant has clearly demonstrated that yields can be dramatically increased.... But such clarifications do not remove the serious human, economic, and social obstacles to establishing these new varieties and practices in the rice paddies of Southeast Asia.'[16] Byrnes later recalled that he began to stop talking about a 'behavioral science approach'. Natural scientists at IRRI did not, he said, believe a social science was possible, 'so I began calling our approach "social technology". As they were technologists more than scientists, they liked the idea of social technology more than behavioural science.'[17] It is probable that many staff at IRRI in 1965 also did not wish to hear about the obstacles in the rice paddies of South-East Asia, nor that the decision of the Office of Communication to concentrate on the Philippines 'was based on the ease of working with research findings most readily adaptable to local situations'.[18] This ready adaptation was a variation on the ecological theme of advocates of multiple cropping, and most scientists at IRRI preferred to leave problems of local adaptation to others. They were after the great breakthrough to the big jump in yield.

In 1966, 1967, and 1968, IRRI's Office of Communication carried on some implicit criticism of the technology-first and big-jump-in-yield orientation of most of IRRI's scientists. They reported on the great difficulties faced by government agencies and farmers in actually utilizing the HYV seeds. In the 1966 *Annual Report* it wrote: 'While the public read glowing accounts of the "miracle rice", as it was called by the press, Institute informational efforts tried to de-emphasize the miracle and to focus the attention of responsible government official and the public on the total range of factors upon which the solution of rice problems depends.'[19]

However, evidence suggests that other people at IRRI, including some members of the board of trustees, were pleased with the glowing accounts and felt that press accounts of a miracle rice seed helped justify IRRI and its mission.[20] The newly released variety IR 8 was the full embodiment of that mission to many people. The Office of Communication, however, said that 'the world-wide publicity generated by the mass media incident to the release of IR 8 tended to obscure' the fact that its successful utilization depended on factors which most countries in Asia usually lacked.[21] In 1967 the Office of Communication reported that its policy was to inform the public that credibility in IRRI's kind of work required a much broader approach than that characteristic of most modern agricultural research:

Mass media representatives, government officials, visitors, and trainees were informed that understanding the people and problems of Southeast Asia depends upon understanding the role of rice in the daily activities, economy, politics, social organization, culture, and basic philosophies of the millions of people in these rice-growing, rice-eating areas. The person who seeks to become sensitive to the life of Asians at the 'rice-roots' level particularly must understand the importance and pervasiveness of rice. If he does, he speaks the 'language of rice', and thus builds rapport and gains credibility.[22]

The Office of Communication reported in 1966 that it had already worked out a methodology to study the effects of field demonstrations on the knowledge, understanding, and attitude of rice farmers in the Philippines. The results of this work pointed to the phenomenon already well known in the US and Asia, of 'farmer-resistance'. The office concluded that this resistance could not be understood except in terms of the 'language of rice', but this conclusion would have appeared absurd to many IRRI scientists: rice had to be understood, they believed, in the language of science. It was assumed that resistance was irrational or resulted from ignorance. These cautionary statements from the office came in the very year in which some of the governments of South Asia decided to make a major investment in IR 8 seed as a partial solution to their food problems. Despite exceptions, there were, it seems from numerous recollections, many people at IRRI and elsewhere who did not feel any need to become sensitive to the life of Asians at the rice-roots level. But already the Office had drafted a paper advising foundations on detailed communication procedures to be followed in establishing new international agricultural research institutes; this statement concerned IRRI, CIMMYT in 1966, CIAT in 1967, and IITA in 1968. Byrnes's advice to scientists and administrators included emphasis on genuinely co-operative projects, emphasis on the achievements of participating countries, and de-emphasis of the role of the institute, stronger personal ties between institute scientists and 'relevant nationals' transcending their official relationships, consistency in policy statements by institute scientists, and modification of 'the behavioral styles of scientists' through open and efficient exchange of information within the institute.[23]

At the end of 1968 the Office of Communication was closed and its functions divided. It had already begun to take responsibility for the rice production training courses, and in the following year was split into the Office of Information Services and the Division of Rice Production Training and Research under the guidance of William Golden and Vernon Ross. Byrnes went to work at CIAT in Colombia. The 1968 *Annual Report*, partially written by Byrnes, focused on 'farmer resistance' to IRRI's technologies and on the overriding need to ensure their adoption through communication 'by actions not words'. The Office of Communication had, the *Report* said in 1968, evidence to support 'the thesis that rice farmers, to a large degree, are resisting the ill-equipped "change-agents" as much as, or more than, they are resisting change in the technology itself'.[24] The view of this book is that 'the technology itself' is never encountered in isolation and that rice farmers experienced the new rice technologies only in combination with their associated requirements and the organizational paraphernalia which required access to scarce resources such as change agents.[25] However, it was common for many people at IRRI to use this mystifying distinction, indicative of an intellectual confusion, between 'change-agents' and 'technology itself' because they failed to recognize the degree to which there were requirements strongly linked to their technologies and that many of these requirements were not being met. Farmers, however, had realized this.

Byrnes maintained his earlier views in the last *Annual Report* of the Office of Communication: if IRRI was to succeed and deal with farmer resistance, the technology itself must be perceived as useful by various audiences in both public and private sectors of agriculture. Should the new technology under demonstration fail to impress, both the technology and the rice specialist responsible for it would find difficulty in gaining acceptance. With training in the fundamentals of the communication process combined with the use of rice technology, the rice specialist should be able to demonstrate simultaneously his own credibility and the efficacy of the technology.[26] Since 1968 was the year in which IRRI felt it was reaping the benefit of years of work, this advice does not seem to have been welcomed.

When he reflected later on IRRI's resistance to this point of view, Byrnes said that IRRI had to choose, at that time, between a path which pushed IR 8 with a propaganda 'which eliminates alternatives', and a path which responded to the needs of development and extension specialists of recipient countries, thus retaining a number of alternatives.[27] The narrow first path was chosen at the time, he said; scientists insisted that they wished to change no one's behaviour, only to help them grow more rice. 'They resisted seeing this conflict between their actions and values because they were reacting against the failure of US-style extension schemes in Asia.'[28] Byrnes speculated that the rice scientists did not wish to be contaminated with activities of lower-status extension workers; it was a reflection of the ranking within the 'holy trinity' of research, teaching, and extension in US land-grant universities, he said. Calls for training a new type of specialist, though celebrated in IRRI's publications, were resisted 'all the way up to the Chairman of the Board of Trustees. IRRI staff paid only lip service to training Asians'; it simply was not understood by scientists, he said, that there had to be different methods and rewards not oriented to the US system. Some senior scientists refused to participate in training, said Byrnes, until they had been at IRRI for a few years. 'Essentially they believed their place was in their labs, everything else was wasting time.'[29]

4.3 The Outreach System and Asian Diversity

When the possibility of big production breakthroughs appeared in the late 1960s IRRI's donors and administrators were anxious to expand the outreach system, although not always for the same reasons. So in terms of media relations, production, and training this expansion had a major effect, absorbing one-fifth of IRRI's total budget. Thus closing the Office of Communication did not seem to affect the selling of IRRI's products. Though there was no more talk of 'the language of rice' that scientists found hard to accept, rice production training was continued, and by 1971 there had been 660 production-specialists trained from forty countries, the majority from South and South-East Asia. The total trained was 1,600 by 1976. IRRI continued to receive full exposure in the media; in 1969 the visit to IRRI by US Vice-President Spiro Agnew, made in conjunction with his talks with President Marcos about the Vietnam War, provided IRRI's biggest US media coverage of the decade. In 1969 IRRI promoted the idea of Minikits, cardboard boxes containing seeds, fertilizers, pesticides, and herbicides adequate to permit cultivation of 50 square metres of new rice varieties. Sent to extension officers for presentation to 'average peasant farmers [sic]', these Minikits were explicitly intended to bypass the slow process of approval for rice varieties by government agencies: 'Sometimes farmers and the releasing authorities do not see the qualities of varieties in the same way, so farmers may reject "new and improved" varieties. Such differences could cause farmers to doubt the credibility of the releasing authorities and thus delay the acceptance of the new technology in general.'[30] Government officials were thus being asked to reaffirm their commitment to acceptance of a more productive technology, yet to admit that their own procedures for screening and distributing might not be consistent with farmers' interests and views. These kits were distributed by the thousands, first in the Philippines, then in Vietnam, and elsewhere. The promise of IRRI's rice varieties was made a part of the propaganda in the Vietnam War at this time. For example, thousands of leaflets airdropped over North Vietnam in 1969 entitled 'South Vietnam Is Experiencing a Rice Revolution' stated, among other things, 'IR 8 is South Vietnam's miracle rice: all Vietnamese can enjoy this rice when peace comes.'[31] The Minikit expanded the seed-diffusion activity IRRI had undertaken from the very beginning: between 1963 and 1971 IRRI had distributed 65,000 packages of its breeding-line seed and IRRI-named varieties to government scientists and extension workers in Asia.

After this first ten years of promotional activity, the time seemed to call for reflection about the constant round of flying visits by senior IRRI scientists, the cycles of conferences and joint research trials, the gradual increase in scientific training done at IRRI or in association with UPLB, the flow of research reports and annual reports, and the stationing of teams of IRRI scientists in Asian countries. Associate Director McCLung admitted at IRRI in 1972 that 'the impression may at times have been gained that [outreach activities] amounted to introducing IRRI developed varieties and related technology to other countries and fostering their adoption'.[32] He is clearly acknowledging a common national criticism of IRRI, a criticism which became even stronger in the next few years and originated in the belief that the prevailing approach was American: technology-first and IRRI-first. Some IRRI scientists began to respond to this kind of criticism, stressing the need to expand the outreach system in a genuinely co-operative manner. True co-operation posed continual difficulties because of the striking differences of power and approach between rice researchers in Asia and IRRI. The expansion of the outreach system had already been fairly successful in terms of recruiting IRRI's clients and demonstrating concerns to the underwriters. Budgets for outreach activities at the end of the 1960s were four times as large as they had been in the beginning: $326,000 in 1963 and $1,365,000 in 1969.[33]

In the late 1960s and early 1970s, the two foundations were vigorously searching for other agencies to help pay for IRRI, allowing it to expand further special programmes and deliver the new technology. The voluminous correspondence points to their excitement in thinking the food problem might be solved but also their growing realization that potential technologies were stalled in delivery. The high cost of expansion, to which they were fundamentally committed as part of the solution to the food problem, worried them; meetings were held with the UN Development Program (UNDP), and the Canadian, Dutch, and German governments, as well as the American government, to increase the number of investors.[34] As early as November 1968 both foundations had met in New York to hear Chandler of IRRI review work in progress and to review its budget: a representative of the Canadian government was also present.[35] Not only at the level of agricultural scientists at IRRI but also in investors' and donors' boardrooms, expansion was seen as desirable, inevitable, and costly. The size of IRRI's Board had to be increased to accommodate new US and Canadian investment.[36] One consequence of this expansion was the founding of the CGIAR; another was the great increase of IRRI special projects favoured by individual donors.

While hundreds of officials of Asian governments had met IRRI officials, the problem of how to build a co-operative relationship with Asian departments of agriculture remained. At IRRI it was slowly recognized that the traditional conception of the role of the agricultural advisers must change. Many early advisers in Asia were American, with only undergraduate degrees and experience limited to the US extension system. Owing to the inadequate preparation of some of the men who were acting as advisers, as well as to a strengthening national feeling and superior education among those who were supposed to be advised, improvement of the quality of the people in the outreach programme became necessary. In fact, according to Evenson and Boyce, the 'retirement' of the agricultural adviser around 1970 could not have been avoided by IRRI.

On the whole, this retrenchment developed independently of the international institutes. That is, the retrenchment was not the direct result of the decision to build the international system. It was probably the other way around. The early success of CIMMYT and IRRI *and* the partial insulation of the international institutions from national political forces provided the incentives to move ahead with the international system.[37]

In effect national political forces caught up with the advisory relationship. Within three or four years, for instance, the number of foreign agricultural advisers in India was reduced steadily, with departures including those like Wayne Freeman, attached to IRRI's work in India. However, in Bangladesh and Sri Lanka IRRI teams remained through much of the 1970s to work with the national rice research programme. For instance Golden, who promoted the training programme in IRRI with Byrnes, moved in 1972 to become IRRI's liaison officer in Sri Lanka and to work with some of his former IRRI trainees, among whom he was highly regarded. He was succeeded by Ross, who was a Minikit promoter and who had USAID contract experience in India. However, IRRI's liaison scientists occasionally were judged so incompetent or irritating that Asian countries asked them to leave, as in Bangladesh in 1976 and in Sri Lanka in 1978.

The new advisers shifted the definition of their role. They were not to be primarily identified with fostering the adoption of new rice varieties from IRRI as they had been in the past. Rather, there was a gradual realization that research networks of co-operating institutions had to be formed if IRRI was to remain at the centre of rice research. Donors in the newly formed CGIAR (1972) were asking for a rationale from IRRI as to what these outposted IRRI researchers could achieve. By the early 1970s the number of IRRI scientists working in national programmes in Asia grew to about half the number of scientists at Los Banos. These were the scientists who nurtured the co-operative research networks.

As the numbers of outreach advisers swelled, they began to request some of the rights and privileges enjoyed by their colleagues at Los Banos. With very few exceptions, these outreach personnel were US citizens. Typically they had no employment security beyond a two- or three-year contract and had no additional benefits of schooling or pension arrangement comparable to those of staff of Los Banos. In effect, they were requesting comparable status and benefits, that is, recognition in the core budget of IRRI. Outreach projects totalled over one-quarter of the IRRI budget in 1976 (where total means core plus outreach funding). But donors to the CGIAR treated the request for core budget from IRRI separately from requests for outreach programmes. The latter were often bilaterally funded annually, as special short-term projects.

In 1976 TAC's Quinquennial Review Team stated that 'the outstanding personnel problem of IRRI is the uncertainty of the continuity of employment of staff outposted to regional programs'.[38] In addition, said TAC, this 'disruptive regime' would be unacceptable in other forms of professional employment at this level, and the 'callous disinterest', on the part of IRRI, was 'counter-productive'. These scientists worked in 'relative isolation from their country-men', and their families had real difficulties with children's education 'unlike H.Q. staff'; in addition they were expected to adopt a low profile and could publish little from the research they were committed to encourage.[39] Two years after TAC's report, IRRI's associate director said that in 1978 donor-resistance was still hindering IRRI's attempts to bring outreach personnel into the core budget. Outreach staff and donors believed, in turn, that there may have been resistance within IRRI itself, that some Los Banos scientists might be trying to impede the re-attachment of outposted staff back into Los Banos, and that some senior Asian scientists at IRRI on the core budget reacted against TAC's suggestion that they and outreach staff be interchanged regularly because it would 'be particularly advantageous for the developing countries to have the opportunities to draw on the experience and ability of their nationals at IRRI'.[40] Few such researchers wanted to be transferred away from IRRI. Keeping outreach staff off the core budget substantially reduced IRRI's personnel costs, supplied donors with quality research teams at lower costs, and reduced competition for recognition and rewards at Los Banos.

Yet, the TAC secretariat said in 1976 that outreach work largely financed by specially funded projects 'is causing a somewhat insidious expansion of the whole system'.[41] The popularity of specially funded projects with the donors arose because donors could excercise more influence over special projects than over the core budget. IRRI conducted sixty-six special projects in 1976, for which it received $2.8 million in addition to its core budget of almost $9 million.[42] In addition to this investment through CGIAR, a separate outreach service was planned by some CGIAR donors from 1973. Some individuals who planned this new International Agricultural Development Service (IADS) felt that institutions like IRRI still limited their influence because their research was conducted entirely on their experimental farms.

When the idea of IADS was finally adopted at the Rockefeller-sponsored Bellagio VII conference in Canada in 1975, 'there seemed to be a consensus that no new institutions apart from the proposed IADS should be created to deal with strengthening of national agricultural research systems'.[43] IADS offered to work in smaller countries: for example Bangladesh and Nepal, not India. It would refrain at first from working in a country in which there existed an international centre, like the Philippines or India. By 1977 IADS had substantial long-term contracts with Indonesia ($8.9 million), and Bangladesh ($1.6 million), countries in which IRRI already had major teams. Three of the IRRI's former associate directors (Wortman, McClung, and Athwal) and one former director (Cummings, as well as his son) worked with IADS from the beginning. One of them, Wortman, was its first president. Others as deeply connected to IRRI as Drillon, Harrar, and Freeman served on its early governing committees. In a very real sense its underwriters viewed IADS as part of the efforts to sell not only IRRI's products but the full range of modern agricultural technology. IADS thus expanded the agricultural development thinking on which IRRI was founded in the 1950s.

Through the late 1960s the outreach budget remained about 20 per cent of IRRI's total budget. That is one dollar in five was to be spent away from Los Banos. In 1978 core staff numbered 70, of whom 21 were visiting scientists; outreach staff numbered 32.[44] There was a high proportion of outreach scientists in 1978 to permanent core scientists (32 : 49). A report by the CGIAR in 1980 showed IRRI had 32 separately budgeted 'off-campus' projects, some of them involving millions of dollars ($1.23 million in Philippines, $3.12 million in Sri Lanka, $2 million in Burma, $2.6 million in Bangladesh, $7.7 million for the international rice testing programme).[45] The networks of researchers for whom the principal connecting link was IRRI, in combination with IRRI's outreach staff, made the number of people available for selling IRRI's technology and its methodologies much larger than the number of scientists at Los Banos. This fact helps to explain why there was tension over methodology of the experimental farm and laboratories and also why scientists at IRRI felt more competition for their privileges and rewards. This competition was recognized as a problem by the LRPC, thus forcing it to deal with improved co-operation between IRRI and national ministries of agriculture. LRPC argued for the advantages of such co-operation:

> Comparison of results across widely differing conditions enables researchers to draw conclusions more rapidly. Some research problems require specific environments for their solution. It may be unwise to create such condition in the laboratory, as with certain insects and diseases.... In many collaborative projects results from one location have limited value without results from other locations.[46]

However, given the centralized and centralizing biases in IRRI's work and institutional structure, this principle of multisite research had, until the late 1970s, few supporting voices. These voices would eventually be strengthened, because research was already under way in the late 1970s which would soon reveal the difficulties in the field which IRRI avoided while relying on its narrow focus and headquarters-oriented methodology. These were the very difficulties which should have been addressed throughout IRRI's development and which a few IRRI scientists had been stressing. A really effective study of consequences of technology would probably have provided a formal understanding inside IRRI of these phenomena. By 1980, however, they were understood as part of the insitute's important oral culture (judging by our conversations at the time), but were outside the official definition of IRRI's programme.

Examples from Thailand and the Philippines illustrate the difficulties which IRRI's earlier simplified approach had not addressed but which had to be accounted for after 1980. Work already underway by B. and K. Rerkasem was showing

> that each species, and indeed each variety within a species, must fit into an existing agroecological niche for acceptance by small farmers. In their study of the persistence of traditional rice varieties in the Chiang Mai Valley, the Rerkasems discovered that, despite government efforts to replace traditional varieties with genetically improved high yielding varieties (HYV's), yield was only one of many important niche parameters. In those areas where farmers grew garlic after rice, a traditional variety yielding large quantities of straw was retained, despite its low grain yield, because the high demand for mulch for the garlic beds made the straw nearly as valuable as the rice itself. In other areas, where farm plots were small and the farmers depended upon income derived from harvesting the fields of larger landlords, late maturation was the key niche parameter. This allowed time for them to complete harvesting the fields for others for wages before cutting their own grain. The HYV's could not satisfactorily fill such specialized niches and were therefore rejected in favor of the traditional varieties.[47]

The Chiang Mai area resisted the spread of HYV rice, but the area just outside Bangkok was deemed by IRRI to be a prime location for its work because these central plains were embedded in a nexus of transport, marketing, communication, and irrigation systems.

Central Thailand had long been exposed to pressures to adopt HYV rice, including calls from foreigners and Thais for reduction of the number of varieties planted. This position was strikingly expressed in the approval in 1960 by S. Dasanada, a senior official of the Rice Department of Thailand, of the strategy of 'eliminating the poor yielders and selecting a few superior varieties'. (S. Dasanada eventually became director general of Thailand's Rice Department and a member of the board of trustees of IRRI in 1968 and 1969.) Rather than refine the logic of prevailing agronomic practices and attempt to refine them, he preferred that the Thais continue the practice of Dr Love of Cornell University, who had designed 'a program for rice improvement [for Thailand] from 1950 which has led to a considerable reduction in number of varieties as well as the establishment of superior varieties with greater adaptability'.[48]

Also in central Thailand, Pontius began research, supported by the Rockefeller Foundation, at Nakhom Pathom to identify those factors which influenced the patterns of adoption and rejection of HYV technology at the household level: 'the farmers' use of various sources and channels illustrates that when communication relationships are the unit of analysis, the factors controlling the availability of effective communication [about HYV technology] are certainly more complex than previously hypothesized'.[49] The selective diffusion process meant, said Pontius, that all locations, and thus all potential adopters, did not have the same opportunity to adopt. Entrepreneurs, officials of the Ministry of Ariculture, farmers, and local extension agents were studied in terms of from whom they received information and how far and how often they travelled to get information about fertilizer, herbicide, insecticide, or fungicide. A number of key leaders of local public opinion, who had particularly intense communication with information sources in the city, operated beyond the local relationships structured by the government and marketing corporations. The variation in adoption was explained by variation not simply of availability of the inputs but more by the disjunction between changing local systems of influence and decision-making and official expectations of how the technology would flow. This disjunction was precisely the subject of research proposed by IRRI's Office of Communication before its dissolution in 1967. Certainly the LRPC referred in 1979 to the complexities of new technology adoption but ruled out, as did the later conference on communication responsibilities, the question of direct linkages between researchers and farmers.

Closer to Los Banos, research was being done at IRRI in 1980 which examined not only the communication relations among key actors and institutions, but also the contradictions between the organizational implications of the research strategies of two IRRI departments—namely irrigation and entomology. Discussing the choice between planting and harvesting simultaneously or in some staggered sequence, Goodell says that several of the main components of the HYV rice 'have built-in requirements for farmers' organizational configurations ... these organizational requirements are integral to the technology itself'.[50] The contradictions in requirements for water, pest control, labour, harvest technology, trucking, and credit lead, says Goodell, to serious bottlenecks which would affect levels of production increase. Moreover, they would put new opportunities beyond the reach of smaller farmers or businessmen and into the hands of larger, more distant institutions (like banks, governments, corporations). So serious are these bottlenecks, she argues, that IRRI should conduct its research in truly interdisciplinary teams and directly in farmers' fields, with the active participation of rice farmers.[51] Otherwise IRRI's technology would be at risk of being irrelevant to many cultivating environments, often because they presumed a form of organization which could not be sustained.

Many Filipino rice farmers had, after all, long been participating in the government's programme of increased production, called Masagana 99, and were beginning to experience some of the consequences of that participation. In work begun at IRRI in the early 1980s, evidence emerged of the risks to health of the heavy use of insecticides in conjunction with HYVs. This is the kind of study of the consequences of the use of technology never contemplated during the days of promoting the Green Revolution. But researchers certainly already knew of the flood of chemicals into Philippines agriculture and of the risks associated with their use.

Loevinsohn, who previously worked at IRRI, reported on studies of four rural populations of central Luzon in the Philippines, the heartland of the production increases brought about by Masagana 99.[52] Over 80 per cent of the heads of households in these villages were engaged in rice farming in the two periods studied: 1961–71 (low use of insecticides) and 1972–84 (high use of insecticide). Among the compounds in common use were carbofuran, endrin, and parathion, classified as 'extremely' or 'highly' hazardous by the World Health Organization. Adult males who sprayed with backpacks, usually without protective clothing, were widely exposed. Loevinsohn found a 27.4 per cent mortality increase in the population, concentrated among adult males (mortality rates for females and children did not increase). Deaths diagnosed as chemical poisoning increased 247 per cent between these two periods. The use of these chemicals and some of the consequences were known at IRRI at the end of the 1970s, in part because IRRI had actively supported Masagana 99. Again, it had not been IRRI's official policy to study the consequences of the technology in these directions.

All of this kind of information was in circulation in 1980, though not necessarily in its published form. Researchers at IRRI were aware of it and of the research on which it was based.[53] There were people already at IRRI who believed IRRI's previous approaches were quite inadequate: in fact in the research groups on cropping systems, integrated pest management, and agricultural economics were those who had contested IRRI's prevailing conception of its role and the classic cluster of disciplines. In doing so they planned and initiated research which challenged IRRI's earlier conceptions, and the contest occurred most poignantly at the time of Internal Program Review—where the Institute's directorate critically reviewed the research plans of each group.[54] The fact that this research was being initiated by 1980 shows that IRRI had learned much from the previous twenty years. And at this time there was a gradual 'de-Americanization' of IRRI: the new deputy director was British, the search for the next director-general resulted in the choice of an Indian scientist for the job. The consequence of such change is beyond the scope of our book. But was IRRI in 1980 quite different from IRRI in 1970? The majority of the scientists at IRRI and the greater part of the budget were still dedicated to the research farm, to the plot–lab methodology, and to a centralized conception of research. This dedication created a tremendous inertia for those who wished to change IRRI. All of the activity of the previous twenty years, including the changes in personnel, had resulted by 1980 in a curiously stable organizational culture at IRRI. The insight of people working at IRRI confirmed some of our perceptions and other observers, before and after 1980, made descriptions of IRRI which suggest that our perceptions were of stable, not transitory, phenomena.

The fundamental distinction at IRRI between those who lived inside the IRRI compound and those who did not paralleled the distinction between those who were from other countries and those who were from the Philippines. The 'culture', that is the feeling of IRRI (the aesthetics, the atmosphere), in the late 1970s was that it was a hybrid between a large outdoor YMCA camp, a paramilitary base, and a multinational corporation. The style of mid-American agricultural universities prevailed. The director-general's own Mormon upbringing—the values of hard work, long hours, sacrifice, and close family life—were all prevalent. Mormonism is characteristic of the culture of a New World agrarian society and, along with other forms of modern Protestantism, this set of values was a marked contrast to the complexities of an Asian Catholicism in the Philippines.

At lunch-time in the late 1970s senior scientists drove home to the IRRI compound and drove back afterwards. Within the small compound people sometimes drove from one house to another. Outside the compound, life for other IRRI employees and staff was quite different: these facts were still visible in 1983 when an anonymous author described the different IRRI groups as 'castes'.[55] The two basic restrictions of caste life—no intermarriage and no interdining—were being followed by each group, the anonymous author said. Also noted was that much time and psychic energy were spent on the articulation and preservation of one group's privilege with respect to another, including frequent discussion of 'minute differences'.[56]

Yet at the same time, the anonymous author noted, there was an overriding application of the rule of 'first-name basis only', extending to a 'false back-slapping heartiness' which did not fit with the values of some non-American members of IRRI.[57] And there was a structural sexism which reinforced the Philippines law forbidding work by spouses of foreign IRRI scientists (most of these spouses were women). The result was that female spouses, some of whom were well qualified to work, were blocked in their own personal development and had more or less to remain at home, 'keeping fit and going crazy', as one of them joked in 1979.

These divisions of status and rank, or of gender and ethnicity, were not peculiar to IRRI or to other international research centres. Institutions like BRRI also had their own divisions and prevailing organizational culture. What is important is that this situation described at IRRI grew over twenty years, and was, by 1980, the organization which directed most of the co-operative rice research networks in the world and appropriated most international financing for rice research. The organizational culture which carried on in the 1980s is thus important for the understanding of rice science and scientists, and surely informed the ways in which IRRI communicated its mission.

4.4 IRRI's Mission Criticized, and the Underwriters' Response

The questioning of IRRI's mission and its goals made its formal appearance after twenty years of vigorous activity in selling IRRI's products and transmitting new methodologies. Long a topic of conversation and a theme in memoranda, the goals of IRRI and its communication responsibilities were discussed in a major official conference and an unofficial 'counter-conference' in 1979. Sharp divisions of opinion surfaced again over the relation of IRRI to national research programmes and to rice farmers. In fact relations to all its originally defined audiences, what we call its communication responsibilities, were evaluated and questioned. At the very same time the LRPC was making its pronouncement about IRRI being in the Third World but not of it: not subject to nationalistic pressures, and independent of financial limitations, political considerations, and staff shortcomings 'that pervade such organizations in the Third World'. Four of the six activities in which the LRPC believed IRRI had a comparative advantage over national research institutions were essentially communication activities. Each closely followed the original purposes in the 1960 Articles of Incorporation: genetic resource conservation and sharing, organization of research networks, training of researchers, and dissemination of rice research findings. The idea of the single definitive centre with a dominant role had changed little, but was being adapted to shifts in power, sources of funding, and political perceptions. These areas of comparative advantage which IRRI enjoyed were added to the other two purposes (basic research and testing methodologies); all of them speak of a special kind of communication, with very little dialogue, in which IRRI was at the centre. There is a striking parallel with the thinking of the RF planners in the 1950s, because being in and not of the Third World, as the LRPC described it, *is* unique for agricultural researchers. The comparative advantage follows logically from IRRI's unique position and the planners foresaw that. There could not be many competitors to IRRI.

The official 1979 conference on communication responsibilities, however, defined types of communication deemed important to IRRI. Rice cultivators were omitted by the proceedings, and Asian governments were to be approached only indirectly through their researchers.[58] This left communication with agricultural scientists working in advanced countries, and communication between international research centres and national research programmes, presumably the most important and complex of all the relationships. While there were memoranda of agreement, signed during the 1970s, between IRRI and various countries in Asia, the conference pointed again to the fact that research institutes like IRRI did not define their relation with national programmes. This relation was carefully *ad hoc*; 'no uniform policy has emerged', the conference summary says, because this politically sensitive relationship depends on factors outside IRRI's influence. The conference carefully limited its examination of the complexity of these relationships and the great variation between various research centres and various countries, pointing to the undeveloped relation of research centre communication specialists to their counterparts in various national programmes. It was recognized that IRRI's 'main clientele is primarily agricultural researchers' who work in departments which have little to do with communication and extension. The extension and communication departments were viewed, in turn, as 'poor in many countries partly because research scientists do not provide technical messages for them to target to production specialists and farmers'.[59] Communication specialists in the research centres were asked to help their counterparts and 'fellow information specialists' to improve their conditions and thus to raise the status of the whole enterprise with respect to agricultural scientists. Finally, communication between centres and donors–investors having grown to such proportions, it was recommended that research centres appoint a full-time staff person to deal exclusively with specialized publics like policy-makers and advisers who influence their future. While communication specialists should 'help to identify and serve these audiences', scientists at the conference said that interaction with donors was difficult because donors knew very little about rice cultivation, because donor-representatives changed frequently requiring constant re-education, and because they 'often ask[ed] questions we are not yet ready to answer'. Though scientists might have preferred to avoid communicating with donors, they were reluctant to allocate this crucial task to inexpert communication specialists.

In contrast with the official conference, held at IRRI, financed by USAID and organized by the Agricultural Development Council in New York, the University of Wisconsin, and IRRI, a counter-conference was organized at the same time at the nearby university by UPLB students. Its findings were that the official conference had promoted a number of 'ideological myths' which were peculiar to international scientific élites in agriculture.[60] Some of the myths examined at the counter-conference were that the most critical constraint on rice farmers is not organization or communication but capital and technology. Besides, another myth states, there are too many farmers and too few scientists, so the International Agricultural Research Centres must produce communication specialists to represent farmer interests to researchers. The ultimate myth, the conference said, is that scientists know agriculture better and know what is needed better than farmers. The disappearance of the farmer from the official discourse was challenged: to leave communication with farmers in the hands of national research programmes, which are in a state of 'notorious disrepair', was proof that international centres were more interested in lateral communication with other élites. The counter-conference decried the emphasis on communication aimed at passive audiences rather than on dynamic, responsive, multifaceted communication with co-equals. There was too little listening and too much broadcasting: it found the official conference 'hypocritical and irresponsible', saying that the political party (of élitism) goes hand in hand with information control. The communication strategists, said its final report, 'have come of age as a political class'.

Discussing these communication responsibilities raises the question of the ways in which IRRI's underwriters evaluated its performance and in which its researchers responded to these evaluations. For example, in the report of the Second Asian Agricultural Survey in 1976, the Asian Development Bank said that 'the most optimistic view which can be taken of the food situation is that the region is not much worse off now' than it was in 1967 when the first survey was conducted. This view is not nearly so optimistic as IRRI's planners in the 1950s or the Green Revolution warriors of the 1960s would have liked. Because IRRI scientists were intimately involved in the second survey of 1976, IRRI foresaw these sobering results and thus could not be surprised by the public reaction of Asian governments in the Asian Development Bank (whose officials sat on IRRI's board of trustees). Director-General Brady used these results in 1977 to reinforce the need for IRRI's work in Asia: 'The ADB statement is sobering. And it is a poignant reminder of the seriousness of our task. But the ADB report is in harmony with other evaluations of the state of the world food supply.'[61]

By anticipating the published report, IRRI could remind even its board of trustees in Asia of the seriousness of its mission, the need to expand its outreach budget and increase the participation of researchers at other locations, as well as to expand its own staff. The implications of the ADB agricultural survey had a definite impact at the highest level of Japanese planning, in part because of the deep involvement of Japan in the ADB. Clearly failure to increase agricultural production would adversely limit Japan's trade and investment plans. This impact is seen in the programme of research initiated by Saburo Okita (who became, in the 1980s, Japan's Minister of Foreign Affairs), on massive new irrigation investment of $54 billion by 1993. This plan was to result in the doubling of rice production in South and South-East Asia: it was promoted for wide international discussion by the Trilateral Commission, a private North American European–Japanese initiative on 'matters of common concern'.[62] During this 'Trilateral Process' the authors discussed rice production in Asia with dozens of influential persons around the world. This suggests that the issue of rice politics of the 1950s was vital twenty-five years later.

The ADB's report went on to give criticism of centres like IRRI and modified approval by stressing that though they had deleterious effects on some national agricultural research efforts, their high costs were still justified:

Existing international centers, while they have used up much of the internationally available funds for research over the past ten years and have hindered the development of national programs by dominating them in some instances, still have contributed more than could have been expected from national centers, and the evidence is strong that their benefits have justified their costs.[63]

All the so-called 'returns to agricultural research' studies by economists have stressed that there was 'serious underinvestment' in agriculture compared to other public investment projects. Just as the CGIAR was created by convincing investors in 1971 that institutes like IRRI were the best link in an 'innovation cycle' which would lead to increased productivity and profitability, so the national research programmes were, even after the 1977 Second Agricultural Survey, seen as the best new link in the cycle. A senior ADB official said in 1978:

> We don't finance much research yet—not because we are sceptical but because we can't really spend much money on agricultural research in the developing countries. The national programmes have a low absorbtive-capacity. But we know there's a pay-off, we've had Evenson do studies for us. So we are now proposing to our Board that we spend a lot of money on national research systems because we know that's where the maximum returns will be. The IRRI-type of thing is fine, and probably necessary, but the national systems will give the big returns on investment.[64]

Important members of the Asian Development Bank and clients or donors of IRRI (Japan, Philippines, Thailand, India) would be happy to hear of the pay-offs. And if it would not undermine regular investment in IRRI, scientists there were, with mixed feelings, satisfied to see new investment in national research programmes. After all, they had long argued that new technologies were 'underutilized' by various countries because their research programmes could not absorb and promote new technologies. New research investment would improve their absorbtive capacity for new technologies as well as for further investments for the agricultural sector as a whole (machinery, chemicals, storage and processing, etc.).

A study in 1979 by IRRI economists confirmed the pay-off on investment in national research systems, but called for careful study of them: it is a curious fact of IRRI's first twenty years that knowledge of research values and national organizations was incidental and *ad hoc*; the matter never received scientific and critical attention:

> The consistently high internal rate of return found in all studies suggests a chronic state of underinvestment in rice research. There has been considerable discussion regarding both the accuracy and meaning of these findings.... Certainly it would be wrong to conclude that because returns are so high it is unnecessary to worry excessively about the allocation of funds. In fact, just the opposite conclusion seems warranted.... In order to achieve efficient utilization of very limited research resources, understanding the structure, organization and administration of research is a matter of great importance.[65]

Others with less direct material interest in the pay-off did not think the returns to investment calculations adequate, though as accountant–economists they accepted the concept of such calculations. For example, in its 1977–8 study of the whole research institute system under CGIAR the US General Accounting Office concluded:

> The existing methods of evaluation of the work of the research centres ... do not necessarily provide a measurement of the actual or potential research

benefits, especially in relation to the costs of such research.... A challenge for CGIAR is the development of effective methods to gauge the potential 'pay-off' of present and proposed research efforts.[66]

What was the response of the underwriters to some of the criticisms levelled at IRRI? Did they direct IRRI to incorporate the study of consequences more deeply into rice research? Oasa reviewed the evidence, and says that the CGIAR's response was only partial, and omitted the systematic study of consequences.[67] He shows that through their membership in the CGIAR, and through the enquiries of the prestigious Technical Advisory Committee (appointed by the FAO but working for the CGIAR), the donors were well informed about the level of response at IRRI to the complex reactions to the progress of the Green Revolution. They knew that the centres like IRRI were being scrutinized by other groups like the US General Accounting Office and the Asian Development Bank. The evidence shows that concern among the underwriters of research was for efficient use of funds and priorities for re-directing new research programmes. Oasa says, 'the CGIAR's post-1976 consolidation period placed the small resource-poor farmer at the forefront of agricultural research'.[68] He offers evidence of discussion of the subject at the CGIAR in 1978–80 between directors of centres like IRRI and donors, and yet the study of consequences by IRRI scientists remained underdeveloped because so little of the budget was allocated to it, and because so few researchers were committed to it.

There have been four potential buyers for IRRI's technology, and thus four audiences for IRRI's messages about rice research. To cultivators IRRI spoke of 'science in the service of man' and increased family income from increased yield per hectare; to Asian governments IRRI listed the national benefits from commercialized agriculture, increased aggregate production, and self-sufficiency; to other agricultural scientists at Asian research stations or international institutes it pointed to a new importance for the role of science in development and the status of scientists and technologists and their institutions; and to big investors, otherwise calculating the comparative profitability of slum improvement in Manila, highway construction in north Thailand, jute-exporting in Bangladesh, or fish-farming in south India, IRRI extolled the virtues of the pay-off from the expansion of investment in all aspects of agricultural research and stressed the economic disadvantages of the political instability that would result from neglecting investment in science and technology for development.

It is a conclusion of our work that values and models of research and social organization have been transferred along with agricultural technology. Researchers arrived in the Philippines to build IRRI with ready-made ideas about mission-oriented research and about multinational corporate activity, ranging from the relation with its workers to the role of its governing board of trustees. They also transferred to the Philippines, and then to other Asian countries, models for the marketing of the product (for example IR 8) and for the image-building of the institute. Even IRRI's own Office of Communication reported an 'excessive concentration' on image-building, and an associate director later acknowledged that many clients perceived IRRI's work simply to be promoting its own technologies. This marketing and image-building was a natural feature of the world which most of IRRI's research staff inhabited; it was an essential part of the commercial agriculture to which IRRI was committed. The axiom of these concepts and models was and is that the life of the user–consumer shall be transformed by the purchase or adoption of the 'product'. That was the dream in IRRI's message.

In what precise sense was this selling function essential to the livelihood of IRRI scientists? Was it that if the 'product' was not successfully sold their image would decline and their underwriters withdraw? Were they simply fighting for survival? No, its role was always more subtle and pervasive; after all, the underwriters could surely employ another marketing team and leave the scientists alone to work because their work was considered autonomous. The fact is that the sale of IRRI's technology and methodologies was one of the causes of a shift upward in the status of agricultural scientists and technologists themselves. In this shift, IRRI's builders and Asian scientists had a mutual interest. When IRRI was first opened, agricultural research had no glamour. Only the most senior Asian scientists could travel on the IRC circuit, while a trickle of younger people studied in North America where agricultural research also had low status compared to other callings like biology or economics. But IRRI came to Asia as part of an attempt to transform the rice production system. It arrived while other new institutions, like the agricultural university, were being created, while existing institutions like rice research stations were being 'improved' and 'modernized', while these various institutions were being linked in research networks such as AICRIP in India, and while new symbols of agricultural research's political prestige, such as the Bangladesh Agricultural Research Council or the Philippines Council on Agricultural Research, were being established. If governments had not bought these new rice technologies this status shift for agricultural scientists would not have occurred.

In this way IRRI was part of a global (and often self-serving) campaign for 'science in the service of man'. Tangible evidence of this campaign was the first UN conference on Science and Technology in Development in 1963 and a long list of subsequent events intended to reward agricultural scientists and give their work a bit of nobility or glamour. This cause was fulfilled in 1970 with the award of the Nobel Prize for

Peace to Norman Borlaug, for research done largely in Mexico at CIMMYT. Both the international research centres and agricultural scientists (and particularly plant breeders) thus gained glory and perhaps a little more power. For example, Borlaug's name was soon used in the creation of a Borlaug Award given by Coromandel Fertilizers Limited in India, beginning in 1971.[69] With a gold medal and a cash award of Rs 10,000, this award was for Indian nationals working anywhere for an 'outstanding contribution' to agricultural research, but particularly intended as an encouragement to younger scientists. The Indian co-ordinator of AICRIP, Shastry, who worked in the 1960s with Freeman and IRRI, won it in 1974. IRRI scientists also won this award: Pathak in 1973, Khush in 1977. When the award was given there was a Coromandel Lecture offered simultaneously, and people of the stature of Borlaug, Hopper, Cummings, Crawford, and Hutchison have delivered the lecture. This is an example of glamour and the upward shift in status of research in the twenty years since IRRI's founding. The value of the status shift of agricultual research in Asia was clearly recognized at IRRI and also at numerous research stations in South and South-East Asia. The Long Range Planning Committee concluded:

During IRRI's first decade of research, the high yielding varieties and their associated technologies that came from its programs and those of co-operating national research centers sparked a revolutionary change in scientific outlook and created hopes for a world free of hunger. Less obvious, but perhaps more important, was the great change in the concept of the role of science and technology in increasing rice production. A spirit of confidence and expectation replaced the timidity, uncertainty, and half-hearted support for research that had characterized the 1950s.[70]

The key to understanding this statement is that timidity and uncertainty characterized the attitude of Asian politicians, or at least American perceptions of those attitudes. This 'revolutionary change in scientific outlook' only really attracted attention when a number of countries and donors were prepared from the early 1970s to invest through CGIAR in the apparent higher economic returns generated by the possibility of increased rice production.

Perhaps partially in response to the views presented at the 1979 conferences, scientists writing in the *LRPCR* viewed IRRI's communication responsibilities with the assumption that research in national programmes had already improved because training was provided to 'key scientists such as plant breeders and entomologists'. But IRRI had failed, they said, to emphasize the training of 'professional agricultural educators' who would act as extension agents for IRRI's products.

Partly because most of the developing nations lack such specialists, many national programs now have a backlog of improved rice technology, much of which never reaches the farmer.... Communication research should be organized to achieve better understanding of the flow of information between IRRI and the national programs so that more effective communication strategies can be developed.[71]

Once again, we are offered the vision of public goods waiting for proper private use, but owing to communication obstacles, unutilized. There is a backlog, not because farmers do not know about new technologies, but because there are no 'specialists'. Communication with national agricultural development and research programmes continued to be the only way to reach 'the farmer'. Whether plant breeders and entomologists are in more intimate contact with 'the farmer' as a result of being provided with IRRI training can be judged in the succeeding chapters on rice research and development in Sri Lanka and Bangladesh. But again there is no formal scientific acknowledgment here that much improved technology is appropriate for only some cultivating environments in Asia and is not intended, in its origins or in government terms of access, to reach every farmer. This is the message derived from the clear patterns of adoption and non-adoption, resistance and change, that can be seen throughout Asia by 1980. The research and the communication from a single definitive centre raised expectations but failed to meet the requirements of local diversity. In this sense at least IRRI could be said not to have been definitive.

NOTES

1. Discussions, K. Ramiah, Bangalore, March 1978; and N. Parthasarthy, Bangalore, May 1979. These two scientists were the first two Rice Consultants appointed to the FAO in the 1950s and were stationed in Bangkok at the headquarters of the IRC.
2. P. C. Ma, 'Introducing the International Rice Research Institute', *IRC Newsletter*, Bangkok, June 1960.
3. A. Colin McClung, 'IRRI's Role in Institutional Co-operation', in *Rice, Science and Man*, Los Banos: IRRI, 1972, p. 37.
4. *IRC Newsletter*, Bangkok, June 1961, p. 21.
5. Sixth Session of FAO Consultative Subcommittee on the Economic Aspects of Rice, Rangoon, February 1962 (doc: CCP Rice/62/15, FAO Rome, pp. 15–16).
6. *IRC Newsletter*, June 1961, p. 22.
7. Ibid.
8. IRRI, *Annual Report 1961–62*, Manila, 1962, p. 12.
9. LRPCR, p. 32.
10. Harry Cleaver, 'The Origins of the Green Revolution', unpublished Ph.D. thesis, Stanford University, 1974, pp. 285–6.
11. Wolf Ladejinsky to Walter Rudlin, 17 January 1963, reprinted in Louis

Walinsky (ed.), *Agrarian Reform as Unfinished Business: The Selected Papers of Wolf Ladejinsky*, New York: Oxford University Press, 1977, p. 331. The letter of the same date is in FFA 006–102.
12. Discussion, New York, January 1979.
13. IRRI, *Annual Report 1963*, Manila, 1964, p. 170.
14. IRRI, *Annual Report 1964*, Manila, 1965, p. 298.
15. Ibid. 299–302.
16. Ibid. 327.
17. Discussion, New York, January 1979.
18. IRRI, *Annual Report 1965*, Manila, 1966, p. 328.
19. IRRI, *Annual Report 1966*, Manila, 1967, p. 271.
20. Numerous discussions, Los Banos, Washington, New York, Ottawa, Dhaka, Cuttack, 1977–9.
21. IRRI, *Annual Report 1966*, p. 271.
22. IRRI, *Annual Report 1967*, Manila, 1968, p. 278.
23. Francis Byrnes, 'Toward an Analysis of the Dynamics of an International Institute or Center for Research and Development in Agriculture', IRRI, Working Draft, May 1967.
24. IRRI, *Annual Report 1968*, Manila, 1969, p. 368.
25. For an account of this situation, but without sufficient focus on the state's apparatus, see Steven Pontius, 'The Communication Process of Adoption: Agriculture in Thailand', *Journal of Developing Areas*, October 1983.
26. Ibid. 370.
27. Discussion, New York, January 1979.
28. Ibid.
29. Ibid.
30. IRRI, *Annual Report 1969*, Manila, 1970, p. 248.
31. Robert W. Chandler, *War of Ideas: The U.S. Propaganda in Vietnam*. Boulder, Colo.: Westview Press, 1981, p. 109.
32. A. Colin McClung, 'IRRI's Role in International Co-operation', in *Rice, Science and Man*, p. 37.
33. IRRI, *Annual Report 1976*, Los Banos, 1977, p. 294.
34. Re: Canada, see D. E. Bell to F. F. Hill, 3 June, 1968; F. F. Hill to W. S. Gaud, 13 and 15 January 1969; F. F. Hill to D. E. Bell and Wm. Myers, 22 January 1969; F. F. Hill to R. F. Chandler, 13 February 1969; F. F. Hill to D. E. Bell and Wm. Myers, 28 February 1969. Re: conversation with Myers and Maurice Strong, see D. E. Bell to F. F. Hill, 19 September 1969; L. S. Hardin to F. F. Hill, 6 December, 1972, FFA PA 65–55. Re: Dutch financing for IRRI work in Indonesia, see J. Bresnan to E. S. Staples, 10 September 1969; D. E. Bell to F. F. Hill, 19 December 1969; F. F. Hill to R. F. Chandler, 6 February 1970, FFA PA 65–55. Re: West Germany, see D. E. Bell to L. S. Hardin, 4 August 1970, FFA PA 65–55. Re: UNDP, see F. F. Hill to Ford staff, 5 February 1968, FFA PA 65–55.
35. E. S. Staples to Files, 6 November 1968, FFA PA 65–55.
36. In January 1969 F. F. Hill offered a seat on the IRRI board to USAID when/if they accepted a request for $2.75m. (F. F. Hill to William Gaud, 15 January 1969, FFA PA 65–55.) This total sum was not explicit in his letter of request to Gaud. However, when Hill forwarded Gaud's 17 January 1969 response

to D. E. Bell and Wm. Myers he wrote: 'I would have preferred a Treasury check in the amount of $2,750,000 to the attached letter from USAID' (22 January 1969; FFA PA 65–55). One month later, in discussing a Canadian (CIDA) commitment for $2.75m, the FF said that IRRI's Chairman of the Board (F. F. Hill) had already rejected a Canadian suggestion of an IRRI board seat, although it had been given to the Asian Development Bank instead: 'it seems to me there is a problem here to be solved. The Canadians, not the Bank, are the real source of the money. They have, it seems to me, a legitimate interest in knowing how the money is spent, joining in discussions of IRRI's future direction, etc.' D. E. Bell to F. F. Hill, 13 February 1969, FFA PA 65–55.

37. James Boyce and Robert Evenson, *National, and International Agricultural Research and Extension Programs*, New York: Agricultural Development Council, 1976, p. 63.
38. TAC, *Quinquennial Review*, 1976, p. 68.
39. Ibid.
40. Ibid., p. 67. Also various interviews, Washington, New York, Dhaka, Los Banos.
41. TAC, *Priorities for International Support to Agricultural Research in Developing Countries*, Rome, 1976, p. 33.
42. US General Accounting Office, Report to Congress, January 1978, p. 15.
43. International Agricultural Development Services, *First Report 1976*, New York, 1977. p. 9.
44. Calculated from IRRI, *Research Highlights for 1978*, Los Banos, 1979, pp. 116–17.
45. CGIAR, Report of the Stripe Analysis of the Off-Campus Activities of the International Agricultural Research Centres, Washington, DC, 15 September 1980 (ICW/80/09). See also CGIAR, *Report of the Review Committee*, Washington DC, September 1981.
46. LRPCR, p. 34.
47. A. T. Rambo and P. E. Sajise, 'Alternative Crops', *Science*, 14 November 1986, 801.
48. Sala Dasanada, 'Method of Reducing the Number of Varieties in Thailand', *International Rice Commission Newsletter*, March 1960, p. 7.
49. Steven K. Pontius, 'The Communication Process of Adoption: Agriculture in Thailand', *Journal of Developing Areas*, October 1983, 104. For a comparative view see Bart Duff, 'Changes in Small Farm Paddy Threshing Technology in Thailand and the Philippines', in Frances Stewart (ed.), *Macro-Policies For Appropriate Technology in Developing Countries*. Boulder, Colo.: Westview Press, 1987.
50. Grace Goodell, 'Bugs, Bunds, Banks and Bottlenecks: Organizational Contradictions in the New Rice Technology', *Economic Development and Cultural Change*, 33, 1 (October 1984), 23.
51. Note that in 1981 an exploratory workshop was held at IRRI on the effective use of social scientists, including anthropologists, in interdisciplinary teams: this event and the ensuing programme of research was due to the initiative of Grace Goodell, James Litsinger, and Gelia Castillo, in IRRI and UPLB.
52. Michael Loevinsohn, 'Insecticide Use and Increased Mortality in Rural

Central Luzon, Philippines', *Lancet*, 13 June 1987, 1359–62.
53. The authors organized an international symposium on agricultural research strategies at the American Association for the Advancement of Science in January 1980. Researchers from IRRI were present, and the event was very well attended by people with professional interest in this subject. Presentations by and discussion with IRRI researchers showed the awareness of these issues.
54. See for example the 'Annual Internal Program Review of the Integrated Pest Management Program and Consequences of Technology', 1 February 1980, IRRI: this programme combined the services of an entomologist, an anthropologist, and an engineer. They sought to understand the logic of farmers' practices and attitudes in order to evaluate farmers' interest in adopting new technology. For a complete account see G. E. Goodell *et al.*, 'Rice Insect Pest Management Technology and its Transfer to Small Scale Farmers in the Philippines', in *The Role of Anthroplogists and Other Social Scientists in Interdisciplinary Teams Developing Improved Food Production Technology*, Los Banos: IRRI, 1982, pp. 25–41.
55. Anon., 'International Research Systems: Part A Caste and Class at IRRI, A Descriptive Case Study', 1983.
56. Ibid. 7.
57. Ibid. 8.
58. Agricultural Development Council, *Communication Responsibilities of the International Agricultural Research Centres*, Los Banos: IRRI, 1980.
59. Ibid. 16. The entire subject is well reviewed in Robert C. Hornick, *Development Communication: Information, Agriculture and Nutrition in the Third World*, New York: Longman, 1988.
60. 'Findings of the Counter-Conference on the Crisis of Communication in Agricultural Research in the Third World' (draft), Los Banos, May 1979.
61. IRRI, *Annual Report 1976*, p. 1.
62. Umberto Colombo, D. Gale Johnson, and Toshio Shishido, *Reducing Malnutrition in Developing Countries: Increasing Rice Production in South and Southeast Asia*, New York: Trilateral Commission, 1977. This publication was based on the Report of the North–South Food Task Force of the Commission, 'Expanding Food Production in Developing Countries: Rice Production in South and Southeast Asia' (1977). Yujiro Hayami, economist at IRRI, played a major role in the Task Force.
63. Asian Development Bank, *Rural Asia: Challenge and Opportunity* (2nd Asian Agricultural Survey), New York: Praeger, 1977, p. 252.
64. Discussion, Senior Economist, Asian Development Bank, Manila, April 1978.
65. Randolph Barker and Robert W. Herdt, 'Setting Priorities for Allocating Research Resources for Rice in Asia' (presented at AAAS Symposium on Science, Values and the Politics of Agrarian Change, San Francisco, January 1980), p. 7.
66. US General Accounting Office, 1978, p. 19.
67. Edmund Oasa, 'The Political-Economy of International Agricultural Research: A Review of CGIAR's Response to Criticisms of the Green Revolution', in Bernhard Glaeser (ed.), *The Green Revolution Revisited*,

London: Allen and Unwin, 1987, p. 26.
68. Ibid.
69. I am grateful to officers of Coromandel Fertilizers, Hyderabad, for informative discussions on the Borlaug Prize in March 1978, and for copies of all the Coromandel Lectures to that date.
70. *LRPCR*, p. 3.
71. Ibid. 36–7.

5

Sri Lanka: Success and Stagnation in an Independent Research Tradition

> The IRRI agronomist, James Moomaw, said 'Sri Lanka should get its rice varieties from IRRI and abandon its own breeding programme.'
> J. W. L. Peries, Deputy Director Research (1961–74),
> Sri Lanka, 21 May 1978

> Sri Lanka has the most sophisticated agricultural research structure of *any* country. They think in terms of land types and land utilization and so the structure of research fits the environmental situation. [Emphasis in original.]
> Dr Howard Zandstra, IRRI Cropping Systems Program,
> 28 April 1978

THE foundation upon which the government's rice production and rice science policies have been built is that Sri Lanka does not grow enough rice to meet its own consumption needs. As far back as reliable records go, Sri Lanka has been an importer of rice in spite of rice being grown in all districts of the island and in the great majority of its villages. In recent decades, even when there has been a doubling in annual rice production, the consuming population has more than doubled in size and so the need for imports has continued. This long-standing dependence on rice imports is illustrated by Table 5.1.

Another dimension of the continuing need to import rice has been the political and social concern to distribute both the imported and the domestically grown rice at prices which the rice-deficit households can afford. While the governmental concern for rice prices may have antedated the development of the tea plantations in the mid- and late nineteenth century, by the late nineteenth and early twentieth centuries the government was importing rice to feed the migrant Tamil labourers on the great tea estates. If the estate workers had to pay more for rice, then their wages would go up and so would the minimum auction price for tea. The companies owning the estates were anxious to maintain their market share and to pay dividends to their shareholders. The companies' success in meeting these two objectives was directly affected by the price of rice. However, if the price of domestically produced rice

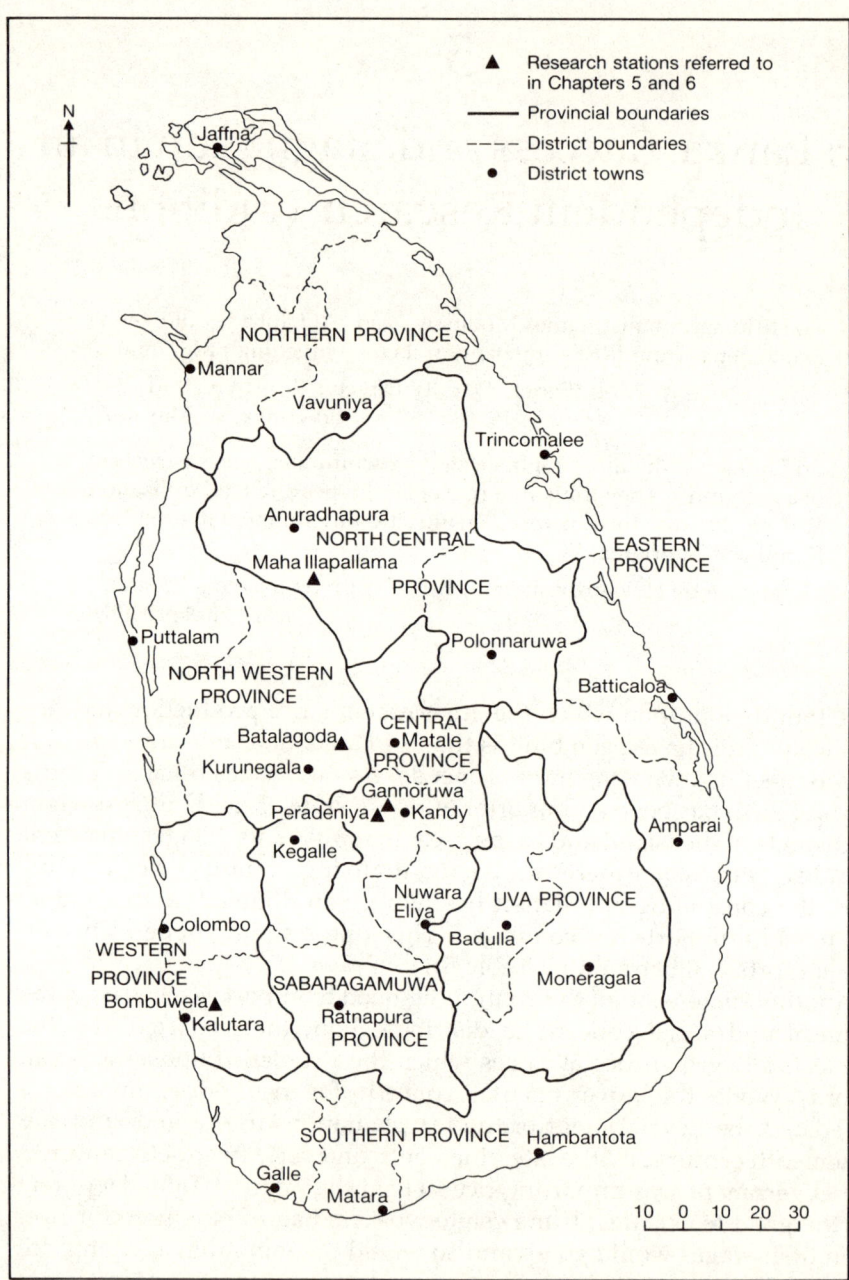

MAP 1. *Sri-Lanka: provincial and district boundaries*
Source: K.M. de Silva, *Sri Lanka: A Survey* (C. Hurst, 1977).

TABLE 5.1. *Rice imports, Sri Lanka, selected years, 1886–1977*

Year	Quantity (tonnes 000)	Value (Rs 000)	Value of rice as % of all imports
1886–7	152	18,093	47.8
1906–7	n.a.	42,383	33.5
1916–17	482	62,826	31.9
1926–7	532	103,224	22.5
1936–7	588	52,932	19.1
1946–7	375	185,712	23.9
1956–7	559	259,540	14.7
1966–7	429	n.a.	n.a.
1976–7	475	n.a.	n.a.

Source: 1936–7 to 1956–7 reported in Youngil Lim, 'Export Industries and Pattern of Economic Growth in Ceylon, Ph.D. thesis, University of California, Los Angeles, 1965, p. 158. For 1966–7 and 1976–7, see N. F. C. Ranaweera, *et al.*, 'Policy Issues Relating to Rice Production in Sri Lanka', in *Rice Symposium 1980* Colombo: no publisher, n.d.

TABLE 5.2. *Net cost of the Sri Lanka food subsidy programme* (Rs million)

Fiscal year	Net cost	Fiscal year	Net cost
1955–6	79.5	1965–6	290.0
1956–7	105.5	1966–7	201.8
1957–8	112.0	1967–8	296.3
1958–9	146.5	1968–9	328.8
1959–60	193.0	1969–70	340.0
1960–1	248.0	1970–1	459.1
1961–2	235.4		
1963–4	375.4	1973[a]	480.3
1964–5	290.0	1978[a]	549.7

Note: [a] Median projected net food subsidy.

Source: Gavin W. Jones and S. Selvaratnam, *Population Growth and Economic Development in Sri Lanka*, Colombo: Hansa Publishers, 1972, pp. 150–5.

was held down, then local growers would not bother trying to cultivate a marketable surplus. They would only grow enough for their own household's consumption. Efforts by the government to reduce the importation of rice and improve the island's balance of trade foundered on the reluctance of the smallholder to grow a marketable surplus.

The resolution of the rice-price dilemma was to have two prices. One was a government-run distribution system to distribute rice in the form of a ration or to sell it at subsidized prices to consumers at as low a price as the government could afford. The second was a purchase price for domestic production which encouraged the growing of a large surplus by the paddy-cultivating household. This latter price was often low, but the government subsidized the costs of production to give the cultivator an attractive net return. However, the gap between the government's selling price and the cost of purchase, whether from domestic or international suppliers, kept widening. By the early 1960s the cost of the rice or food subsidy programme was growing out of control, as may be seen from Table 5.2.

The issues surrounding the historic rice deficit and both the colonial and independent governments' efforts to provide cheap food and yet encourage production will run as a constant theme throughout this and the following chapter.

5.1 The Different Physical Environments and Social Organization of Rice-Growing

In addition to historic deficits, a second fundamental condition of rice-growing which had a continuing influence on the government's rice production and rice research policies was the extreme variation in the physical settings in which rice was grown. Within the 65,600 square kilometres of the island, there are mountain ranges towering over 2,000 metres, heavy tropical rain forests, long stretches of dry scrub bush and lush alluvial and coastal lowlands. The diverse topography and rainfall regimes have provided the environments within which cultivators have developed locally successful rice-growing systems. The result has been a bewildering number of agricultural systems, many with their own favoured rice varieties. Indeed the number of viable rice varieties collected on the island acts as an index of the number of ecological/cultivating niches. In 1977, the Central Agricultural Research Institute at Gannoruwa held in its cold storage over 1,700 different varieties of rice seed that had been collected from different parts of the island.

The environmental variety and the different cultivators' adaptations created complications for the government's agricultural research. From its creation in 1921, the Department of Agriculture had begun to examine the variations in crops and agricultural practices in the different parts of the country. Eventually systematic analysis based on elevation, terrain, soil type, together with rainfall and other hydrological characteristics, led to the division of the island's 65,000 square kilometres into twenty-four agro-ecological regions. Moreover, many of these regions are also internally complex, as for example the northern portion of the mid-country region lying within Kandy District, which has been further subdivided into twenty-one minor agro-ecological zones to describe more accurately paddy cultivation conditions. While Kandy District, with other mid-country districts, is unusually varied with its small valleys running in different directions, it does illustrate the great variation in the agricultural ecology.[1] The clearest reflection of this diversity is in the district statistics on types of crops grown. The extent of the land in different crops can be seen in the district land-use figures in Table 5.3; while taken from 1962 data, it still indicates the varied mix of crops for both earlier and later decades.

Paddy area ranges from under 10 per cent in Kegalle to over 80 per cent in Mannar and Amparai. The plantation crops of tea, rubber, and coconut make up over 70 per cent of the acreage in six districts, dropping to less than 10 per cent in six other districts.

TABLE 5.3. *Distribution of agricultural land at 1 July 1962 (%)*

District	Paddy	Temporary crops[a]	Plantation crops	Grassland	Other crops[b]
All Sri Lanka	28.3	4.2	57.9	0.8	8.8
Colombo	15.8	0.7	79.4	0.6	3.5
Kalutara	19.0	0.9	73.9	0.3	5.9
Galle	22.5	1.0	68.7	0.1	7.7
Matara	26.7	1.3	56.7	0.1	15.3
Ratnapura	11.2	3.6	71.2	0.1	13.9
Kegalle	9.8	1.2	79.7	0.2	9.1
Kurunegala	25.1	2.7	67.4	0.4	4.4
Puttalam	15.8	3.0	76.1	0.5	4.5
Kandy	14.1	3.9	69.3	1.3	11.5
Matale	21.8	9.9	53.2	0.3	14.8
Nuwara Eliya	12.1	7.7	73.5	1.6	5.1
Badulla	16.6	13.2	59.3	2.2	8.7
Moneragala	23.6	13.6	27.8	0.5	34.5
Jaffna	48.7	8.3	22.3	1.6	19.0
Vavuniya	79.0	8.0	6.9	0.2	5.9
Mannar	82.2	1.4	8.6	0.4	7.4
Anuradhapura	75.5	9.2	7.8	0.3	7.2
Polonnaruwa	68.5	9.8	9.0	7.9	4.9
Trincomalee	75.6	7.2	7.7	0.4	9.1
Batticaloa	76.1	3.0	16.4	1.9	2.6
Amparai	83.4	2.8	7.6	0.3	6.0
Hambantota	33.0	6.4	40.3	2.4	17.9

Notes: [a] Temporary crops are hill paddy, pulses, root crops, oil seeds, fibre crops, vegetables and other crops.
[b] Fruit trees and other perennial crops.
Source: Data bank of the Agrarian Research and Training Institute, Colombo.

The variations in the agro-ecological regions and the crops grown, seen in Table 5.4, were only the most obvious sign of the agricultural variation that ramified through the local economic and social structure. Even setting aside the great tea plantations with their industrial-style organization together with similiar rubber and coconut estates, the smallholder sector revealed great diversity.

In the mountainous districts in the centre of the island—Kandy, Matale, Nuwara Eliya, Badulla, Kegalle, and Ratnapura—the holdings were and are concentrated in the valleys and lower slopes of the mountains with the higher land occupied by the plantations. Confined by terrain and by the land claims of the plantations, the villagers tried to meet the needs of their children and heirs by dividing and subdividing the paddy fields. Over generations as the population grew, the paddy fields were terraced higher and higher up the slopes and divided into smaller and smaller fields.

The smallholding size forced villagers both to intensify their exploitation of all available land and to turn outside the village to find income. The two rainy seasons made it possible to double-crop the paddy lands, often with a higher-valued vegetable crop alternating with rice. Families with a shortage of land competed to lease additional fields. Home gardens surrounding their houses were planted to pepper vines, cardamon plants, tea bushes, nutmeg and clove trees—all producing high-value crops. Where the family's resources in the village were still inadequate, family members sought employment outside of the village, working on a neighbouring tea plantation, joining a public works gang

TABLE 5.4. *Size (1962) and distribution of paddy smallholdings in up-country districts (1946)*

District	Mean size of smallholdings, 1962 (ha)	Owners with under 1 ha of paddy, 1946 (%)
Kandy	0.80	91.0
Matale	1.18	82.5
Nuwara Eliya	0.85	84.0
Badulla	0.84	88.7
Ratnapura	1.13	82.4
Kegalle	1.06	78.0
All Sri Lanka	1.09	71.2

Note: Smallholdings include both paddy land and highland.

Sources: Investigation carried out by the Internal Purchase Scheme 1946 and reported in Farmer, *Peasant Colonization*, pp. 85, 86 and *Ceylon, Census of Agriculture, 1962*, 1965, i. 30.

or migrating to the cities in search of a living. Early statistical evidence of landlessness and tenancy are seen in Table 5.5; this pattern held through 1980.

One of the societal consequences of this pressure on land resources was to increase socio-economic stratification within the village. Families with sufficient paddy and high land to meet their daily needs and provide a surplus for saving and investment improved their status while landless and land deficient neighbours fell behind. The poorer families did not send their children to school for as many years; they were less able to arrange good marriages; the doors to higher-paying jobs outside of the village were closed to them.[2]

Second, for those villagers despairing of making a living from the land and turning away from the village the important connections were no longer with the landowning élite but with the townsmen: traders, government officials, and, increasingly, politicians who could offer contacts and employment in the outside world. For those villagers who turned away from agriculture in the village, the long-standing neighbourly organization of work in the paddy fields began to lose its relevance. The older focus of the village life on its fields becomes blurred and diffuse as other extra-village resources were sought out to provide a living for many village families. A third consequence of these changes is that they contributed to the decline in the societal importance of the village while other forms of social organization became more significant, such as the extended family, schoolmates, work-place contacts, and local political organizations.

In other parts of the island, such as the dry zone, different agricultural conditions prevailed and other social and economic changes were under way. The dry-zone districts ring the central mountains to the north,

TABLE 5.5. *Landlessness and tenancy in up-country districts, 1946 (%)*

District	Landless agricultural families	Holdings cultivated by non-owners
Kandy	19.4	45.7
Matale	38.3	45.3
Nuwara Eliya	41.8	38.0
Badulla	8.8	42.2
Ratnapura	32.2	77.4
Kegalle	20.5	64.9
All Sri Lanka	26.3	

Source: *Census of Ceylon, 1946*, i. pt. 2 and *Report on Survey of Landlessness*, Sessional Paper 13 of 1952 cited in Farmer, *Peasant Colonization* pp. 83 and 87.

east, and south—Puttalam, Mannar, Vavuniya, Trincomalee, Polonnaruwa, Anuradhapura, Batticaloa, Amparai, Moneragala, and Hambantota. (Until the 1963 Census, Polonnaruwa was joined to Anuradhapura, Amparai to Batticaloa, and Moneragala to Badulla.) All of these districts were sparsely populated, with villages centred on a water source and separated by tracts of scrub jungle. The majority of rural families owned paddy fields, and some families had very large holdings, particularly in Batticaloa and Hambantota districts, where tenants and labourers were used to work the fields. All of this can be seen in Table 5.6.

In addition to their rice land, most families had access to an unirrigated highland area which was traditionally cultivated with pulses, oil-seed, and vegetable crops by burning off the scrub and farming it for two years or more before letting it revert to jungle. The combination of rice and highland crops was sufficient to meet the subsistence needs of the dry-zone families, together with the products of their home gardens and animals as well as the fish that were taken from the tanks, streams, and lagoons. Few families were forced to look outside of the village for their living. All of this can be seen in Table 5.7.

Consequently in the dry zone the village remained the principal focus of the family's economy while social relations within the village were of paramount importance. Descendants of the founding families who retained a claim to a home site, irrigated paddy land, and highland were of the highest status.[3] In Batticaloa and Hambantota, where large holdings had been amassed, owners or their agents held the dominant

TABLE 5.6. *Size (1962) and distribution of paddy smallholdings in the dry zone (1946)*

District	Mean size of smallholdings, 1962 (ha)	Owners with under 1 ha of paddy, 1946 (%)
Anuradhapura	1.77	41.1
Vavuniya	2.47	21.3
Batticaloa	1.43	6.1
Trincomalee	1.66	18.7
Mannar	1.82	20.8
Puttalam	1.35	52.8
Hambantota	1.49	13.3
All Sri Lanka	1.09	71.2

Note: Jaffna district has been omitted, for it is distinctive among the dry-zone districts.

Sources: Investigation carried out by the Internal Purchase Scheme, 1946, reported in Farmer, pp. 60 and 62; *Ceylon, Census of Agriculture, 1962, Peasant Colonization* i. 30.

TABLE 5.7. *Landlessness and tenancy in dry-zone districts, 1946 (%)*

District	Landless agricultural families	Holdings cultivated by non-owners
Anuradhapura	21.9	23.7
Vavuniya	25.6	26.8
Batticaloa	57.8	38.2
Trincomalee	20.0	34.0
Mannar	11.1	21.2
Puttalam	27.2	25.3
Hambantota	34.0	44.3
All Sri Lanka	26.3	39.2

Sources: For tenure, *Census of Ceylon, 1946*, i, pt. 2; for estimate of landlessness, *Report on Survey of Landlessness*, Sessional Paper 13 of 1952 reported in Farmer, *Peasant Colonization*, pp. 57 and 66.

economic interest.[4] There was little evidence of changes similiar to those occurring in the up-country villages, where many of the villagers were being forced to look beyond the village for their livelihoods. In the dry zone, dominance of the leading agricultural families, the 'feudalistic' pattern of resource management, prevailed with little indication of change.

Yet a third variation in the pattern of agriculture and social relations is associated with the densely populated, low country districts of the wet zone—Colombo, Kalutara, Galle, Matara, and Kurunegala. The villagers in the coastal strips of the first four of these districts were influenced by the string of cities and towns stretching along the coastal railway and roads. For these villagers there was an alternative to agriculture in the manufacturing, handicrafts, and service occupations found in and around the cities.[5] However, the pressure of numerous families on the agricultural resources was as intense as anywhere on the island, as can be seen from Table 5.8. The size of the paddy fields was small, even away from the coastal strip. Moreover the highlands were occupied by coconut plantings, many consolidated into large plantations, rubber estates, and low-country tea estates. Throughout these five districts the plantation crops occupied between 55 and 80 per cent of the cultivated area, most of which was in holdings greater than 20 acres. The possibilities for smallholder families to provide for their own needs from agriculture were limited, for many of the tenant families, difficult, and for the landless impossible, as may be inferred from Table 5.9.

The societal consequences of this third agricultural situation, which dates back long before the Sri Lankans gained a measure of self-government in 1931, was to force villagers to supplement their agricultural resources by other income-earning activity. The coastal

TABLE 5.8. *Size (1962) and distribution of paddy smallholdings in the low country, wet zone (1946)*

District	Mean size of smallholdings, 1962 (ha)	Owners with under 1 ha of paddy (%)
Colombo	0.73	71.8
Kalutara	0.89	70.2
Galle	0.83	75.2
Matara	0.98	70.8
Kurunegala	1.41	63.8
All Sri Lanka	1.09	71.2

Sources: Investigation carried out by the Internal Purchase Scheme, 1946, reported in Farmer, *Peasant Colonization*, p. 85. And *Ceylon, Census of Agriculture, 1962*, ii, table 2, calculated.

TABLE 5.9. *Landlessness and tenancy in low country, wet zone districts, 1946 (%)*

District	Landless agricultural families	Holdings cultivated by non-owners
Colombo	14.2	41.1
Kalutara	22.0	47.3
Galle	20.0	38.8
Matara	20.2	62.2
Kurunegala	12.1	52.7
All Sri Lanka	26.3	60.8

Sources: *Report on Survey of Landlessness*, Sessional Paper 13 of 1952. *Census of Ceylon, 1946*, i, pt. 2. Both reported in Farmer, *Peasant Colonization*, pp. 84 and 89.

districts stretching from Colombo south to Matara were the site of most manufacturing on the island. Weaving, rope-making, pottery manufacture, furniture-building, boat- and cart-building, all manner of small carpentry and metal-working shops, as well as the processing of fish, coconut products, including arrack, combined with the mining of plumbago, gems, and even coral all helped provide alternatives for the land-poor villagers. For the more prosperous there were opportunities to send their children to the best schools in the country with the expectation that they would be able to enter government service or the professions. Consequently many of the villagers in these districts had turned their attention away from the paddy fields or the smallholdings of tea, coconut, and cinnamon to the local workshops or to the factories, stores, and offices of the towns. The important figures in these communities were the wealthy and educated who had contacts in the world of commerce and government. Often large landholdings went with such positions in business and politics, but the significant resource was the connections into the outside world, not the land. The powerful families acted as patrons to a network of family members, dependants, and clients.[6]

The variations in land use and agricultural organizations in these three important agro-ecological zones only suggest the extent of the variations which exist. In parts of the southern dry zone of Hambantota and Moneragala there are yet other physical conditions with distinctive patterns of land use and agricultural systems. To the far north in Jaffna, with yet different agro-ecological conditions, the smallholder vegetable gardener was and is more characteristic. These differences in the physical environment and the associated variation in social organization of agriculture laid down a second major limiting condition within which the Department of Agriculture and successive governments had to work in devising rice research strategies.

5.2 The Foundations and Beginnings of Rice Research to 1931

From the inception of systematic plant research on the island there has been a concentration on plants with an economic potential.[7] Soon after the British occupied the seaboard of the island, they set up gardens in Colombo to experiment with the cultivation of tropical crops that would have commercial value in Europe. With the conquest of the Kandyan Kingdom and the taking over of the central highlands, they gained the royal gardens at Peradeniya where the Kandyan court had cultivated medicinal and other plants. There trials were conducted which identified rubber, improved varieties of cinnamon, coffee, cocoa, tea, and chinchona as crops which would flourish in the island.[8] By the end of the nineteenth century, Peradeniya Gardens had developed a research unit which included a systematic botanist, a mycologist or plant pathologist, an entomologist, and an agricultural chemist. The work of the successive botanists and entomologists laid the foundation of present classification of plant and insect species on the island. The studies of these men such as Wills, Peyt, and Trimen are still referred to in modern botanical and entomological publications. Throughout the early history of crop experimentation and research the preoccupation was with identifying economic crops, improving their yields, and providing protection from pest and disease. These were export crops designed for overseas marketing and for production on carefully managed estates. Very seldom was there any interest in improving crops for domestic consumption. The nearest thing to concern for domestic supply was the effort to grow cotton at Maha Illuppallama in the northern dry zone between 1903 and 1908.

However, during the World War of 1914–18, when the German U-boats sank hundreds of British freighters, there was no shipping available to carry rice from Bengal, Burma, and Thailand to Colombo harbour. The acute shortage of rice set off a clamour to do something to improve rice production on the island to feed both the city population and the large number of Tamil labourers employed on the great tea estates. It was recognized that attention must be paid to peasant agriculture and that the exclusive concentration on plantation crops should end. Out of this concern came the decision to set up a Government Department of Agriculture which came into existence in 1921. Since it incorporated the Botanical Gardens at Peradeniya, it was headquartered at Peradeniya and took over the research establishment which had been in existence for many years in the gardens. The research organization simply continued the earlier divisions of agricultural chemistry, economic botany, entomology, and mycology. To carry out effective field trials experimental farms or stations were set up in some of the different agricultural zones of the island—Maha Illuppallama in the northern dry zone, Batalagoda in the wetter mid-country, Bandarawela in the high wet zone, Angunakilapelessa in the southern dry zone, as well as others. These stations were directed from Peradeniya, carrying out the instructions on varieties to be tested, types of growing trials to be run, irrigation and manuring levels to be followed, and so on. As the former Director of Agriculture stated, it was 'research by proxy' for all of the field work was done by the subordinate station staff while the research officers lived and worked in Peradeniya with only occasional visits to the stations.[9] Since there were many fewer links between the research staff and the field stations and, more important, between the stations and the peasant cultivator, the research was judged by research officers in the 1970s to be of little relevance to the improvement of rice cultivation. In contrast, for the plantation crops the growers (particularly for tea but also for rubber and coconut) had national associations which represented their interests to government and to the Department of Agriculture. Moreover, since these growers were predominantly British or well-connected and educated Sinhalese, they often had personal access to the Department of Agriculture and the research officers to help in the solution of their problems. In contrast, the small Sinhalese paddy cultivator had few connections outside his village let alone into the distant and forbidding world of government and the research section of the Department of Agriculture. Under these circumstances the research and plant-testing objectives of the Department of Agriculture were set by the senior staff of the department with little awareness of the needs of the cultivators.

In spite of the departmental isolation, the policy directive to improve food supplies and the production of rice was clear. In the 1920s under the leadership of Lord and Illiffe, both British plant scientists, research began at Peradeniya on rice. They began with a systematic collection of rice varieties from the different agro-ecological regions. The most promising country varieties were then planted at Peradeniya and selections were made from the best-performing plants of the best varieties for both the Maha (September to March) and Yala (April to August) growing seasons. These initial 'pure-line selections' were then distributed to the eleven field stations throughout the island for further field tests and a more refined selection under the varied regional growing conditions.[10] In this way a number of very good traditional varieties were identified and systematically screened for the highest-performing plants. It was recognized that there were distinct age classes of rice varieties in spite of nearly all of them being photosensitive, that is, they would only flower under certain conditions of light intensity and duration, which meant that the Maha planting would flower and the grain would set in late February or early March whether the seed was sown in late September or in October. The research staff identified, selected, and made available several varieties—Ma wee; Muthu Samba; a 5-month variety—Devaraddiri; a 4-month—Vellaialunkara; and a 3-month—Pachaiperumal. Most of these pure-line selections were capable of yielding two tonnes to the hectare in the department's field stations. Performance in the cultivators' fields was lower, ranging between 1 and 1.26 tonnes per hectare under normal growing conditions.[11]

5.3 The Politicians' Concern for the Rice Grower and the Place of Research, 1931–1950

In 1931, with the creation of an elected State Council where fifty of the sixty-one members were elected on a universal adult franchise, D. S. Senanayake took over the chairmanship of the Executive Committee on Agriculture and Lands. In 1935, Senanayake who in 1948 became the first prime minister of independent Sri Lanka, published a book entitled *Agriculture and Patriotism*. He wrote of the increase of population saying that 'the problem of how to sustain these increasing numbers cannot much longer be ignored'. He called attention to the heavy bill for importing food: 'It is a remarkable fact that we in Ceylon, while repeating in season and out, that ours is an agricultural country and that her prosperity is inextricably bound up with her agricultural progress, should yet be apparently content to pay a bill in a year of depression

(1933) of nearly eighty-seven million rupees for the import of our food and drink.' He observed that the concentration on the profitable export crops of tea, rubber, and coconut had left food production in the hands of a land-starved and debt-ridden peasantry. He concluded that agriculture's 'status is a matter of national concern calling for deliberate and far-sighted policies, not only to conserve the national and human resources involved in it but to provide the national security, promote a well-rounded prosperity and secure social and political stability'.[12]

To create an agriculture capable of providing a living for many more people, to reduce the food import bill, to improve the condition of the cultivator as well as meeting more general national goals was the new government's responsibility, Senanayake asserted. The state should be the initiator and constant supervisor of policies to improve agriculture. It was not a responsibility that would be left to the individual, to private colonization companies, or to 'the invisible hand' of market forces.

A new, interventionist set of policies must be devised by the government for the food-producing rural areas. In broad terms, the new policies were to replace the 'feudalistic' management of rural resources by the village élite sustained by the colonial administration with a 'technocratic and bureaucratic' management of rural resources led by the government. The new policies for the village-based smallholder sector were to replicate in adapted form the practices and organizations earlier devised to encourage plantation agriculture—land laws, labour regulations, credit arrangements, improved market access, and crop-specific agricultural research establishments. But what had been done earlier to improve the profitability of tea, rubber, and coconut production on the large estates was now being undertaken for a much more diffuse array of objectives for the multitude of smallholders growing rice and other food crops.

Given the determination of Senanayake, of his political colleagues and of the bureaucrats in Agriculture and Lands to improve the production of food and the condition of the small cultivators, the most readily available resource was land. Rather than concentrating on improving the farming practices of the food growers, rather than building organizations to provide better access to agricultural inputs, rather than offering high guaranteed prices to all producers, or other such strategies, the government turned to the opening up of new land for cultivation. Land had been accumulated by the colonial government under earlier land laws, most notably the Crown Lands Encroachment Ordinance of 1840 and the Waste Lands Ordinance of 1897, which had asserted that all undeveloped land was the property of the Crown. In the dry zone it was estimated that as much as 85 per cent of the land was claimed by the Crown as late as the 1930s.

From this large reserve of land the government would systematically transfer land to the small cultivator. In particular the transfer would be encouraged in the sparsely populated dry zone where the government would undertake to provide irrigation and other supporting services to landless Sinhalese colonists from the crowded wet zone. The largest of these colonies was the Minneriya scheme, which had been developed during the preceding decade. 'The special attraction of Minneriya lay in its role as the granary of the island in ancient times.'[13] The ancient water storage tank, whose origins in the third century AD were attributed to the kings of the old Sinhalese capital of Polonnaruwa, was restored and the jungles cleared to reduce the dangers of malaria.[14] Building on earlier experiences and inspired by the vision exemplified and articulated by Senanayake, a renewed peasant colonization programme was undertaken.[15]

An essential part of this new programme was the recognition that land use must be secured for the colonist in perpetuity and that the past experience of land gradually being alienated and fragmented until the cultivator had been reduced to tenancy on uneconomic smallholdings was to be avoided. To prevent the repetition of the subdivision of land, the Land Development Ordinance of 1935 was promulgated. The chief feature of the ordinance was to give land allotments to colonists and their successors on a perpetual lease, subject to continued proper cultivation and the passing on of an intact economic holding to the successor.

To make these malaria-ridden districts attractive was a formidable task. Senanayake believed that 'science had to be brought in to support the innate skills of the peasant cultivator who [also] had to be helped with financial assistance, technical guidance and attention to his health needs'.[16] As part of the assistance for colonists in the dry zone the government undertook to supply tools, bullocks, and seeds. These paddy, pulse, cotton, and other seeds were probably the selected pure-line seeds chosen under the guidance of the Peradeniya Research Centre. There was no research station entrusted with breeding plant varieties for the conditions of the dry zone and the specific growing conditions in the irrigation colonies. The old research station at Maha Illuppallana, which was well located to serve colonies such as Minneriya, had been closed down in an economy move in 1919 and was not reopened until 1943. The closest government institution to the colonies, both agro-ecologically and physically, was the farm at Batalagoda, near Kurunegala. This farm was set up for pure-line seed multiplication in 1939 and was primarily concerned to supply Kurunegala District, one of the major rice-growing areas of the island. In the 1930s, the supplying of seed varieties played a minor role in the government's total programme for the development of the dry zone. Far more significant than agricultural technology were the engineering of the large irrigation works; the clearing, stumping, levelling, and bunding of the land; and preventive measures to limit the toll of malaria.

Thus from at least the 1930s, government was taking the initiative to improve agriculture through a series of policies which centred on land transfer but which included irrigation and the application of science to the agricultural and health needs of the cultivator. The subsequent structure of the scientific and technical assistance in agriculture rested on the foundation of plant-breeding and seed selection for higher yields and greater disease and pest resistance. Thus, plant-breeding with rice as the most important crop became an integral part of the agricultural development strategy from the time that the first Sinhalese took responsibility for the Department of Agriculture in 1931.

This approach to extending the cultivation of the dry zone through irrigation, preparatory clearing, and supplying basic social services was very expensive. Using post-World War II figures, it was estimated that average cost only for construction of irrigation works for nineteen projects undertaken between 1947 and 1953 was Rs 2,272 per irrigated hectare.[17] Added to the construction costs were the land-clearing and preparation costs as well as the cost of providing basic services and support until the colonist could harvest a crop. These were estimated by B. H. Farmer to average Rs 8,150 per colonist in 1951, though the costs were sharply cut back by 1953 to Rs 3,775 per colonist. Such high costs combined with the limited acreage that could easily be brought under irrigation restricted the numbers of colonists that could be settled on the land. The tabulation of figures for the longer period 1935–70 reveals that only 74,265 allottees were set up in major colonization schemes while 580,104 were provided with allotments under the village expansion schemes in the same period.[18] The facts of high cost and limited number of settlers were already known by the time the effects of the Japanese entry into the war reached Sri Lanka in 1942.

The coming of war once again emphasized the importance and urgency of improving rice and other staple food production on the island. Shipping was concentrated in the Atlantic region, which together with the subsequent Japanese miltary action soon led to drastic cuts in food shipments to Sri Lanka. Thailand and Burma, which had supplied much of the imported rice, were occupied by the Japanese, production in Bengal was disrupted, and shipping in the Bay of Bengal was subject to attack by Japanese submarines and occasionally aircraft. During the war, rice imports were halved from the level of 1939.[19] The restricted rice supply was rationed, prices were controlled, and the government undertook emergency measures to increase local production. In 1942, district officials were authorized to alienate for food production any available land without reference to earlier priorities.[20] Every tea plantation over 35 acres was required to turn 24 per cent of its total acreage into food production. An Internal Purchase Scheme was introduced which guaranteed payment of Rs 2.50 for every bushel delivered; the following year, 1943, the price was raised to Rs 6.

Paddy area, which had changed little from the earliest estimate in 1921 to 1946, was expanded rapidly in the late 1940s and subsequently. The statistics are not reliable but suggest an increase in land used for paddy from around 324,000 ha in 1921 to 364,000 ha in 1946, a 12 per cent increase. Between 1946 and 1962 the area increased by around 89,000 ha, a 24 per cent increase.[21] The increase took place in most districts as owners of uncultivated lands were encouraged either to bring them into or to restore them to paddy cultivation. The irrigation programmes, both major and minor, were also emphasized, though wartime and post-war scarcities of personnel, equipment, and construction materials slowed the increase of irrigated area and the spread of double-cropping.

At the end of 1943, a Paddy Advisory Board was appointed to advise the government on how to increase production rapidly to meet a larger proportion of the country's rice requirements. At its inaugural meeting on 28 June 1944 it recommended that, since the extension of cultivated acreage would be both slow and expensive, first priority should be given to increasing yields. If an average yield of 2 tonnes/per hectare could be achieved, Sri Lanka would approach rice self-sufficiency. Key to the improvement of yields was the availability of high yielding varieties, capable of converting fertilizer into grain.[22] The pure-line seed selection programme to find high yielding local plants had continued and produced some additional good varieties, such as Podiwee A 8, a 5–6 month variety with a station yield of 2.5 tonnes per hectare; Murungakayan 301, a 4 to 4½-month variety identified and selected at Maha Illuppallama which became the dominant variety in this age class. To supplement the locally available varieties, rice plants from abroad were imported in the 1940s, crossed, tested at Peradeniya and the field stations, and released to the cultivator. Those varieties such as Ptb 16, Mas, and Remadja, mainly based on genetic materials from Indonesia and India, were capable of yielding up to 3 tonnes per hectare with fertilizer application in the fields of the experimental stations. However, in the cultivator's fields with low levels of fertilizer application none of the pure-line varieties or imported and crossed lines were averaging much more than 1.5 to 1.75 tonnes to the hectare.

5.4 The Successes and Limitations of the Rice-Breeding Strategy from 1950

The limitations of searching out existing higher yielding varieties and adapting them to some of the agro-ecological regions of Sri Lanka were recognized. It was believed that a clearer analysis of the desirable plant characteristics and more systematic crossing to produce the desired characteristics were needed if yields were to be raised. While it is not clear from the recollection of participants or from the published documents when the new phase of rice-breeding began or who exactly initiated it, it is clear that during the mid-1940s there was a radical shift in research orientation from the selection of high yielding local or imported varieties to an active breeding programme to develop new plants.[23] From 1946 systematic rice-crossing experiments were started at Maha Illupallama and the associated rice farm at Batalagoda.

This pioneering work was initiated and carried forward by three men. The genetic theorizing was the work of Chandraratna. Drawing on the work of the Japanese on rice and the British on wheat, Chandraratna espoused the view that an analysis of 'the yield components' of rice pointed to four constituent elements for improvement. Firstly, the yield potential per unit of land sown was affected by the number of plants in the field, which was in turn constrained by the space requirement of the individual plants; hence plant shape was critical. The three other components were the number of pannicles per plant, the grains per pannicle, and the weight of the grain. Since the hybridizations of the *japonica* group with the *indica* group of rices that were being attempted at the Indian Rice Research Centre at Cuttack were consistently running into the sterility barrier, Chandraratna concluded that the improvement of the yield components should be sought within the *indica* group of plants. Moreover, he was convinced that *japonica* was not suited to the higher temperatures of the tropics. From the pure-line seed selection that had been carried out at Maha Illuppallama since 1946 under the supervision of Abeyaratne, one 4-month variety of rice, Murungakayan 301, was distinguished from among the local *indica* varieties as having unusually large pannicles. In addition Murangakayan 301 (M. 301) had more tillers, that is additional stalks bearing heads from a common root, and would utilize fertilizer for grain formation better than any other local pure-line variety while being resistant to blast, a common fungus. The problem was to find a suitable *indica* parent that would further improve the number of grains per pannicle and respond to fertilizer. Chandraratna, attending a meeting of the International Rice Commission in Bandung, Indonesia, saw the remarkably long-eared varieties of *indica* that had been bred by the Indonesians and brought home a selection of seeds, including Mas M. 24. These seeds were then crossed with M. 301 and sister lines, such as M. 302, in the breeding trials that were now centred at Batalagoda in the newly designated Central Rice Breeding Station.

The plant-breeding programme was being conducted by Weeraratne, as the research officer in charge at Batalagoda from 1952 onward.[24]

While Weeraratne was under the general supervision of Abeyaratne at Maha Illuppallama, the breeding trials were his responsibility. By 1956 the cross of M. 302 and Mas M. 24 was proving itself in the fields at Batalagoda. The parent stock of M. 302 was yielding 2 tonnes/ha and Mas M. 24 3 tonnes/ha, but the new crosses H2 and, more important, H4, were producing yields of 6 tonnes/ha at the Batalagoda station. After comparative trials over several seasons H4 and subsequently H5 were released to the cultivators for the 1958 growing season.

The new varieties, H4 and H5, could utilize more fertilizer for grain development than any other plant. It was reasonably blast-resistant and a non-photosensitive variety of a 4½-month age class. As such it fitted neatly into the Maha crop season with the rainfall patterns of the north-east monsoon which brought the heaviest and most reliable precipitation to the low country and mid-country where the greatest rice acreage was concentrated. Planting could begin with the onset of reliable rain in late September or early October; there was usually rain for the flowering phase in late January; and harvesting could take place in February, 4½ months after germination. With fertilizer applications of 45 kg of nitrogen per hectare and with adequate and timely water, H4 could regularly yield 4 tonnes/ha in the farmer's field while on the agricultural station it was yielding 6 tonnes/ha. H4 more than doubled the yield of the earlier pure-line varieties when grown in the farmers' fields.

The greatest problem faced by the government was the limited adoption of the high yielding varieties. In the ten years after the release of H4 and its sister lines the proportion of paddy lands planted to the new seeds varied widely from district to district. Cultivators in some areas enthusiastically adopted the new varieties, planting as much as 90 per cent of the sown area in the Maha season.[25] In other areas the total acreage planted to the H lines was only about 15 per cent of the sown area, as seen in Table 5.10.

The reasons for the variations in use of the new seeds differed from locality to locality. Where there was assured water and good drainage with long hours of sunlight, H4 was very productive. Throughout the dry zone it was widely adopted. However, in some areas of the wet zone, H4's growth was deformed by infestations of blast, a generic term for a type of fungus with a broad spectrum of subtypes.[26] In other fields, soils with high levels of iron or low levels of phosphorous stunted the growth of H4. In general, however, it was in the low-lying, poorly drained rice lands, which were concentrated in Colombo, Kalutara, Ratnapura, and Galle districts, that H4 did less well and the local varieties continued to be planted.[27]

Beyond the constraints imposed by soil types and rainfall patterns, many cultivators apparently decided not to plant the H line of seeds

TABLE 5.10. *Adoption of H4 and sister lines by district in 1966–1967 seasons (% sown area)*

District	Maha 1966–7	Yala 1967
Colombo	35.8	11.8
Kalatara	16.6	13.7
Galle	14.0	26.5
Matara	36.1	43.7
Ratnapura	26.9	8.1
Kegalle	41.9	53.4
Nuwara Eliya	74.2	88.6
Kandy	70.8	72.3
Matale	44.7	25.8
Badulla	68.0	68.0
Kurunegala	56.3	43.8
Puttalam	68.3	83.7
Anuradhapura	61.5	34.5
Polonnaruwa	70.7	54.7
Hambantota	90.7	81.4
Amparai	89.0	48.1
Moneragala	61.6	86.1
Jaffna	42.6	94.2
Vavuniya	89.7	99.0
Trincomalee	87.5	39.0
Batticaloa	83.2	48.5

Source: Adapted from District Agricultural Extension Officer reports as tablulated in W. B. Medagama and S. H. Clarles, 'Seed Paddy Production . . .', in *Rice Symposium 1980*, table 2. For cultivation statistics from 1962 to 1967, *Statistical Abstract of Ceylon, 1967–1968*, Colombo: Department of Census and Statistics, 1970, tables 61–76.

because of the incompatibility of the new varieties with the established cultivating practices and social arrangements of production devised for the traditional varieties of rice. The planting of H4 was associated with a shortening of the Maha cultivation season to 4 or 4½ months from 5 or 6 months required for the maturation of the traditional varieties, thus increasing the time available for planting a second crop and altering the annual schedule of cultivation and labour requirements. In addition, H4's capacity to turn moderate applications of fertilizer into grain greatly encouraged the use of nitrogenous fertilizer, making the H4 cultivator more dependent on suppliers. The increase in volume of paddy for each field resulting from higher yields and increased double-cropping created problems of threshing, storage, and marketing for the grower, again increasing his dependence on merchants and government purchasing agencies. None of these cultivator problems had been appreciated by the rice scientists or the policy-makers.[28] There had been no systematic enquiry into the diversity of traditional agrarian practices and the rationality that underlay them. No cultivators had been associated with research policy-making. No farmers discussed their problems with the breeders. The rice plant was bred to meet the criteria identified by the analysis of the potential of the rice plant and not by an analysis of the cultivator's practices and practical wisdom. The plant was changed by rice scientists, and hundreds of thousands of farmers were expected to adapt to the plant's requirements.

The broken runs of national statistics on percentage of cultivated acreage sown to H4, transplanting, rates of fertilizer application, and tractor ploughing all indicate the transformation of the cultivating practices at the national level. While there are proportionate increases in each reported activity, the actual increase is much greater for the area cultivated in each season. Over the ten-year period Maha 1957/8 plus Yala 1958 to Maha 1967/8 plus Yala 1968 there was a 64 per cent increase in area sown, as can be seen in Table 5.11.

One of the more fundamental changes facing the cultivators was the time schedule and labour requirements that came with H4. The principal rice cultivation season in the wet zone coincides with the north-east monsoon, October to February. Field preparation starts when the soil has been softened by the first rains.[29] Broadcast sowing or transplanting follows so that some four months later H4 is ready for harvest, usually sometime in February. Harvest now takes place when there are still showers. Threshing is more difficult and the grain must be thoroughly dried before it can be stored, both more laborious processes than with the later-maturing traditional varieties. By May, field preparation for the Yala crop should be underway.

Since the onset of the south-west monsoon, May to September, is often accompanied by heavy rains, many cultivators germinate the seeds in specially protected and watered seed-beds ready for transplanting when the seedlings are 20–25 cm tall. Once rooted in the field the seedlings are able to withstand storm damage and survive flooding much better than later broadcast-sown plants. Preparation of the seed-beds, watering them, if need be, preparing the fields, and transplanting consume much more labour than simple field preparation and broadcast sowing. Each cultivating household must divert more family labour into the fields or else hire labour to sustain the double-cropping regime made possible by H4.

Another fundamental change was in the requirements for fertilizer and increased need for credit. All synthetic fertilizer is imported into Sri Lanka and, until the establishment of the Ceylon Fertilizer Corporation in 1964, was distributed through a network of private firms.[30] The nationally aggregated figures report the volume of urea and similar

TABLE 5.11. *Change in area and cultivating practices from Yala 1957 to Yala 1968*

Season	Area sown (ha)	Pure lines (% area)	H4 (% area)	Fertilizer application (kg/ha)	Average yield (kg/ha)
Yala 1957	160,037	16.3	—	29.4	—
Maha 57–8	300,050	27.4	—	—	1,759
Yala 58	210,988	15.5	—	30.8	1,795
Maha 58–9	321,532	19.9	—	—	1,747
Yala 59	202,074	14.4	—	50.5	1,877
Maha 59–60	335,712[a]	33.5	—	—	1,824
Yala 60	215,555	17.3[b]	0.4	35.5	1,860
Maha 60–1	356,562	29.7	2.6[c]	—	1,815
Yala 61	217,691	13.6	—	49.6	1,823
Maha 61–2	379,186	35.5	10.6[c]	—	1,920
Yala 62	225,180	—	—	62.3	1,905
Maha 62–3	405,167	—	22.0[c]	—	—
Yala 63	227,411	—	—	75.2	—
Maha 63–4	410,512	—	—	—	—
Yala 64	231,493	—	—	94.7	—
Maha 64–5	398,753	—	—	—	—
Yala 65	190,663	—	—	72.7	—
Maha 65–6	425,277	—	—	—	1,815
Yala 66	213,158	—	—	63.0	1,749
Maha 66–7	426,790	63.0	58.7	—	2,064
Yala 67	236,977	49.0	28.3	81.1	2,122
Maha 67–8	464,518	65.2	59.0	—	2,400
Yala 68	241,182	54.1	27.4	119.0	2,265
Maha 68–9	—	68.0	55.9	—	—

Notes: [a] Total areas reported as cultivated for any given season.
[b] The area sown to pure-line seeds is not reported from Maha 1959–60 to Maha 1961–2; instead the number of kg of improved seed issued is reported.
[c] Estimate.

Sources: Administrative Report of the Director of Agriculture from 1958 to 1970. Rate of fertilizer application calculated from the same source and recorded in A. P. A. Fernando, 'Agricultural Development of Ceylon since Independence (1948–1968)', Ph.D. thesis, University of Leeds, 1972, p. 125. Extent under pure-line seeds and old improved varieties from Maha 1966–7 reported in W. W. B. Medagama and S. H. Charles, 'Seed Paddy Production: Highlights over the Last One and [a] Half Decades', *Rice Symposium 1980*, table 1.

fertilizers distributed increased by 350 per cent from 1958 to 1964.[31] In 1958 about one-quarter of the paddy land was given synthetic fertilizer. Ten years later the rate was over 100 per cent, which is explained by the double or triple applications during the same season. (See Table 5.11.) While the proportion of the fields treated increased fourfold, the absolute area sown to rice increased by 60 per cent, so that there was a greater increase than is suggested by the proportionate figures. With the increased use of fertilizer as well as with the higher labour requirements, the cultivators had greater need for cash. Obviously the need for rural credit did not begin with the release of H4, but the demand increased with the spread of the new varieties.

In sum, the linked adjustments of labour demands, cultivation schedules, fertilizer, and credit requirements were seen by the government as impeding the adoption of H4 as well as holding back the increase in rice production. Specific remedies to these impediments were devised which, in aggregate, led to a massive and many-sided intervention by government agencies in paddy growing.

The most widespread and fiscally most expensive intervention was in the supply of credit. Supplying credit had begun decades earlier with the development of the irrigation colonies in the dry zone. But increasingly credit had been extended to cultivators outside of the colonies as part of the government's subsidy programme to keep down the purchase price of rice while helping the cultivator to gain a reasonable return on his crop.

Studies of agricultural credit in six villages in Matale District around 1940 and a survey in Badulla District in 1948 revealed that these cultivators were deeply in debt to local traders. The rates of interest were described as exorbitant, which resulted in 'an entire season's product ... [being] barely enough to repay the advances already taken from the trader'.[32] Some relief was provided by the government from 1947 with the first scheme for providing institutional credit to small paddy growers.[33] Initially the credit was administered through the Land Commissioner using the co-operative societies as the village-level institution. In 1952, the Department of Food Production took over responsibility which was subsequently assumed in 1957 by the Department of Agrarian Services. The volume of credit supplied rapidly increased, as is represented in Table 5.12.

The terms for extension of credit were that it be applied to production purposes only and that part of the loan could be given in kind: certified seed, fertilizer, agrochemicals, barbed wire, etc. Loans limits were set at Rs 432 per hectare, up to a maximum of 15 hectares, and the interest rate was 5 per cent. It was soon apparent that the cultivator's cash requirements for both production and consumption far exceeded the government's ability or, at least, willingness to supply loans. The Central Bank of Ceylon's survey of rural indebtedness showed that in 1950 only 30 per cent of rural households were in debt. By 1969 it had risen to over 50 per cent of households of which 8 per cent came from private sources—moneylenders, family, etc.[34] Obviously these figures do not unambiguously demonstrate a connection between the cultiva-

TABLE 5.12. *Government credit for paddy production* (Rs Million)

Year (in 3-year intervals)	Loans granted	Recovered	Cumulative balance outstanding
1947–8	4.3	2.5	1.8
1950–1	6.7	4.1	9.8
1953–4	11.7	11.3	18.3
1956–7	21.9	27.7	20.6
1959–60	14.0	13.8	24.1
1962–3	10.7	9.4	—
1964–5[a]	27.5	16.5	54.2

Note: [a] A 2-year interval.
Source: D. M. B. Marapone and J. E. D. Karunaratne, 'Institutional Credit Support', p. 2.

tion of the H lines and the increase in the cultivators' indebtedness, yet we know from village-level studies of paddy farmers that they were becoming more deeply enmeshed in the market economy, more concerned about their cash flow, and more heavily dependent on credit to meet production costs associated with the fertilizer responsive varieties of paddy.

With the great increase in fertilizer use, 350 per cent from 1958 to 1964, a fertilizer subsidy for paddy was introduced in 1962 to hold down prices, and the government-run Sri Lanka Fertilizer Corporation was set up in 1964 and assumed responsibility for the low-cost distribution of fertilizer. In addition to low-cost fertilizer and credit, the cultivator needed greater crop security before he was willing to invest in seed and fertilizer purchases. The paddy-crop insurance scheme was started on a trial basis in 1958 and was extended in 1962 and 1963. In short, by addressing one after the other of the production problems originating with the planting of the new seeds, successive governments presided over a transformation of smallholder agriculture after 1958.

Beyond the concern for paddy production itself, the left coalition government of S. W. R. D. Bandaranaike and, after his assassination in 1959, the government of his widow Sirima Bandaranaike, toyed with the idea of a radical restructuring of rural society. The principal exponent of such a restructuring was the new Minister of Agriculture and Food, Philip Gunawardena, leader of the revolutionary Lanka Sama Samajist Party. Gunawardena and his principal assistant G. V. S. de Silva, an economist who had studied agrarian conditions in both Kandy and Matara districts,[35] believed that the traditional conservatism and class structure was a bar to increasing productivity. They saw the need to 'prepare rural Ceylon for cooperative and collective ideas', and, further, 'to carry the class struggle into the villages'.[36] For Gunawardena, the 'feudal remnants' who blocked rural development were most clearly operating in the landlord–tenant relations.[37] One of Gunawardena's earlier actions was to prepare a new bill designed to protect the tenants from the landlord's exactions, the Paddy Lands Act No. 1 of 1958.

The new Act was much broader in its scope than its title suggests, for it undertook to reorganize agrarian relations. The preamble of the Act itemized the concerns of the minister and the government.

An Act to provide security of tenure to tenant cultivators of paddy lands; to specify the rent payable by tenant cultivators to landlords; to enable the wages of agricultural labourers to be fixed by Cultivation Committees and agricultural labourers to be appointed as tenant cultivators and collective farmers; to provide for the consolidation of holdings of paddy lands, the establishment of collective farms for paddy cultivation, and the regulation of the interest on loans to paddy cultivators and the charges made for the hire by paddy cultivators of implements and buffaloes; to make provision for the establishment of Cultivation Committees; to specify the powers and duties of such Committees; to confer and impose certain powers and duties on the Commissioner of Agrarian Services; to abolish the liability of proprietors within the meaning of the Irrigation Ordinance, No. 32 of 1946, to pay remuneration to irrigation headmen; to control the alienation of paddy lands to persons who are not citizens of Ceylon; to repeal the Paddy Lands Act, No. 1 of 1953; and to provide for matters connected with or incidental to the matters aforesaid.[38]

Each of these provisions marked a further penetration by the government into the fabric of agrarian relations. Terms of tenancy and rent were regulated. Wages were to be fixed. Interest on loans and charges for hiring buffaloes and tractors were set. Fragmented paddy lands could be consolidated. The old irrigation headmen who claimed a share of the harvest from each cultivator were abolished, and their functions were taken over by new headmen paid by the state. Elected cultivation committees were provided to act as the instrument of the government in the paddy-growing villages. All of the foregoing provisions were to be monitored and further expanded by the newly created Department of Agrarian Services. In short, the state now officially recognized the limiting conditions for rice-growing were not only the availability of seeds and modern inputs but an interconnected set of practices concerning moneylending, tenancy, wage labour, and the very structure of property holding itself; and furthermore, that the means of easing these problems were increased state regulation of the relations of production, the creation of self-help cultivators' committees—the cultivation committee—with the longer-term possibility of moving towards collective agriculture.

The agrarian strategy of restructuring the relations of production was not in fact carried very far in spite of much enthusiasm for learning from China and Chairman Mao among the left-wing radicals. The Paddy Lands Act's amended provisions and halting implementation undercut the vision of its preamble, and the reality of 14,500 registered complaints of eviction of tenants by the end of 1959 revealed the extent of landlord resistance.[39] As an instrument for increasing rice production and bringing about the modernization of the villages, the edge of the Paddy Lands Act had been seriously blunted. Rather it was the progressive technology of rice plant-breeding and its entailments that seemed to be spreading change through the paddy-growing villages.

While successive governments devised policies to increase rice production, the rice researchers were refining and developing their research strategies. Within the Department of Agriculture and among the plant scientists there was conflict emerging, though it too was related to productivity and the deeper questions of the ways in which research issues and the relations of research scientists to the cultivator were defined. There were three different positions upheld within the Department of Agriculture, two which had an effect on contemporary policy and one position which was to subsequently emerge and affect policy after 1974. These three positions may be characterized as: first, a cautious elaboration of the earlier pure-line seed selection approach through crossings to meet specific local growing conditions; second, advocating the rapid diffusion of two or three high yielding varieties which had been bred to realize the full potential of the rice plant; and third, the explicit recognition of the diversity of agro-ecological zones and cultivation practices which required the adaptation of the rice plant rather than the modification of the environment and changing the practices of the cultivator.

The first position was maintained by M. F. Chandraratne, Director of Agriculture 1956–9, Leslie Peries, Deputy Director Research 1961–74, and Hector Weeraratne, plant breeder at the Central Rice Breeding Station, 1952–76. Their general position was to integrate the breeding programme which had produced H4 into the pure-line seed selection practices. They planned to cross a series of new varieties for each different age class, subject them all to rigorous field tests, select the best plants of the best varieties, and offer proven varieties to the cultivator. They looked for incremental gains in yield by offering a broad spectrum of tested new varieties to the cultivator. Moreover they understood that cultivation practices of the villagers would limit the gains in yields of the new plants. As Peries stated, they would be content to get the 1.5 tonnes/ha cultivator to reach 2.5 tonnes/ha, the 2.25 tonnes/ha cultivator to reach 3.75 tonnes/ha, and the 3.75 tonnes/ha man to reach 6 tonnes/ha. To secure such yield gains across varied cultivating conditions and practices required a range of varieties and not just a few varieties with a broad spectrum of response to fertilizer, etc. The goal of this policy was to raise productivity throughout the island and throughout the different socio-economic classes of cultivators. It was an approach which sought to build on the existing capability of the cultivators, whether their holdings were large or small, on rain-fed or irrigated land, whether rich or poor. Every rice cultivator was to make his contribution to the drive for national food self-sufficiency. Basically it was a conservative position attempting to extend the proven practices of the research division and accepting the existing socio-economic structure of agriculture. It was also a self-protective and increasingly nationalist position, for it gave preference in research to local pure lines and cross-bred varieties, limiting the importation and recommendation of foreign varieties, particularly the International Rice Research Institute's first release, IR 8.

The second policy position was supported by Ernest Abeyratne (Head of Maha Illuppallama Research Station, 1946–61; Deputy Director Extension, 1961–74; and subsequently Director of Agriculture, 1974–7) and other research officers in the department who persuaded the Minister in the United National Party (UNP) government of 1965–70, M. D. Banda, of the merits of their position. The emphasis of this position was that dramatic gains in yield and production could be obtained by releasing high-potential crosses, whether locally bred or imported. Moreover, Abeyratne saw that he could strengthen his own position in the department if the Extension Division was the instrument for distributing the new seeds introduced independently of the Research Division. He also recognized that there would have to be a major improvement in input supplies and services to the cultivator and that rapid advances in the extension of irrigation, particularly in the dry zone, would have to be made. Less emphasis was placed on local plant-breeding and research, and much greater concern was focused on modernizing and improving agricultural organization and services. Such a widespread shift in policies would enhance the importance of the Extension Division as well as transforming the Department of Agriculture into a more interventionist and politically important organization. These men accepted the idea of supporting the best cultivators and allowing the natural demonstration effect plus the 'trickle-down' effect to carry the new varieties and practices to the smaller and less productive cultivators. It was a more radical position and one which drew inspiration from the reported achievements in Taiwan and Japan and was responding to the urgings of the international agencies, the Rockefeller Foundation and, after 1962, IRRI.

Out of the encounter between these two approaches and through direct local experience with trying to improve production came a third concept which is most closely associated with the subsequent Deputy

Director of Research, C. R. Panabokke, 1974–80. It will be simplest to characterize this policy in the following chapter, where the context which encouraged it is discussed.

The issue between the improvers and the modernizers, to give them neutral-sounding labels, was first joined over the question of whether varieties for early, large-scale distribution to the cultivators should be imported, and, specifically, whether IR 8 should be imported. By the mid-1960s the limitations of H4 and its sister line, H5, were becoming apparent. Neither performed well in the low country wet zone or in the wetter areas of the central highlands. Moreover, if the levels of nitrogen greatly exceeded 36 kgs/ha, H4 was prone to lodging. It also lodged extensively if rains came in the weeks before harvesting. With variations in soil conditions and cultivation practices H4's performance appeared to fall short of the advertised virtues of IR 8. Given the difficulties of realizing the potential yield of H4 in many parts of the island, Abeyratne and others urged the import of IR 8, which was being offered by IRRI as a much higher yielding variety—at least 25 per cent better than H4 at fertilizer levels of 107 kg/ha—and which was stated to be of wide adaptability to many soil conditions.

However, there was opposition from 'the improvers'; Hector Weeraratne, who had visited IRRI where the breeding of IR 8 was under way, had brought back an early example as well as the parent genetic material to Batalagoda. Running his own trials on IR 8 and experimenting with the parent varieties, Dee-Geo-Woo-Gen, a dwarf *indica* from China, and Peta, from Indonesia, Weeraratne discovered that IR 8 and the parents were very susceptible to the Sri Lankan varieties of bacterial leaf blight. Weeraratne, supported by the Director of Research, argued that it would be a great mistake to release IR 8 and any other imported variety that had not been thoroughly tested under local growing conditions. Moreover, the plant pathologists, led by the Director of Research, who was himself a pathologist, pointed to the danger of major imports of seed paddy when there were no facilities for controlling the spread of alien pests or diseases that might travel with such large volumes of seed paddy. Abeyratne, who had just assumed the Deputy Directorship of Extension, was eager to provide an internationally praised variety to the cultivators and was sceptical of the dangers of bacterial leaf blight in the dry zone. The attraction of a major dramatic increase in rice yields and the prospect of economizing on foreign exchange by improving local rice production were compelling, and 9,000 tonnes of IR 8 seed was imported from the Philippines in 1966.[40]

The results were disappointing. IR 8 was selectively distributed under government auspices, but additional seed was distributed indiscriminately by an international charitable organization, Oxfam, to cultivators in many different locations. Within a few seasons it was apparent that

IR 8, as predicted by Weeraratne, was very susceptible to bacterial leaf blight (BLB) and that it would only perform well in the dry zone where lower humidity checked the spread of BLB. However, even in the more favourable climate of the dry zone, IR 8 revealed other limitations. It was well known that to approximate the potential yields of IR 8 a new set of inputs and cultivation practices were needed. The distinguishing characteristics of IR 8 were that it was a non-photosensitive variety of 'dwarf' stature, that is, under 100 cm, which was capable of converting high levels of nitrogen into more grain. The short stature and many subsidiary shoots or tillers, each of which bore a pannicle, as well as the erect, stiff leaf structure which helped support the stems, gave an extremely high yielding plant. However, its short stature was accompanied by a shallow root network which failed to provide adequate nutrients as fields dried out. It needed both high levels of nitrogen fertilization, 107 kg/ha, and plentiful water to reach optimum yields. Since the nitrogen also fertilized the weeds it was necessary to control the weeds, many of which would grow taller than IR 8 and would shade the plant and so reduce yields. Since the plant was short, earlier practices of impounding the water to drown slower-growing or shorter weeds could no longer be used. The alternatives were hand-weeding or the use of weedkillers.

Moreover, the numerous tillers of the plant required that each plant be given ample space to put out its secondary stems and leaves to avoid mutual shading. The normal practice of broadcast sowing was no longer possible; either row sowing or nursery beds with row transplanting were necessary. Since seed-beds had other advantages such as earlier germination, screening against germination failure, and easier protection of the shoots, transplanting became the preferred means of spacing the plant. Transplanting did require much more labour, which usually led to the hiring of labour or participation in labour exchange. With regularly spaced plants it was then possible to do hand weeding, though this was so labourious it was an unpopular practice. Harvesting of the shorter, many-stemmed plant required more labour, while the numerous smaller pannicles required longer and more careful threshing to recover as much grain as possible. Since most threshing was done by driving water buffalo over piles of cut stalks which had to be constantly turned, picked up, with the grain collected from the threshing floor, IR 8 took longer to thresh and so was more costly for the hire of buffalo and labour.

Finally, it was discovered that IR 8 had a shorter dormancy period than the established varieties. Stored paddy would start to sprout in eight or nine months, which caused problems in holding over seed or keeping it as a food supply. The older varieties could be used for seed even after two or three years and in good dry wooden storage chests could be used for food as long as seven or eight years after harvest. These foregoing characteristics which had direct implications for the cultivator's labour time and costs further limited the popularity of IR 8.

In the same year, 1966, that IR 8 was being imported, IRRI approached the government to enter into an agreement for the posting of an IRRI staff researcher in Sri Lanka. The IRRI staffer was expected to negotiate with the relevant plant scientists, the Director of Agriculture, and the government to participate in the various programmes centred in Los Banos. The first of these representatives was James Moomaw, 1966–8, who was well remembered by the Sri Lankan plant scientists for his abrasive arrogance.[41] Moomaw reportedly urged the Sri Lankans to phase out their rice research programme on the grounds that IRRI with its greater scientific resources would supply all the new rice varieties that were needed on the island and indeed throughout wet-rice cultivating Asia. Moomaw categorically asserted that IRRI's new variety, IR 8, would not only flourish in all soil conditions but would yield more rice than any locally bred variety. He acted as though irrigated paddy fields provided identical growing conditions around the world. Variations in soil types, nutrient availability, levels of insolation, or in disease and pest genotypes were discounted. All that was needed was ample irrigation water and massive doses of fertilizer for the IRRI rice varieties to outperform the local varieties. Moomaw is reported as saying that anything less than 100 bushels per acre for IR 8 was a crop failure, though he himself was often unable to reach these yields on his own test plots in spite of massive doses of fertilizer. Moreover Moomaw horrified the local entomologists by recommending that IR 8 be protected from pests by introduction of pesticides into the irrigation water every twenty days whether needed or not. As H. E. Fernando, an entomologist, observed, such wholesale applications of pesticides would destroy the whole ecosystem in the rice fields. Understandably Moomaw did not last very long and he was replaced by William Golden (1968–72) whose personal style combined with an interest in local rice-growing and the achievements of the rice scientists made him a more effective representative of IRRI.

The governmental decision to enter into an agreement with IRRI and to import IR 8 had little effect on the research programme within the department. Here 'the improvers' under Leslie Peries had a free hand to continue their favoured approach to research. Weeraratne at Batalagoda concluded that the IR 8-type dwarf varieties were not well-suited to Sri Lankan cultivation and with the support of the Director of Research started a breeding programme to produce a high yielding, semidwarf or intermediate plant size. The problem of lodging was to be handled by a shorter plant size than the pure-line varieties or the H lines and by increased straw strength. The intermediate stature of the plant would be associated with a deeper root system which would draw moisture and nutrients from further down in the soil than the dwarf varieties such as IR 8. A plant derived from an H4–Podiwee A8 cross, H8, was crossed with an Indonesian variety, Engatek, to produce BG 11–11, the second major success for the Central Rice Breeding Station under Weeraratne. BG 11–11 could produce 6.5 tonnes/ha at the field station and was capable of 5 tonnes/ha in farmers' fields, a 15–20 per cent improvement in yield over H4. It lacked the very high yield potential of IR 8 but responded very well to intermediate levels of nitrogen fertilizer of 62–71 kg/ha, rather than requiring the 108 kg/ha of IR 8. It was moderately resistant to bacterial leaf blight and was the popular 4–4½-month class variety. Equally important it had vigorous early growth and with its taller stature was able to outgrow and hence shade out weeds. The new variety was given its trials and released in 1968.

The successes of Batalagoda were the most spectacular in the rice-breeding programme, but ever since Peries took over as the first Deputy Director of Research in 1961 other stations had been encouraged to turn their varietal testing and selection programmes into small-scale breeding programmes. This policy was an extension of the earlier moves towards decentralization of the late 1940s and of the posting of research officers in the field stations. Peries believed that a range of varieties would be needed for particular local environments and to be compatible with cultivation practices. This logically and necessarily led to the practice of giving each plant breeder a wide measure of discretion in identifying optimal plant characteristics for the surrounding area. To keep a check on the research officers' directions and performances and to speed up the testing of new varieties, Peries introduced the Co-ordinated Rice Varietal Testing programme (CRVT) in 1968.[42] Each breeder from Batalagoda, Bombuwela, Ambalantota, Maha Illuppallama, or other stations where rice was bred nominated his best varieties in each age class for an eight-station co-ordinated seasonal test. Separate trials were conducted for each age class—3, 3½, 4, and 4½-month varieties. Under an agreed cultivation regime with set levels of fertilization and care the various breeders' nominations were grown and compared with control plots of H4 and IR 8: Every conceivable and measurable aspect of each variety's performance was measured, such as germination rate, tillering, culm length, pannicle weight, and, finally, yield. The senior research officers, including all the rice breeders and the entomologists, pathologists, and others from Peradeniya, visited each of the eight stations and personally inspected the performance of all varieties in the test. Combining the visual inspection with the accumulated statistical data, the officers evaluated and ranked each plant breeder's nominations. This elaborate procedure served a double purpose: first, it greatly stimulated rice-breeding in all the stations, for it threw the breeders into competition with each other and measured their successes comparatively; and second, by subjecting all varieties to comparative testing in eight locations simultaneously, scientists could gather information on pest and disease resistance, capacity to stand up under adverse conditions, as well as performance under diverse conditions, in one season. In this respect it was a more efficient and rigorous way of testing varieties than running trials at one station for eight years.

Complementary to the CRVT was the earlier programme of systematically collecting plant material for a genetic bank which could be drawn upon by the rice breeders in the different stations. Starting in the early 1960s, genetic material was collected, classified, and, after 1967, placed in the newly built cold-storage facilities at the Central Agricultural Research Institute (CARI) at Gannoruwa. (By the mid-1970s over 1,500 varieties of the rice plant had been identified and stored. Samples of local varieties had been sent regularly to IRRI and to Fort Collins in the United States since 1962. As of May 1978, IRRI held 1,743 viable accessions from Sri Lanka which had been mainly supplied from Gannoruwa.)

The newly released BG 11–11 was put through an early sequence of seasonal trials and outperformed IR 8 at most locations. Since the Minister of Agriculture, M. D. Banda, remained sceptical of the virtues of BG 11–11 after the experience of IR 8 and its susceptibility to blast and BLB, they insisted on a full-scale trial with 60 hectares of H4 and 60 hectares of BG 11–11, rather than basing a conclusion on the findings of small test plots of one-twentieth of a hectare or less. In this test H4 yielded 2.7 tonnes/ha while BG 11–11 produced 4 tonnes/ha under identical cultivation practices. Moreover, BG 11–11 demonstrated that it had reasonable resistance to blast and BLB.

The rice-breeding programme and the rigorous testing which produced BG 11–11 were to produce a number of other improved varieties. In the four years following the release of BG 11–11 in 1968, Batalagoda was able to breed and test successfully a number of high yielding, intermediate-statured plants in different age classes. A 3½-month variety, BG 34–6, which was well suited as a minor or Yala season planting alternating with BG 11–11, was released. It increased the farmers' yield figures fourfold over the most widely grown pure-line varieties, such as Vellai Perumal, which improved yields by 40–50 per cent over H7, which had been released in the early 1960s. With a yield potential of 6 tonnes/ha, BG 34–6 was regularly reported as producing 4.5 tonnes/ha in the field. Soon after, BG 34–8, the first high yielding 3-month variety, was released. It was hailed by plant breeders as a miracle rice, for it had a '20 kg-a-day' performance. Within 90 days from germination it could produce 4.5 tonnes per hectare. Good farmers were reporting yields of 5.5 tonnes/ha while, in the field stations, they were getting yields of 7 tonnes/ha in 90 days. Colin McClung, the then Associate Director of IRRI, hailed BG 34–8 as a remarkable achievement and utilized it in tangible evidence of the capabilities of Sri Lankan plant scientists and the rice-breeding programme. It was widely borrrowed and utilized in the breeding programmes in other countries and at IRRI.

The BG 34–8 variety was widely adopted and is one of the major varieties now being cultivated. The advantages of a high yielding variety of this age class are many. It can be used as a second or even a third planting in well-watered lands. The short maturation period means that from seed-bed sowing to harvesting and threshing will be about 100 days and that with careful management and ample labour it would be possible to have a second planting of rice in the fields within 110–20 days of the transplanting of BG 34–8. Another valued advantage is that on rain-fed lands when the rains are delayed and the growing season is truncated BG 34–8 can be sown to mature before the last dates of high-probability rainfall needed at flowering time. If there was no photo-insensitive 3-month variety available, cultivators would have to plant a longer age class and take a chance on erratic late rainfall or abandon the hope of cultivation during that season. Similar situations emerge when there are storms and flooding with damage to early planted varieties or when a pest infestation such as stem borers destroys the new shoots. In these situations it is possible to plant a 3-month variety and have a reasonable chance of taking a good harvest off the land.

While other varieties in other age classes were bred and released, none marked as dramatic an advance as BG 34–8. These newer varieties were bred as intermediate-stature, non-lodging, fertilizer-responsive varieties. They were also systematically screened for resistance to blast, bacterial leaf blight (since 1967), and other common diseases and pests. From sometime in the early 1960s the laboratory screening programme for new varieties was enlarged to include tests for milling characteristics; acceptable size, shape, and colour of grain; satisfactory amylase content (which directly affects the cooking results); parboiling characteristics, and others. It was recognized that it was a waste of effort to produce a high yielding, pest- and disease-resistant variety if its processing and cooking characteristics were not acceptable to the consumer. As yet there was no systematic testing of the nutritional levels provided by different varieties. The target was rather to produce a reliable, high yielding variety of an acceptable quality rather than of high quality.

Looking ahead to the problems of the mid-1970s, the breeding objectives which had been centred on yields with a gradually enlarging series of secondary desiderata were suddenly called into question by the outbreak of a plague of brown plant hopper in Batticaloa and Amparai Districts in 1973–4. Brown plant hopper (BPH) was an endemic pest in the island but had never been known to damage crops extensively. The reasons for the outbreak were not clear. Whether the cause was the evolution of a new variant form of BPH, whether variant forms of BPH came in with imported seed, whether the hybrid varieties of rice were more susceptible to attack, or whether increasing levels of nitrogen application or weedkiller and pesticide use had created a favourable environment for a massive expansion of the BPH population was not known. The effect of BPH plague, which spread in following seasons with lesser effect to other districts, was to force the breeders to turn their attention to pest and disease resistance. When plant damage and crop loss due to these causes was assessed it was found that gall midge and stem borer were also causing substantial damage; in 1976 and 1977, respectively, the toll of these two pests increased.

From this time the breeding programme not only sought to maintain and improve yields but to also breed varieties that were resistant to the principal pests and diseases. It was recognized that this would be a constant breeding objective, for the pests were changing, evolving in parallel with the human-directed development of the rice plants; it was discovered, for example, that there were a number of different types of BPH in different environments around the island.

The search for resistant varieties led to the importation for breeding and testing purposes of varieties from other countries. IR 26, which had been developed as a BPH resistant variety in Los Banos, was imported, but its resistance soon broke down. It was discovered that the biotype of BPH in the Philippines was different than those of Sri Lanka and India, and so IR 26 performed poorly in both India and Sri Lanka. Fortunately a variety from South India, Ptb 33 from Pattambi, was found to have resistance to the types of BPH found in Sri Lanka. Crossing was undertaken and a resistant high yielding variety was tested and released in 1979. Similarly, the search for resistant varieties to gall midge led to the import of promising types from Warangel in India. The search for a resistant variety to stem borer attacks had, as of 1980, not identified promising parental stock.

The broadening breeding desiderata were further complicated by the recognition that soil types, hydrology, and nutrient levels all had a large impact on plant performance—yield, processing, and cooking characteristics, as well as disease resistance. While in concept this was known to the plant scientists, it was not until about 1971 that the recognition of the need to breed varieties for a wide range of edaphic conditions began to enter the planning and specification of breeding objectives. There is no single event or act marking the growth of this awareness in the thinking of the principal plant scientists. The most compelling reason for its gradual acceptance was, however, clear. In spite of a broad spectrum of high yielding varieties of different age classes being available through the Extension Division of the Department of Agriculture, the cultivators in many parts of the island refused to adopt either the H4 and associated lines, which are known as the old improved varieties, or the BG series, known as the new improved varieties. The Department of Agriculture estimated that only 60 per cent of the rice land was planted with either the old or the new improved varieties (OIV and NIV). In the irrigated tracts of the dry zone the adoption rate was 90 per cent or better; in the low-country wet zone the OIVs were between 10 and 15 per cent, NIVs another 10 per cent, leaving 75 per cent of the rice area still being planted to the pure-line selections or traditional varieties. It was estimated that as many as 80,000 hectares in this region were still under the unimproved varieties. In the mid-country and high country no one knew what the adoption rate was. Farmers occupying different elevations with different soils and different water retention properties made their selection to suit their own particular needs. The second and related fact which was being brought home to the plant scientists was that even when the improved varieties were being grown the yields often fell far below expectations and the known achievements of some farmers. Thus low adoption rates and yields, far below the improved plants' known performance levels, forced the agricultural scientists and the government once again to consider the root causes and what they should do about them.

NOTES

1. Information provided by Dr Somasiri, Land and Water Division, Department of Agriculture, May 1978.
2. A number of studies report on this change in the last thirty years. For earlier years, see discussion in *Report of the Kandyan Peasantry Commission*, Sessional Paper 18 of 1951, pp. 90–103, 142–7, 200–7. A recent study is Barrie M. Morrison, et al. (eds.), *The Disintegrating Village: Social Change in Rural Sri Lanka*, Colombo: Lake House Investments, 1977.
3. For an excellent description of a dry-zone village, see E. R. Leach, *Pul Eliya: A Village in Ceylon*, Cambridge: Cambridge University Press, 1961.
4. These summary observations are based on Morrison's and students' field work in Batticaloa and Polonnaruwa districts as well as upon Rudolph Wikkramatileke's *Southeast Ceylon: Trends and Problems in Agricultural Settlement*, Chicago: Department of Geography, University of Chicago, 1963, pp. 89 ff.
5. The best study of the coastal strip and its immediate hinterland is M. P. Moore and G. Wickramasinghe, *Agriculture and Society in the Low Country (Sri Lanka)*, Colombo: Agrarian Research and Training Institute, 1980. For the interior villages, see G. Obeyesekere, *Land Tenure in Village Ceylon*, Cambridge: Cambridge University Press, 1967.

6. For an excellent historical discussion of the social and economic position of one group in this region, see Michael Roberts, *Caste Conflict and Elite Formation*, Cambridge: Cambridge University Press, 1982.
7. Much of the information in this section was provided through the courtesy of Dr E. F. L. Abeyratne, Director Department of Agriculture, 1974–7.
8. On the nineteenth-century significance of economic botany and the world-wide exchange of planting materials, see Lucille H. Brockway, *Science and Colonial Expansion: The Role of the British Royal Botanic Gardens*, New York: Academic Press, 1979.
9. Interview, E. F. L. Abeyaratne, 27 May 1978.
10. L. Lord, 'The Selection of Pure-Line Strains of Paddy, their Testing and Distribution', *Tropical Agriculturalist*, 68, 5 (May 1927), 309–18.
11. Interview, D. Senadhira, Research Officer in Charge, Central Rice Breeding Station, Batalagoda, 16 May 1978.
12. All quotations from *Agriculture and Patriotism* are taken from H. A. J. Hulugalle, *The Life and Times of Don Stephen Senanayake*, Colombo: Gunasena, 1975, pp. 117–119.
13. Vijaya Samaraweera, 'Land Policy and Peasant Colonization, 1944–1948,' in K. M. de Silva (ed.), *History of Ceylon*, Peradeniya: University of Ceylon, 1973, p. 457.
14. S. Arumugam, *Water Resources of Ceylon*, Colombo: Water Resources Board, 1969, pp. 242–4.
15. For a full discussion of these policies see B. H. Farmer, *Pioneer Peasant Colonization in Ceylon*, London, Oxford University Press, 1957.
16. Hulugalle, *Senanayake*, p. 110.
17. Farmer, *Peasant Colonisation*, pp. 316–20.
18. M. P. Moore, 'The State and the Peasantry in Sri Lanka', Ph.D. thesis, University of Sussex, 1981, p. 114.
19. L. A. Wickremeratue, 'The Economy in 1948', in K. M. de Silva (ed.), *Sri Lanka: A Survey*, Honolulu: University Press of Hawaii, 1977, pp. 134–5.
20. Samaraweera, 'Land Policy', p. 460.
21. P. C. Bansil, *Ceylon Agriculture: A Perspective*, New Delhi: Oxford and IBH Publishing, 1971, pp. 250–5. See corrections in *Economic and Social Development in Ceylon, 1926–1950*, Colombo: Ministry of Finance, 1951, pp. 7–8.
22. D. Senadhira, M. P. Dhanapala, and C. A. Sandanayaka, 'Progress of Rice Varietal Improvement in the Dry and Intermediate Zones of Sri Lanka', paper presented to the Rice Symposium 1980, Colombo, 25–6 September 1980, in *Rice Symposium 1980*, Colombo: n.p.
23. Interviews, M. F. Chandraratna, Director Dept. of Agriculture, 1956–9, on 31 May 1978; J. W. L. Peries, Deputy Director Research, 1961–74, on 21 May 1978; and with E. F. L. Abeyaratne, former Director Dept. of Agriculture, on 27 May 1978.
24. Interview, D. Senadhira, Research Officer in Charge, Central Rice Breeding Station, Batalagoda, 7 March 1977.
25. Figures reported for Maha 1966–7 in W. B. Medagama and S. H. Charles, 'Seed Paddy Production: Highlights Over the Last One and One-Half Decades [sic]', in *Rice Symposium 1980*.

26. Interview, Irwin Gunawardena, plant botanist, Central Agricultural Research Institute, Gannoruwa, 28 February 1977.
27. Interview, Paul E. Peiris, plant breeder, Bombuwela Rice Research Station, 10 March 1977.
28. Interviews, Irving Gunawardena, Central Agricultural Research Institute, Gannoruwa, 28 February 1977; U. Tennekoon, agricultural instructor, Wariyapola, 17 May 1978; and L. T. P. de Soysa, district agricultural extension officer, Kalutara, 23 May 1978.
29. This illustrative sequence of cultivation is based on observations in a village in the Udunuwara division of Kandy District. See Barrie M. Morrison, 'Meegama', in Barrie M. Morrison et al. (eds.), *The Disintegrating Village: Social Change in Rural Sri Lanka*, Colombo: Lake House Investments, 1979, pp. 71–113.
30. R. A. P. Malasekera, 'Distribution of Fertilizer in Sri Lanka', paper presented to Seminar on Economic and Social Consequences of the Improved Seeds, 19 April–20 May 1973, mimeo.
31. Senadhira, Dhanapala, and Sandanayaka, 'Progress of Rice Varietal Improvement', p. 8.
32. *Report of the Kandyan Peasantry Commission. SP XVIII-1951*, p. 143.
33. D. M. B. Marapone and J. E. D. Karunaratne, 'Institutional Credit Support for Paddy Production in Sri Lanka', paper presented to Seminar on Economic and Social Consequences of the Improved Seeds, 19 April–20 May, 1973, mimeo.
34. Ibid.
35. See Martin E. Gold, *Law and Social Change: A Study of Land Reform in Sri Lanka*, New York: Nellen, 1977, pp. 27 ff.
36. Ibid. 30, quoting de Silva.
37. The following paragraphs are largely based on the excellent discussion in M. P. Moore, 'The State and the Peasantry in Sri Lanka', Ph.D. thesis, University of Sussex, June 1981, pp. 126 ff. and Gold, *Law and Social Change*, pp. 27–42. See also R. J. Herring, 'The Forgotten 1953 Paddy Lands Act in Ceylon: Ideology, Capacity and Response', *Modern Ceylon Studies*, 3, 2 (1972), 99–124.
38. Paddy Lands Act No. 1 of 1958, p. 3, quoted in Moore, 'State and Peasantry', p. 130.
39. Gold, *Law and Social Change*, p. 36.
40. Interview, J. W. L. Peries, May 21 1978.
41. Interviews, H. E. Fernando, Head, Central Agricultural Research Institute, 28 February 1977; C. R. Panabokke, Deputy Director Research, Dept. of Agriculture, 20 May 1978.
42. The contribution of William Golden, the second IRRI staff researcher posted to Sri Lanka from 1968 to 1972, to the development of this pioneering varietal testing programme is ambiguous. The Sri Lankan senior staff asserted that the programme was entirely of their devising while IRRI staff cited it as an experimental national programme shaped by Golden and subsequently taken up and expanded by IRRI as their International Rice Testing Program.

6

Sri Lanka: The Quest for a New Research Strategy, 1971–1980

> On the research stations it is too easy to be a little self-contained enclave and not connect to real problems. Scientists must have an appreciation of the whole system and the whole problem. We [the scientists] have a moral responsibility to speak up about what is really happening.
>
> C. R. Panabokke, Deputy Director Research, Dept. of Agriculture,
> Sri Lanka, May 1978

BY the middle of the 1970s the research strategy which had led to the release of H4 in 1958 and the semidwarf BG 11–11 in 1968 could be assessed. Together with their sister lines, these new improved varieties, as they were called, had helped stimulate a great increase in yields and production. From the three-year average yield for the Maha season of 1951–2, 1952–3, and 1953–4 to the average Maha yields in 1973–4, 1974–5, and 1975–6, there had been a 51.6 per cent increase in yields across twenty-two districts over the twenty-two years.[1] This marked improvement in yields had not been brought about solely by the application of a seed-led strategy of agricultural innovation as recommended by the Rockefeller Foundation and IRRI. Rather it was the result of innovative seed-breeding coupled with the prior and independent institutional changes such as the creation of the Cooperative Agricultural Production and Sales Societies after 1947, the Paddy Lands Act of 1958, and innovative programmes in crop insurance begun in 1958 and other such changes designed to improve the security of the small rice cultivators.

Whatever the mix of causes of the increase in yields during the preceding decades, by the early 1970s the rate of improvement in yields had slowed, and the limitations of the official strategy were evident. From the base of the Maha crop seasons of 1951–2 to 1953–4 the national average yields had mounted steadily from 1.57 tonnes per hectare for fifteen years through to 1966–7 to 1968–9 when yields reached 2.33 tonnes. Seven years later, the average yields for Maha 1973–4 to 1975–6 were only 47 kg higher at 2.44 tonnes/ha. It was also apparent that the general improvement in yields concealed much variation from district to district. Districts such as Polonnaruwa had absolute increases of over 1.6 tonnes to the hectare over the twenty-two years between 1951-2 and 1973-4, yet other districts, such as Moneragala or Vavuniya, had no increases or only very small increases.[2] It was evident to the new Minister of Agriculture, Hector Kobbekaduwa (in the 1970 Sri Lanka Freedom Party (SLFP) government) and to the senior officials of the department that something more would have to be done to stimulate further increases in yields and production. The limitations inherited from the 1960s would have to be overcome.

The concern for continued growth of yields and production was affected by two internal developments and one external one. First, the country's balance of payments and budgetary deficits were being seriously affected by the continuing costs of importing sufficient rice to supply the rice ration and to meet the needs of the rapidly growing population. From 1946 to 1971, the national population had nearly doubled from 6.7 million to 12.7 million. To feed many more people the imports of rice had grown from an average of 375,000 tonnes in 1946-7 to an average of 430,000 in 1970-1. It was obvious that in spite of the 50 per cent increase in yields that the country was no closer to meeting its rice requirements in 1971 than it had been twenty-five years earlier.

The second internal development occurred in 1971, the year after the election, when the militant Maoist party, the Janatha Vimunthi Peramuna or JVP, attempted a revolutionary seizure of power. For over a week communications with parts of some rural districts were cut off. The government was badly frightened, and both police and army were vengeful in their repression. Thousands of persons were rounded up into detention camps. The government's analysis of the insurrection was that it drew its energy from the alienation of the educated, unemployed rural Sinhalese youth from the densely populated districts of the low country and mid-country.[3] The government concluded that a much more energetic policy of agricultural development was needed, including a redistribution of land to the landless. In the words of Director of Agriculture from 1974-7, Dr Ernest Abeyratne, we 'were sitting on the edge of a volcano'. Some of the leaders in government recognized that they had to create a network of rural institutions to encourage small cultivators to exercise greater control over their production. They had to 'head off a revolution' in rural Sri Lanka.[4]

Pushed by the awareness of the extensive rural unemployment and political alienation and by the costs of continuing to import hundreds of thousands of tonnes of rice each year, the SLFP government of Sirima Bandaranaike undertook vigorously to support rural development. Building on the prior work of the preceding UNP government of Dudley Senanayake (1965-70), when the attention of the planners had begun to shift back to rice production, the SLFP government issued a new five-year plan in November 1971, which identified agricultural as the sector of greatest potential growth.[5] Rs 3 billion were allocated for agricultural development of which over half was committed to the support of 'traditional agriculture' producing for domestic consumption—mainly paddy production. In addition to the projected investments in agriculture a series of new institutional reforms and innovations were carried out. The Land Reform Law No. 1 of 1972 created a Land Reform Commission in which was vested amounts of agricultural land in excess of the ceiling for individually owned land. In the same year the Agricultural Productivity Law No. 2 of 1972 created the Agricultural Productivity Councils to supervise the cultivation of land that was not reaching the norms of productivity. Also in the same year, 1972, the Agrarian Research and Training Institute (ARTI) was created in Colombo with the assistance of the FAO and the UNDP. Part of ARTI's mandate was to undertake 'research into the agrarian structure and to examine the institutional factors basic to the development of agriculture'.[6]

The third change was the growing concern of international aid agencies to deliver the results of research to the farmers. Institutionally, this meant trying to shift research testing off the experimental farms and research stations and into the fields of practising farmers. For example, IRRI began its major off-station research in Sri Lanka in 1973 when Nyle Brady replaced Chandler as director. The concern to reach the farmers also led to much greater attention being paid to extension work. Subsequently the World Bank provided funds in 1973 and 1974 for Sri Lanka to expand and improve its extension services assisted by consultants from the Bank. The shift of attention to the delivery of relevant research to the farmers meshed neatly in Sri Lanka with the long-standing concern of the department with variant soil types and the identification of agro-ecological regions. By 1975-6, the extension workers were the officials who came closest to understanding the real problems of the cultivators and who could give direction and relevance to the research and plant-breeding at the research stations.[7] The way in which their understanding was communicated is the subject of this entire chapter.

6.1 The Central Agricultural Research Institute (CARI) in the Mid-Country and Highlands Regions of the Wet Zone

In these regions paddy cultivation has traditionally been the crop around which the villages' agricultural as well as social organization have been structured. Yet it was evident from both local-level studies of academic researchers and from district-wide surveys of ARTI that by the

1970s rice-growing was declining in importance as a source of household income for the small cultivators.[8] It is in the context of the declining economic significance of rice and the growing importance of other cash crops such as spices, tobacco, vegetables, as well as the enlarging significance of off-farm employment, that the rice research strategies and the work of the extension division must be assessed.

The links between the paddy cultivators and the research stations in the mid-country and highlands region of the wet zone are maintained largely through the District Agricultural Extension Officers for Kandy, Kegalle, Matale, Nuwara Eliya, Badulla, and Ratnapura. The research officers at both the Central Agricultural Research Institute at Gannoruwa and at the Central Rice Research Station at Batalagoda said that cultivators never came directly to the station for advice or assistance.[9] From the cultivators we learn that while there was an informal network spreading out from the stations through the research, office, and field workers to some villagers, the extent of information and planting material that moved in this network is impossible to discover. Generally the contact between the research staff with their findings and the rice cultivator is maintained by the staff of the District Agricultural Extension Officer (DAEO).

The staff available to the DAEO varies from district to district though the grades and general functions are much the same in all districts. In addition to the DAEO himself, who is usually a graduate in plant science or agriculture-related studies, there are one or more additional DAEOs who also hold appointments as officers in the department.

The next level is occupied by Agricultural Instructors (AIs), who are divided into two groups: the headquarter subject-matter officers and the field AIs.[10] All the younger AIs are graduates of the agricultural training centre of Kundasale, which gives science graduates of the secondary schools an additional two-year training in agriculture. The headquarter subject-matter officers are commonly the most experienced of the AIs, who have benefited from periodic additional training at Kundasale or at the In Service Training Institute at Gannoruwa. Specific 'subject-matter' responsibilities were allocated, in Kandy, in the following manner: paddy, subsidiary food crops, vegetables, tobacco, and plant protection; an administrative assistant monitored the Training and Visit Programme, and generally assisted the DAEO. In other districts there were AIs specializing in horticulture and beekeeping, machinery, young farmers' clubs, farm women's extension, and other subjects relevant to the district. The field AIs are stationed in the Agrarian Service Centres or in other suitable centres distributed throughout the district. In Kandy District there are forty-four AI territorial ranges which are grouped into four agro-ecological zones. One centre acts as a zonal focal point for extension staff in that agro-ecological zone. The AIs are expected to supervise the range extension office, identify the seed, fertilizer and agro-chemical requirements of cultivators, and ensure that supplies are available through the extension service, the Multi Purpose Co-operative Societies, or some other agency. The AI is responsible for ensuring that all agricultural statistics are properly kept and reported. He is also responsible for conducting the periodic sample-cutting surveys which are aggregated to form the district paddy yield and production figures for each season. While he has many other administrative responsibilities, such as endorsing loan applications, ensuring the collection of crop insurance premiums, etc., his most important task is the regular and systematic visits to the cultivators, ensuring, as well, that his subordinate staff are making similar regular scheduled visits.

The responsibility of contacting and helping the cultivator largely rests with the next level of extension staff. In each AI's range there are on the average three Krushi Karma Viapthi Sevakas (KVSs) who are graduates of farm schools or of secondary schools with some additional and recurrent short training programmes. In all there are 140 KVSs stationed in Kandy District. All the field KVSs have a schedule for visiting contact farmers and farm groups in the villages in different parts of the AI range. Ideally each KVS has responsibility for contacting 600 to 700 households of farmers. To meet his share of the farmers the KVS is expected to be on circuit Wednesday, Thursday, and Friday and in the course of two weeks meet thirty-five to forty contact farmers in a fixed schedule which, for example, will bring him back to the same farmer every second Thursday at 11 a.m. At the appointed time and place the KVS is expected to call on designated contact farmers. Aside from helping the farmers with their specific problems, the KVS should have general information and advice appropriate to the crop and its growth stage. So in a paddy area served by small tanks where the plants are reaching flowering stage he might advise that water be released from the storage tank if there is no rainfall during the following week and that a final light application of nitrogen fertilizer be given. He will also report on the spread of various pests and plant diseases and urge that a field watch be established and that appropriate agrochemicals and sprayers be readied. In his visit in the following two-week cycle he will repeat and update his advice; thus throughout the year his 'message' will change to meet the changing agricultural conditions. If there are problems about which he lacks information and the Agrarian Service Centre lacks resources, the problem will be reported by telephone to the DAEO or discussed during fortnightly visits by headquarters staff to the AI's range office. More difficult problems will be reviewed by the DAEO staff and raised with appropriate research officers at Gannoruwa. Thus it is principally through the extension staff that the problems and needs of the paddy cultivators are expected to be made known to the research staff and the appropriate responses, technological or otherwise, conveyed to the cultivator.[11]

To the cultivators in these surrounding agro-ecological zones the CARI at Gannoruwa contributes little direction. Most of the needs of the cultivators are met by the existing resources of the Extension Division. When specifically asked what contacts were made with CARI, the staff at the DAEO's office had difficulty in remembering occasions. In 1977 there had been an attack by a variety of caterpillar on paddy seeds, and the assistance of CARI research staff had been sought and effectively provided. A chemist at the research station had provided advice on alternate mixes of chemicals for insecticides when the recommended variety was in short supply, and a soil chemist had identified nutrient deficiency in one local area. When new varieties of potatoes and tomatoes were being tested in the extension field trials, research staff came to three of the dialogues being held between a group of contact farmers and the extension staff. No other assistance could be recalled by the officers in the DAEO in Kandy. The Deputy Director of Agriculture (Extension), Earl Jayasekera, candidly acknowledged that CARI did not relate well to farmers in the highlands region. He suggested that the emphasis on basic and on applied national-level research problems which kept the officers' attention concentrated in their laboratories and the research station's test plots turned them away from concern with the highland cultivators' particular problems.[12]

Even at a more general level it was difficult to see how the CARI research programme would be of help to the mid- and high country wet-zone cultivators. The extension staff who were interviewed, four KVSs, three AIs and the headquarters staff, all took their lead from the priorities given by the farmers. This attitude was summed up by stating that 'the farmer is the deciding factor'. The priorities of the farmers were very clear. For paddy, the cultivators wanted varieties that would give them reliable and moderately high yields across a wide range of soil types and growing conditions. The paddy cultivators were asking for a 5½-month paddy variety that would do well in the 'half-bog' soils which are common where local drainage channels converge. BG 35 was available, but a better variety was needed that could be planted and have vigorous early growth before the onset of the Maha season rains. They preferred lower yielding but pest-resistant and water stress-tolerant varieties to higher yielding varieties that needed many expensive inputs and constant attention. Guided by a similar consideration they wanted inputs available in the most economical and usable form. So they wanted fertilizer and other agrochemicals to be available in smaller quantities, 10–25 kg for fertilizer, so that it could be used with little waste on the small fields operated by most cultivators.[13] With agrochemicals for treating brown plant hopper costing Rs 80 for a half-litre bottle, the cultivators wanted to be sure of sprayers that were efficient and readily available at low rental costs. Thus if the paddy variety needed inputs, they wanted those inputs to be as economical and efficient as possible given the small-sized holdings and low cash sales of most cultivators.

For vegetables, which were of even greater interest than paddy in some of the AI ranges, there was a very heavy demand for better-quality seed potatoes and for onions to meet the market demand. In the AI ranges near Kandy City, the greatest interest was in seed for the household's mixed vegetable gardens rather than in paddy. In Four Gravets, for example, most paddy cultivators had very small plots and other jobs, but the women left at home were very interested in growing vegetables for household use and for market sale. Also regarding vegetables, commercial cultivators were enquiring about storage and marketing services.

Given these concerns, communicated through the extension workers, CARI appeared to have little relevance to these small cultivators in and around Kandy District. The distant Central Rice Breeding Station at Batalagoda was breeding different age classes of paddy that were resistant to brown plant hopper and shortly expected to release several which would be of use to the mid-country and highlands wet zone cultivators. However, there was a clear disjunction between the concerns of the leading research station in the country and the preoccupations of the cultivators trying to generate an income from their land and labour.

6.2 Maha Illuppallama and the Northern Dry Zone

The level of economic development, the agro-ecological conditions, and the organization of agriculture in the dry zone are very different from those found in the highlands of the wet zone. In the districts of Anuradhapura, Polonnaruwa, Batticaloa, Trincomalee, Vavuniya, Mannar, and Jaffna as well as parts of Puttalam, Kurunegala, Matale, and Amparai smallholder agriculture is the principal economic activity. These are the farming districts of the island. For many cultivators, perhaps a majority, high-value cash crops such as chillies, cowpea, green gram, or aubergines, which are grown on high land, receive the farmers' first attention with the onset of the monsoon. While the area planted to paddy exceeds the area in highland crops, the paddy is usually planted only once the highland crops are sown. (See figures on major crops grown by district in Table 5.3.) The preference for highland crops comes from their capacity to produce a marketable crop even when there is a shortage of rain; irrigated rice, on the other hand, will fail to grow or fill out its grain if water is short. As the name dry zone suggests, rainfall levels are lower and are much less reliable, and so paddy growers are dependent on the numerous small water storage tanks or the larger linked sets of irrigation canals to provide reliable water for the rice crop.

The cultivators in the northern dry zone with their larger holdings and different systems of irrigation—major and minor—are served by the Agricultural Research Station at Maha Illuppallama and to a lesser degree by the Central Rice Breeding Station at Batalagoda.[14] As described earlier, the institutional arrangements for farmer–research officer contact was made through linkage provided by the Extension Division. Begun in 1975 in Anuradhapura District, 'the training and visit programme' provides for contact with select cultivators in each village on a regular fortnightly schedule by the KVS (Krushi Karma Viapthi Sevakas), the local-level extension worker. To assist the AIs out in their ranges and the KVSs, the DAEO's office has a small group of specialized AIs who in Anuradhapura were made up of six officers in the following areas: paddy cultivation, dry farming, plant protection, subsidiary crops, farm machinery, and home economics. If the expertise available in the district office is not able to meet the problems raised by the farmers, they telephone the research station at Maha Illuppallama. Normally the contact with the research station is held over until the monthly 'Research–Extension Dialogue', which is usually held at Maha Illuppallama.[15]

These monthly 'dialogue' meetings form the second stage of the research–farmer link. When they were first set up they were held every two weeks, but by 1978 they fell off to once a month. The extension staff represented by the DAEO personnel and, sometimes, all the AIs in the district meet at the research station to report problems and seek solutions. Research officers provide assistance in identifying a paddy virus and working out dosages of available agrochemicals. From time to time these dialogues include field trips to such places as Walagambahu, where a project on multiple-cropping under small tank irrigation is being held. The research staff at Maha Illuppallama have never accompanied the extension staff into the field to look at a specific problem, except for special projects, such as the Mahawelli Ganga Diversion Project, which will provide irrigation water for a linked set of major tanks in the southern central part of the district and for an earlier dry farming project which was reinvigorated in 1976. The flow of information from the farmer through the Extension Division to the research station appears limited in its utility. There were no clear examples of farmers' problems that were identified through the Training and Visit Programme (T & V Programme) and communicated through the Research–Extension dialogue being used to determine specific research programmes at Maha Illuppallama. Earlier specific research goals had been identified with the emergence of problems in the fields of the research station as well as reports of cultivators' problems coming through the Extension Division. One example cited was the virus susceptibility of the locally grown variety of okra. The station was able to breed a variety of okra, MI 5, which was released and was reported to be doing very well. A similar example was the persistence and spread of leaf curl in chillies which was caused by a complex of a mite and virus infestation which defied control by any set of known agrochemicals. Maha Illuppallama bred a new variety of small chillies, Birds Eye Chillies, or MI 1, which resisted both pest and disease.

More generally the research station's emphasis has been directed to serving the northern dry zone in three different ways. First, existing varieties of paddy and, more important, the rain-fed upland crops such as kurakkan, sorghum, other pulses, and grain legumes were to be improved. Breeding lines for sorghum were introduced from the International Crop Research Institute for the Semi-Arid Tropics (ICRISAT). Similarly genetic materials from India, Thailand, and elsewhere were crossed with local varieties to produce improved strains. Second, the potential of new cultivars was tested in the fields of the station before being recommended to the farmers. Examples of these were cowpea, which was developed as a substitute for green gram which was very susceptible to a local virus, and soya bean, which had been tried out earlier and now reintroduced in an improved strain. The latter's very high protein content and high yields in the dry zone made it an attractive crop; in 1981 it was being processed into milk both for direct sale and for mixing with cow's milk. Third, the research station had been working on problems of optimum land use for the dry zone since the time that Ernest Abeyratne was director of the station and Hardy had been visiting. They had been trying to find ways of integrating a cultivation system composed of tank-irrigated paddy land; chena (shifting) cultivation of vegetables, pulses, and other rain-fed crops, and homestead production of coconuts, papaws, and mango with small-scale animal husbandry and poultry. The concept, which echoed the earlier arguments made by Bradfield and the multiple-cropping specialists when IRRI's focus was being defined, attempted to use the land, water, and labour resources to get the maximum sustainable returns for the farm household.

The actual problems of effectively integrating these varied activities had not been tried out in practice until the creation of the Walagambahu Cropping Systems Project in 1976. This project and its proposed extensions into the Tank Modernization Project, the Integrated Area Development Programme, and the Dry Zone Agricultural Research and Development Project all proposed to modify the timing of cultivation, improve water management, and diversify both crops and other farm-income earning activity. The initial project was based at the small tank village of Walagambahu, which lies in the jungle ten miles from Maha Illuppallama. Supported by Canadian funds from the International Development Research Centre (IDRC), channelled through IRRI, a research-action group from the research station was based in the village. The tank which serves the village is approximately eight hectares in extent and irrigates some 16 hectares of paddy land lying down-slope from the bund and sluices of the tank. The prevailing practice was to delay cultivation until the Maha seasonal rains filled the tank to capacity. Once it was clear how much irrigation water was available a decision as to the extent of the paddy to be cultivated was made. The delay of cultivation from the usual onset of rains in late September or early October until December had two effects. It wasted the rain that fell on the paddy fields directly and, with the cultivation of a 5-month or longer age variety, it pushed back harvest into the peak of the Yala season in April. While the April rainfall was commonly no more than 10 to 13 cm it did wet the grain and make harvesting, threshing, drying, and storing more difficult. More important, the lateness of the Maha crop precluded the cultivation of a Yala crop, whether of rain-fed rice, pulses, or vegetables. The principal advantages of the delay were, first, to ensure that only an area of paddy was planted that could be serviced by the available stored water and, second, to till and to control weeds more easily by flooding the fields before sowing. Bearing in mind that the principal cash crops and a large fraction of the food crops were being grown as chena cultivation on highland, the decision to economize on time and effort for paddy cultivation made sense.

Once the project started the research staff recommended that cultivation commence with the onset of the Maha rains in late September or early October. These were sufficient to soften the sun-dried clay soils of the paddy fields and to make ploughing, though not puddling, possible. As there was enough moisture to germinate the broadcast-sown paddy, a 4 to 4½-month variety could be started which would receive reasonably reliable rainfall in January for flowering and grain-setting and be ready for harvesting in February, the month of lowest rainfall. If there was a shortage of water in January, supplementary water could be released from the tank. The key to the system was to have a high yielding, pest-resistant, medium-tall or semidwarf variety of paddy that was competitive with weeds. Since the early 1970s, BG 11–11, a 4 to 4½-month variety, had been available and performing well. With the initiation of cultivation coinciding with the onset of the rains and with a short age variety there was a great saving in the stored tank water. This in turn made possible the cultivation of a second short age variety of paddy, such as the 3½-month variety

BG 34–6, or the 3-month variety BG 34–8, which could be started before the summer month of highest rainfall probability, April, and which could draw on the stored water to carry it through its flowering phase. Under conditions of high rainfall and with little draw on the stored tank water it has been possible to grow a quick maturing, low water-using crop of green gram, cowpea, or soya bean. Thus by timely cultivation, use of the newly available short age varieties of paddy, it has been possible to triple-crop land which previously bore only one paddy crop. Even under adverse conditions of low rainfall it was expected that one paddy crop and a vegetable crop could be obtained. A further advantage in keeping up the water level in the tank rather than planting a belated, long age variety crop and depleting the water level was that the tank could be used for keeping fish more easily, watering cattle, and maintaining a high water table level to supply the household wells. One major drawback of the Walagambahu trial was that the cultivation schedule had to be uniformly kept by all paddy farmers. This required a higher degree of co-ordination than they had customarily practised. At present leadership is taken by the research staff, but there is no functioning village organization or anyone with the authority in the village to organize the schedule in the future and regulate the release of water from the tank. Another drawback was the difficulty of spacing the peak labour demands of the earlier, short age paddy crops with labour requirements of the more profitable chena crops. The onset of the Maha rainy season makes September and October the preferred months for planting the finger millets, cowpea, and chillies which were grown by the Walagambahu cultivators in Maha 1977–8. Since the chena has always provided the major cash income of the farmer in the tank villages of Anuradhapura, farmers are reluctant to shift labour into intensified paddy cultivation.[16] In 1977, after family incomes were recorded and analysed, it was found that of the average gross family income of Rs 312 per month, Rs 152 came from chena cultivation, Rs 75 from two seasons of paddy cultivation, and Rs 85 from off-farm income sources.[17] By this accounting the chena income was twice as much as two paddy crops which were grown with highly subsidized fertilizer and free agrochemicals. At the 1977 price differentials between paddy and the chena crops and with more realistic cost of inputs, including management costs of water, it is not clear that the village household's limited labour supply would be best used in the labour-intensive cultivation of paddy. Another technical problem raised by the cultivation of the paddy field with the onset of rains is that it creates major difficulties in field preparation. The average size of holding is a little over one-fifth of a hectare made up of 3.2 parcels per holding. Each parcel is separately bunded for water control on the field. Being of such a tiny size even two-wheeled tractor preparation is awkward, and buffalo ploughing of the lightly moistened clay soils is very difficult. More relevant is that neither tractors nor buffaloes are freely available, and tractors remain too expensive for a single smallholder to purchase. The whole technology for field preparation needs to be improved. The present thinking is to settle for a minimal land preparation, lightly loosening the soil by hand tools such as a Swiss hoe or long-handled five-prong cultivator to a depth of one inch or so. No other obvious answer to early field preparation is apparent.

Ultimately these various drawbacks or problems provoked by the effort to intensify the paddy cultivation and to economize on the use of the limited water can be understood as emerging from the attempt to shift part of an existing, integrated system of cultivation without shifting the whole system. It seems very doubtful if major innovations in the use of paddy tracts can be brought about without a complementary set of innovations in highland or chena cultivation. The cultivator over time will make those choices which will give him the most secure and highest income for the least outlay of resources and labour. The insistence on placing paddy in the centre of the thinking about the Walagambahu project may be mistaken, given the much greater income provided by highland cultivation. It would seem most plausible to begin with the improvement of the highland cultivation and time management and then fit paddy cultivation around the chena cropping system. The persistence of the researchers in beginning with paddy may have been encouraged by the project-funding being administered by IRRI, which takes as its basic concern rice-based cropping systems and consciously refuses to build on the existing balance in the local cropping system. However, these comments on the possible influence of IRRI are only speculative.

With Maha Illuppallama's leadership the extension staff in the neighbouring district of Kurunegala are planning to extend the Walagambahu approach, even though its impact and implications have not yet been evaluated and criticized, to the northern part of the district where there are many small tanks. This planned early enlargement of the approach is known as the Intensive Agricultural Development Programme and is envisaged eventually to reach all of the functioning two thousand or more tanks in Kurunegala District. Closely related is the Tank Modernization Project, which is reviewing the water-collecting, storage, and dispersal of the tanks to see if available technology and improved organization might deliver more water. Still in the planning stage is a project for the development of rain-fed agriculture in the dry zone which will draw on the experience gained at the Dry Land Research Project at Hyderabad in India and on the resources of ICRISAT. It is intended to focus on the highland cultivation, introducing improved weed control and tillage practices in ways which will augment the existing local practices.[18]

In review, three broad phases of Maha Illuppallama's historic rice research can be distinguished. It began with a rain-fed agriculture support phase (1950–67) which concentrated on variety improvement including a substantial programme in both rain-fed and irrigated paddy development. The following phase, 1967–74, concentrated on irrigated rice and included studies of irrigation techniques and water management. This emphasis was encouraged by the planning for massive irrigation of the dry zone particularly, under the Mahaweli Ganga Diversion Scheme. Finally, the present phase from 1974 on has been primarily concerned with building a closer link with the existing agricultural systems in Anuradhapura and selectively using the research capability of the station to improve and to diversify agricultural production in the northern dry zone.

In this development the research station has been drawn into increasingly closer relations with the cultivators in the surrounding area, although the initiative for this often came from agencies outside of Maha Illuppallama—the World Bank in the case of the Training and Visit Programme and the Research-Extension Dialogue, the International Development Research Centre and IRRI in the case of the cropping systems project at Walagambahu. None the less the effect has been to commit the station to a closer relationship with the surrounding agricultural systems. No longer is it sufficient to improve the planting materials for local growers, to pioneer and to acclimatize the cultivation of new crops, or to provide back-up plant protection advice. By means of the connections with the cultivators which are now institutionalized through the Extension Division and through its cropping systems and tank-related research projects, Maha Illuppallama is clearly committed to broadening its contacts and services. Symptomatic of this change is the appointment of the first agricultural economist to the permanent staff at Maha Illuppallama.

6.3 The Central Rice-Breeding Station, Batalagoda, and the Mid-Country

A third agro-ecological region identified by the Department of Agriculture is the intermediate zone with its semiwet lowlands. It is more precisely located on the recent agro-ecological maps as being the region with a high probability of between 90 and 140 cm of rainfall, that is a lower rainfall than the wet zone's but much more than the dry zone's.

Its terrain ranges from gently undulating to steep hills with narrow valleys. It lies mainly in Puttalam, Kurunegala, Matale, Badulla, and Moneragala districts though it includes portions of other districts as well. The mid-country cultivator facing lower rainfall than in the wet zone and with less buildup of either minor or major irrigation schemes than in the dry zone is a cautious, rain-dependent cultivator. Since much of the mid-country paddy area receives little water, the great majority of cultivators who have upland fields devote most of their time and resources to upland cultivation, particularly coconut cultivation. As one research officer remarked, given a choice between picking coconuts or cultivating paddy most farmers would opt to pick coconuts. It is a more reliable crop, a surer source of income, and requires much less labour than paddy cultivation. In spite of this preference paddy is cultivated on the lower slopes and bottom land of the undulating countryside. In the district of Kurunegala, 25 per cent of the cultivated area is paddy while 67 per cent is under plantation crops, predominantly coconut. Puttalam, the adjacent district in the mid-country zone, has only 16 per cent of its cultivated area under paddy and 76 per cent planted to coconut and other plantation crops. In contrast, Anuradhapura District in the adjacent dry zone to the north has 75 per cent of its cultivated area in paddy. The use made of the arable land reflects the rainfall and water management systems of the mid-country cultivator. The different forms of cultivation for paddy are also a manifestation of that adaptation. In 1978 it was estimated that in the whole of Kurunegala District there were 68,000 hectares of paddy land of which 8,900 hectares were under major irrigation (principally from tanks of 45 or more hectares of storage area), 28,000 hectares under minor irrigation from an estimated 4,000 small tanks, and 30,000 hectares were rain-fed. The irrigated area lay in the drier northern and central areas of the district.

In discussion with the extension staff in the District Office, in several Agrarian Service Centres, and in the farmers' fields it was obvious that the problems of the farmers varied by the water regime and the proportion of land planted to paddy, coconut, and other upland crops.[19]

For example, in the 2,400 hectares of paddy land serviced by one of the major tanks, Magalla, which was fed by water from a perennial river, there was a reliable and ample supply of irrigation water to sustain a regular practice of double-cropping. When the project was started in 1947, the average size of landholding was one hectare for paddy and one hectare of highland. Now the size of the paddy holdings varies widely—some villagers have 16 hectares, the temple has over 8 hectares, while there are small holdings of about half a hectare. Since the land is fertile and well watered, most cultivators concentrate their attention on the paddy fields rather than on any other crop. A cautious estimate of the average yield was 4 tonnes to the hectare for each of two seasons.

There were other farmers who harvested 7.5 tonnes to the hectare in each of Yala 1977 and Maha 1977–8. One farmer reported that even though the costs of field preparation, transplanting, and fertilizer had gone up, the new guaranteed purchase price for paddy of Rs 2 per kg had helped him net Rs 12,000–15,000 per hectare per year. He estimated that his costs were of the order of Rs 2,500–3,000 per hectare but that he grossed on the average Rs 8,600–9,900 per hectare per season. Owning and cultivating 2.5 hectares, he had netted somewhere in the range of Rs 32,000–36,000 over the last two cultivating seasons. The cultivator's claims were treated with some scepticism by the Additional District Agricultural Extension Officer. However, the point remains that paddy is profitable, even though there may be dispute as to how profitable it really is, as is demonstrated by the recent interest in renting paddy land on the part of non-cultivators such as merchants, agricultural officers, schoolteachers, and the like. A hectare of paddy land was renting for Rs 1,200 per season, and two merchants were each renting over 20 hectares in small parcels of 1 to 2 hectares each. The particular advantage of this for the renter was that he could use his commercial connections to hire, or even buy, a tractor and to purchase fertilizer in a lorry-sized lot. In one case, the owner of a 2-hectare holding was hired to cultivate it using his family's labour at Rs 6–7 per day for four months for himself and Rs 200 for his family's labour for the season. He received about Rs 2,000 in rent and about Rs 900 in wages as a risk-free income. In 1978, it was also possible for the owner of paddy land in an irrigation colony to qualify for a cultivation loan from the Bank of Ceylon at Rs 2,200 per hectare. Since in previous years the bank had forgiven defaulting borrowers, this owner had applied for a loan of Rs 3,600 on his 1.6 hectares which he had put in his pocket without any intention of repayment. Thus in this one unusual year his gross income was Rs 7,500. The renter, on the other hand, renting over 20 hectares and with his own tractor for tillage, and with contracted labour and an ample supply of fertilizer, stood to net better than Rs 100,000 per season. While none of these figures should be taken as being reliable, they serve to explain why more and more of the land under this major irrigation scheme is being taken up as a commercial enterprise by non-farmers.[20]

Conversely, they also reveal the major problems faced by small owners under the major irrigation schemes. First, there is an acute shortage of power for tillage in the district, of both buffaloes and tractors. In three different Agricultural Instructor's ranges, we heard complaints about the lack of buffaloes and the costliness and unreliability of tractor cultivation. The cultivator who owns more land than he can prepare by hand, roughly one-tenth of a hectare, needs some form of power to prepare the field for sowing. Keeping buffaloes is a constant bother, or even impossible where pasturage is short, while a four-wheeled tractor costs a minimum of Rs 100,000. Second, while fertilizer is available it is not physically easy to purchase and transport to the paddy field. We were told of farmers who ended up making five and six trips to the Co-op outlet to get fertilizer and then found the bags were torn, that basal fertilizer was not available, and so on. Finally, even if the farmer did get a good crop then he was faced with the problem of bagging it and transporting it to the Co-op for grading and sale. In Maha 1977–8 gunny bags for sacking the rice were in very short supply. The Agrarian Service Centres received shipments which permitted them to distribute one bag per hectare to cultivators. Since each bag held forty kg of paddy, a man with a harvested and threshed crop of 4 tonnes/hectare was in real difficulty to get his grain off the threshing floor and into the Co-op for sale. As the commercial renters have the resources to buy and operate tractors, to purchase fertilizer by the lorry-load direct from the fertilizer depot, to buy and hold gunny bags, and to arrange for transport of a large volume of paddy, it makes sense for the small owner to rent out his land and save himself from the many difficulties of arranging inputs and marketing. Clearly the problems are connected with the expense and difficulty of bringing the needed inputs to the field and, conversely, of removing the product. It is more a delivery system weakness than an actual scarcity. Or expressed in more general terms, it could be regarded as a structural problem created by the need for high cost and scarce inputs among farmers who lacked the income and means for organizing delivery and marketing.

In parts of the intermediate zone outside of the reach of major irrigation schemes but within Kurunegala District there was less emphasis on paddy, and consequently the supply and marketing problems were less acute. In the minor irrigation areas the chief problems were connected with water, water management, and the availability of high yielding, short age varieties that did not require costly inputs. In addition, the paddy holdings were smaller, the income levels lower, the water less reliable, and hence the yields lower.[21] More specifically in the Wariyapola range, coconut is the predominant crop. There are better than 4,800 hectares under coconut and only 1,500 hectares under paddy, of which 600 is under minor irrigation and the balance is rain-fed. The paddy land that is serviced by the small tanks has declined in extent since 1970. The largest tank used to be 20 hectares in extent and serviced 32 hectares of paddy land, but it has not been well maintained and is gradually silting in, with a corresponding decline in the serviced area. The Irrigation Department, which assumed responsibility for all the tanks in 1970, has not regularly cleaned, deepened, and maintained the storage capacity of the tanks. At present, when meetings are called to organize land preparation and schedule irrigation, the farmers do not bother to come to the meetings. Since there is no longer enough water for all the fields and since paddy cultivation is more time-consuming and less profitable than coconut and other crops, the farmers find little reason to take the time to come to what were described as 'useless meetings'. Indeed some cultivators were reported as saying that the government's decision to take away the maintenance of the tanks from locally organized irrigation committees, to entrust it to the Irrigation Department and then to charge the farmers Rs 15 per irrigated hectare was a device to raise revenue rather than to help the cultivators.[22] Supporting these observations were the many comments on the spread of Salvinia and other water weeds in the tanks which both accelerated the rate of siltation and choked the irrigation channels with a tangle of weeds. There were related reports from cultivators of irrigation gates being cut, the planks for adjusting the water levels in the channels stolen, and the padlocks sealing the sluice levers broken.

Another recurrent problem both in the Wariyapola area and around Karandagolla, which is a small rain-fed paddy area dominated by rubber and coconut plantations, was the difficulty of field preparation. With rain-fed cultivation there is an optimum time for tillage. As the Wariyapola cultivators have very small holdings—it was estimated that 98 per cent had less than 0.4 hectares—they had difficulty in arranging for tractor tillage at the time of high demand. Tractor operators were reluctant to come unless there was a reasonable day's work in the neighbourhood. When they did come the charges were commonly at the rate of Rs 370 per hectare with charges going as high as Rs 490 per hectare for combined first and second ploughing.[23] According to some reports the number of buffalo available for ploughing had declined because of the growing demand for meat in Colombo and the emphasis on using imported strains of meat cattle for dairying rather than the multipurpose buffalo which could both plough and produce milk. When buffalo were available charges were around Rs 335–70 per hectare for first and second puddling.[24]

No problems were reported concerning seed availability though the District Agricultural Extension Office was often hard pressed to arrange for the supply of short age varieties when late rains or problems of tillage had pushed back sowing times. None the less the DAEO believed that it was generally able to meet this problem. The most commonly grown varieties in the tank-irrigated paddy lands during Maha season were BG 11–11 and BG 90–2. For Yala, when water was more uncertain, the traditional variety Heenati was widely used although more venturesome cultivators had planted BG 34–8. With sufficient water in the tanks Maha yields were around 3 tonnes/ha whereas in straight rain-fed cultivation average Maha yields were close to 1.5 tonnes/ha. Fertilizer supply was seen as a recurrent problem with the subsidized fertilizer being distributed through the local Multi Purpose

Co-operative Society. Since the MPCS was also entrusted with the sale of rice, subsidized flour, sugar, cotton cloth, and agrochemicals, in addition to being the purchasing agent under the Paddy Marketing Board, it was small wonder that when cultivators needed fertilizer they often had to join long queues to wait for service or, alternatively, that the overburdened MPCS staff could not always keep the recommended mixtures in stock. Another specific difficulty to which every extension staff member referred was the difficulty in marketing paddy. In 1978, the depots of the Paddy Marketing Board in both Kurunegala and Anuradhapura districts were literally overflowing. Large aluminum and steel sheeting walls had been erected and the excess paddy was dumped into these and then covered with tarpaulins. Each depot we saw had at least two large temporary outdoor storage bins, and therefore the local purchasing agents in the co-ops were extremely demanding about cleanliness, dryness, absence of broken grains, good grain colour, etc., before agreeing to buy stored paddy from the cultivator at the guaranteed price of Rs 2 per kg. This left small cultivators, who lacked dry on-farm storage capacity as well as the means of bagging and transporting, dependent on private grain dealers who would come to the threshing floor and buy all the paddy at Rs 1.70 per kg. With lower purchase costs the grain traders were able to distribute grain in the markets of Kurunegala, Colombo, and Kegalle below the government costs which slowed down the movement of purchased paddy out of the storage depots of the Paddy Marketing Board. The cumulative difficulties of water, fertilizer, and marketing were judged to produce a major disincentive for the small cultivator whether on tank-irrigated or rain-fed land.

The alternative crops for these cultivators, besides coconuts, were upland vegetables and legumes, which were easy to cultivate, required fewer inputs, and were readily marketed to dealers for sale in Kurunegala and Colombo. Unfortunately there was little advice available on the cultivation of upland crops from the KVSs and Agricultural Instructors. In some Agrarian Service Centres they did stock vegetable seed and agrochemicals, but such stocks were a minor concern compared to the support given to paddy growing. What is more important, however, was the ministerial, departmental, and divisional fracturing of responsibility for dealing with the cultivator attempting to make optimum use of his land, labour, and other resources. In addition to the Ministry of Agriculture and Lands, with its separate divisions of Research and Extension as well as the affiliated Paddy Marketing Board, Ceylon Fertilizer Corporation, and others, there were the Ministry of Food, Co-operatives, and Small Industries that supervised the MPCs and tobacco purchasing; the Ministry of Plantation Industry with responsibility for rubber, coconut, and tea; the Ministry of Finance that had responsibility for the disbursement of rural credit through the People's Bank and the Bank of Ceylon; the Ministry of Irrigation, Power, and Highways which looked after not only irrigation but land development and various irrigation colonies; and other offices of government. If the cultivator wanted assistance with crops other than rice he had to deal with the Coconut Cultivation Board, the Rubber Control Department, Department of Minor Export Crops, Livestock Development Board, National Milk Board, horticulture division of the Department of Agriculture, and numerous others. There was no ready access to an agency which would help him plan and organize his land, labour, and other resources to get an optimum return. Each agency attended only to its fractional responsibility. Lack of co-ordination of services was a major obstacle facing all cultivators in the most rational choice of crops and efficient use of their resources.

How were rice researchers responding to these obstacles? The Central Rice Breeding Station at Batalagoda, which together with the district extension staff became responsible for meeting the particular problems of the rice cultivators in the intermediate zone, also had its role narrowly circumscribed. While the rice paddy area in absolute terms is greater in Kurunegala District than in any other district, its relative importance to the smallholder is limited. Batalagoda's research and breeding programme was principally aimed at developing pest-resistant varieties. In 1978, the senior officer in charge of the station, D. Senadhira, judged they were two seasons away from releasing an improved paddy variety that was strongly resistant to brown plant hopper. Success was also expected in breeding varieties resistant to leaf roller. The balance of the breeding programme was designed to improve yields in the paddy fields where the improved varieties (IVs) had not spread. Around 30 per cent of the paddy area was still being planted to traditional varieties which produced under conditions in which the IVs failed. The present breeding strategy has been to collect the traditional varieties that do well in different growing conditions and run laboratory, station, and farmers' field trials to select the most promising. These varieties would be systematically selected, much as with the earlier pure-line selection programme, and crossed to try to push up their present yields of 1–1.5 tonnes/hectare by 25 per cent. Already Batalagoda has identified varieties which do well in soils with high iron levels, which had led to a bronzing and low yields of the existing IVs. Another example is the improvement of the rain-fed upland rice varieties. At Maha Illuppallama they concentrated on breeding drought-resistant varieties while Batalagoda went for a strategy of escaping drought with a 3-month age variety that would germinate under adverse conditions. They were working with a variety ideal for upland conditions which farmers in Hambantota had bred that came to be called Ambalantota 62–355. While it has susceptibility to blast, under rain-fed upland conditions it will yield 2–2.5 tonnes/ha with no weed control, no basal fertilizer, and only one application of urea, 3–4 weeks after germination. In a similar manner the breeding for cold-tolerance has concentrated on short age varieties that can be started in January, when the worst of the cold is past, and that can be harvested in June. The cold delays the maturation of the plant which had forced growers to move planting dates forward into the coldest part of the year in order to harvest before the heaviest rains of the following October and November. Thus H4, normally a 4- to 4½-month variety, takes 6 months to mature and BG 94–1, a 3½-month variety, takes 5 months from sowing to harvest. Different strategies were being tried to identify plants that would perform well under specific kinds of environmental stress.

While the varietal development programme goes forward to help improve pest- and disease-resistance and to breed varieties that will yield well under different kinds of growing conditions, Batalagoda has not attempted to meet the problems of local cultivators in the same way as Maha Illuppallama or Bombuwela have done, perhaps less due to the station's reluctance than to the inadequacies of the information flow from the cultivators through the extension staff. The Training and Visit Programme was only started in 1977, and because of staff shortage and the inability to prepare a training programme for AIs and KVSs it did not work well. There was no preparation of instructions to be given contact farmers and hence limited contact with little flow of relevant information from the field into the DAEO. Partly because of this lack of information the holding of a regular Research-Extension dialogue seemed pointless. Extension staff had met once at Maha Illuppallama and once at Batalagoda during the past year. From the point of view of the research station the most useful flow of information came through the results of the Extension Field Trials (EFT). These trials were organized to test the performance of new paddy varieties against established lines. In one of the EFT sites visited, LD 125, BG 90–3, BG 374–1, and BG 90–2 were being grown in parallel plots with second plots receiving higher levels of fertilizer. Information of yield levels, pest-resistance, etc. were recorded by the AIs and passed on through the DAEO to Batalagoda staff. This was the only regular communication channel from the field to the station. If farmers were troubled by pest attacks, the local AI supported by the subject-matter specialist on plant protection working in the DAEO would recommend treatment. No one could recall going to Batalagoda to consult the entomologist there.

The only new departure in research and in the station's relations with the local cultivators has come through the multiple-cropping project administered through IRRI. The project is centred at Katupota and is intended to be a study of a rice-based cropping system. But it was openly acknowledged that the farmers were far more interested in coconut- and vegetable-growing than in paddy. Indeed it was emphasized by the Director of Batalagoda that the economy of the cultivators throughout the intermediate and wet zones was not based on rice. Two comparable paddy tracts and surrounding upland lying within half a mile of each other were selected. The principal objectives were to investigate ways of increasing the paddy yield in Maha and to determine what cropping patterns would use the land and rainfall of Yala most efficiently. The existing rice-growing practice in these tracts had been to cultivate only a single, long age, photosensitive rice crop a year such as Wanni Heenati, Kalu Heenati, Dahanal, or Ma-wee. In the infrequent Yala season when rains were good a second rice crop was planted and matured on the bottom land where moisture levels are higher and water retention better. Under good conditions yields had never exceeded 1.25–1.5 tonnes/ha. The cultivation practices and water management were at a low level. These practices were poor because of 'the risky nature of returns from the paddy crop under rainfed conditions and [because] most of the farmers are not by any means solely dependent on their paddy lands, since they are either owners of an adjacent piece of coconut land or partly involved in some other income-earning occupation [sic]'.[25] The cropping system strategy worked out by the research staff at Batalagoda was to identify the highest yielding varieties for each local soil and hydrological condition. The terrain in both paddy tracts consisted of undulating land with sandy, porous soils on the crests and clay loams in the bottom lands. Trials in Maha 1976–7 indicated that BG 34–8 and BG 94–1 performed better on the crests and slopes while BG 90–2 did well in the valley bottoms. During the subsequent Yala season the cropping strategy followed was to continue with a second short age variety paddy on the moister bottom lands while planting the most promising Leguminosae crops on the slopes and crests. The results impressed the farmers who conceded that they might obtain better results and higher income by alternating rice and legumes on the drier land. During Maha 1977–8 the same varieties were used, but innovations in cultivation practices were introduced, such as row sowing, higher seed rates, different levels of fertilizer application, and so on. Throughout the season day-to-day observations were made on labour use, cultivating practices, and the performance of the crop.[26]

6.4 The Rice Research Station at Bombuwela and the Low-Lying Wet Zone

In the south-western quarter of the country is the low-lying coastal plain which receives some of the heaviest rainfall with the greatest regularity of any part of the island. Covering Colombo, Kalutara, Galle, Matara, and parts of Ratnapara and Kegalle districts, it is the most densely populated large tract in Sri Lanka and one where a substantial area of paddy is grown every season. However, while there is a large total area of paddy land in this agro-ecological zone, paddy cultivation plays a relatively minor role in the household economy of the small cultivators.

In recent village studies, as well as in a survey of small cultivators in Colombo district, rice-growing was found to be extraneous to the economy of most households and either a subsistence crop or a secondary form of family income for the remainder of the households.[27] Looking at the sample survey for Colombo District which lies in the centre of this agro-ecological zone, we find the same situation. The paddy area reported in the 1962 Census of Agriculture was only 15.8 per cent of the total hectares of cultivated land in Colombo District.[28] This figure was well below the national percentage of 28.3 or the average for all the twenty-two districts of 38.9 per cent of the agricultural land. With such a small proportion of the cultivated area in paddy and with a much larger area under coconut and rubber, the significance of paddy in the rural economy of Colombo District is not likely to be large. Moreover the median size of lowland holdings among 144 sampled farmers was 0.4 hectares (with a mean of 0.73 hectares), while the mean of highland holdings was nearly 2 hectares. Thus even on the average there were much larger holdings of highland than of lowland, which strengthens the view that paddy cultivation is not as significant in the rural household economy as other crops.

Beyond the limited importance of paddy, there are a number of constraints on improving paddy cultivation and increases in yields. In the sampled group 65 per cent of the cultivated area in Maha season was planted to traditional varieties. While the proportion of planted to improved varieties in Yala season was slightly higher (60 per cent planted to traditional varieties), the ARTI research officers concluded that 'poor soil and drainage conditions appear to govern the pattern of varietal use'. While cultivation practices have improved slightly with the wider use of fertilizer and weedkillers, most practices remain unchanged. For example, broadcast sowing was used on 94 per cent of the cultivated area, which included both improved and traditional varieties. The main reasons for continuing broadcasting were the higher cost in labour required for transplanting. In addition, cultivators spoke of the tedium of transplanting, lack of funds, unsatisfactory water supply, and boggy soils. Other improved cultivation practices were unevenly used. The result is that yields among the sampled cultivators are low: 1.82 tonnes/ha in Maha and 1.3 tonnes/ha in Yala. The cultivators who planted the improved varieties in Maha reported yields of 0.96 to 0.23 tonnes/ha more than cultivators sowing the traditional varieties. In Yala season, however, the yields of traditional varieties averaged 0.35 tonnes/ha more than the improved varieties.[29] One season's yields are not significant statistically, but they can be discouraging to the small cultivator lacking resources to try again and wait for the averages to work in his favour. For the sampled cultivators the totality of conditions in which they had to operate—environmental, tenurial, cash and labour resources, relative prices—served to discourage experimentation with improved varieties and use of improved cultivation practices. Consequently yields remained low among these cultivators in Colombo District.

In these circumstances how did the Rice Experimental Station at Bombuwela adapt its research goals to meet the needs of the cultivators? The relations between the experimental station, the Kalutara District Agricultural Extension Office staff, and the cultivators were closer and more directly supportive than those in any other district visited. The Bombuwela station, which is responsible for the wet zone, low-lying area in Colombo, Kalutara, Galle, Matara, and Ratnapura districts, is located about seven miles inland from the coast in a low-lying area of bog and half-bog soils that are characteristic of much of this area. In 1977 and 1978, it was the smallest station visited. The staff included three research officers, specializing in agronomy, plant protection, and plant nutrition, plus one experimental officer, a soil chemist, and one plant breeder. New facilities were being added, including a plant-breeding and soil-testing laboratory as well as a series of small test plots enclosed with concrete walls and served by sluices for testing paddy tolerance to submergence. Additional staff will be appointed when these facilities are finished. The experimental station was responsible for breeding paddy varieties to meet the difficult growing conditions in these districts. The staff recognized that paddy formed only a small part of the cultivated area and that in some areas along the coastline near Kalutara, Galle and Colombo cities vegetable-growing had displaced paddy as the preferred crop. The station staff also recognized that the very small holdings meant that the majority of cultivators could only afford to be part-time farmers and lacked the finances to invest in improved cultivation practices.[30]

Given these limitations, the rice plant characteristics valued by the farmer were those that provided for the survival of the crop, such as ability to grow in the wet, often iron-toxic, and occasionally saline soils; capacity to withstand flash floods and prolonged submergence; pest-resistance and reasonable yields at low fertilizer application rates. The second desired quality was good processing and consumption characteristics, such as long dormancy, good storability, easy milling, and palatability. It was estimated by the station head, J. Jayawardena, that between 60 and 70 per cent of the paddy grown was retained for home consumption. The present improved varieties of paddy were only grown on about 30 per cent of the paddy land in Kalutara District since the cultivators found that the traditional varieties better met their requirements. Consequently, the Rice Experiment Station was engaged in a two-pronged breeding programme. The first goal was to improve the existing traditional varieties through resurrecting the former programme of pure-line seed selection. Traditional varieties preferred for the bog soils were very mixed; thus it was necessary to identify and to establish pure lines which would give the best results season after season. The second goal was to breed a photosensitive variety that would flower in August whether planted in March, April, or May. The difficulties of breeding and maintaining a range of non-photosensitive age classes are growing in complexity and take a larger proportion of the resources of the station. One alternative is to revert to a few lines of photosensitive plants which will flower under specific light conditions no matter when they are planted. If such a variety could come to maturity in a minimum of three months, it would help minimize the effects of iron toxicity and salinity which accumulate over the maturation period of the plant. Thus to develop a rapidly maturing plant would be one strategy of coping with adverse soil conditions. Such a variety, if it was also photosensitive, could be planted after the season of greatest flood hazard, yet in time to catch the August rains needed for the grains to fill. There are, of course, other breeding programmes such as those to develop deep-water varieties, thrips-resistant varieties, and other improved lines. These include a comparative study of bronzing-resistant varieties, which is being co-ordinated by IRRI. These breeding programmes have been largely oriented to meet the needs of the rice cultivators in the service area of the Bombuwela station.

At a more specific level the research officers attended a fortnightly Research–Extension Dialogue where they reviewed the cultivators' problems as reported to the extension staff. In late 1977 and early 1978, the experimental station undertook what they call 'adaptive research'—that is, bringing in new varieties, recommending agrochemicals and cultivation practices, and testing to see that they perform well in the growing conditions of the low-lying wet zone. When a new fertilizer was offered the station worked out the best and most economical dosages for nursery-bed and field applications. When some Matara District farmers devised a form of thrips protection by soaking seed in a mixture of insecticides, the research officers followed up by experimenting to identify the effective component and the best concentrations to reduce the hazard of thrips. In a related manner, the plant protection officer set up a trial 'pest watch' system in close co-operation with extension officers and cultivators. Light traps for insects were built and hung in the fields and a 'bug count' done to keep track of the incidence of different insects that attack paddy and vegetables. It was hoped that this, together with observation of the fields, could permit development of an early-warning system to head off large-scale pest infestations. The foregoing are only examples of the specific way in which the station and extension helped cultivators meet their specific problems.

Besides responding to regional problems the Rice Experiment Station assists the DAEOs in preparing a fortnightly bulletin of advice for paddy cultivators.[31] This is then relayed through the extension network down to the contact farmers, advising them of the appropriate steps to take in the cultivation cycle and what protective measures to follow. The extension staff spoke with appreciation of the willingness of the research officers to go into the field and look at problems in the farmers' fields. In the course of reviewing the substance of the relationship between research, extension, and the cultivators, it was acknowledged that, in fact, the Rice Experiment Station had very little at present to offer the cultivators. Consequently the extension staff were hard pressed to find significant plant varieties, agrochemicals, or even advice and information to pass on to the cultivator. It was recognized that unless there was material of genuine significance to the farmers being passed on, the contact farmers and others would not bother to meet the AIs and the KVSs on their scheduled fortnightly meetings. The punctual meetings and the elaborate network for transmitting information would be useless if there were no matters of real, practical importance to the cultivator being communicated.

As reported at Bombuwela, the lack of substantial information remained the major flaw in the whole of the Research–Extension–Cultivator Dialogues throughout all the agro-ecological regions. If the cultivators were actually engaged in initiating relevant research instead of being the passive recipients or, at best, the testers of research findings arrived at elsewhere, the research problems would be focused more obviously on the issues that concern them. Substance for communication would be vitally present, which it is not at present.

In spite of this criticism of the Research–Extension–Cultivator flow of research findings and information, it is clear that the Bombuwela Rice Experiment Station and the Kalutara District Agricultural extension programme is in advance of other stations in addressing the specific local concerns of the rice cultivators. However, as a very small station with instructions to concentrate on rice, it does not address the paramount concerns of the local cultivators. As noted earlier, the small cultivators are much more concerned with the yields and profitability from their vegetable-growing and from their highland cash crops than they are with rice-growing. Accordingly Bombuwela is of marginal value to most farmers in these low-lying wet-zone districts.

Looking back over the activities of each of these research stations, we can see that substantial changes in research focus were evident. Through the increased attention to the specific conditions of the farmer, as understood and reported by the Extension Division, the stations were slowly beginning to piece together an alternative research strategy, research strategy which was concerned with the optimum use of land, water, and labour to produce the largest income for the small cultivator within a specific environment. Rice was being dethroned as the staple crop of choice in favour of a variety of crops that took fewer resources and produced a higher net income for the cultivator. The emergent research policy built upon the earlier analysis of agro-ecological regions and restated the arguments that the multiple-cropping advocates had made twenty years earlier to the RF.

6.5 The Relevance of IRRI to Sri Lankan Agricultural Research in 1977–1978

The most important IRRI programme in the eyes of the Sri Lankan researchers was the International Rice Testing Program (IRTP) in which Dr Ikehashi and other IRRI scientists supplied seed varieties collected from around the world for testing in different countries. In 1976 the 700 lines in a first shipment were sent for testing of resistance to iron toxicity. Subsequently, 400 varieties were sent to be tested in phosphorous-deficient soils, and varieties were also sent for saline-resistance testing. The IRRI scientists spoke highly of the well-run tests in Sri Lanka and commented that the island with its diverse growing conditions and many problem soils made it ideal for evaluating newly collected lines of rice. The results of these tests in Sri Lanka and in other countries were then tabulated, and the characteristics and potential of each variety of rice were slowly discovered.[32] The best varieties for different growing conditions were thus identified and could be used as either planting material or as a gene pool in hybridization projects.

A second closely related IRRI programme that was highly valued was in genetic evaluation and utilization. This programme provided for the simultaneous testing of newly hybridized varieties developed at IRRI and in other countries' breeding programmes in test plots in many different locations in many different countries. Applications of irrigation water, fertilizer, plant protection chemicals, as well as general cultivating practices, were standardized as nearly as possible so that the only variations were the soil and climate. The alternative used before had been to test the variety over a number of growing seasons to see how it performed under different climatic conditions. Only after it had performed well for four or more seasons would it be given to the seed farms for multiplication. Now, through the elaborate and extensive comparison of the performance of a new variety described above an earlier assessment of its potential could be made. The Sri Lankans were very familiar with this method of testing new varieties since their own Co-ordinated Rice Varietal Testing (CRVT) programme, begun in 1968, was identical, though only run within the island. In fact, Leslie Peries, Deputy Director Research 1961–74, claimed that Henry Beachell of IRRI had copied the Sri Lankan CRVT programme when setting up the IRTP in Los Banos.[33]

The third valued programme was the cropping systems project which was centred at the Maha Illuppallama Research Station. From IRRI's funding there had been an interest in multiple-cropping systems which would introduce crops with lower water requirements than rice in the dry seasons so that the alternation of rice and other crops would keep the fields producing for as many months as possible.[34]

With different soils, topography, and climatic conditions in the different regions, each cropping system had to be tailored to the local environment. In Sri Lanka the cropping-systems project was testing earlier planting, faster harvest and field preparation, and quicker replanting to get two and, possibly, three crops where only one crop of wet rice had been grown earlier. The village selected for study, Walagambahu, was expected to have its first three-crop year based on one low-water demanding, short-maturation period rice, BG 11–11, a second wet-rice crop and a third 2-month variety of sorghum.[35] The role of IRRI in the project had been to act as the project initiator and as the administrator of the funds which came from the IDRC in Ottawa.

In addition to these three projects which were judged to be valuable by the senior Sri Lankan research administrators there were four which were judged unfavourably. The International Rice Agro-Economic Network, led by the Agricultural Economics Department at IRRI, was primarily concerned with the farm-level constraints on production.[36] The weak methodology and trivial nature of the findings of the constraint project induced one Sri Lankan to characterize it as 'bull shit'. Other IRRI projects in Sri Lanka were not regarded with much more favour, such as the soils project, which was designed to measure the level at which iron, sulphur, and other elements became toxic to the rice plant. Even though headed by a Sri Lankan scientist, Felix N. Ponnamperuma, it was regarded as being of little relevance in Sri Lanka. The same was true of the International Network for Fertilizer Efficiency (INFER) project, headed by S. K. De Datta of IRRI, which was set up to study the variation in form and timing of fertilizer applications. The specific test designs were prepared in and sent from Los Banos to Sri Lanka where the tests were run and the results reported back to IRRI. Through many seasons of trials it was expected that the accumulated data would reveal patterns about more efficient fertilizer utlization.

Sri Lanka: Quest for New Research Strategy

This was judged to be a project of doubtful value to Sri Lanka or indeed to any other participating country, for there were so many other variables involved in the plant's uptake and utilization of fertilizer. Finally there was the IRRI-sponsored engineering project to develop small machinery such as mechanized rice threshers, seed cleaners, and mud-land seeders. Sri Lanka was co-operating in this project despite its reservations about the value of such intermediate-level mechanical devices that fell between the traditional means and the modern Japanese or Taiwanese equipment for carrying out these operations. In sum, the Sri Lankans participated in these doubtful projects because they valued the association with IRRI for the International Rice Testing Program and the genetic evaluation and utilization work and to a lesser degree the cropping-systems work.

Beyond these specific in-country projects was the training for junior researchers in Los Banos, generally acknowledged to be the most important single programme of IRRI for Sri Lankans. But beyond the value of any specific project or training programme was a major gap between IRRI's definition of plant science and the definition of the Sri Lankan Department of Agriculture. There is no better way of conveying this difference than to paraphrase a senior IRRI scientist commenting on the Sri Lankan approach. 'The Sri Lankan approach to research is based on a fundamental analysis of land types and land utilization. The structure of plant research is built upon this analysis of the environmental situation.'

The paramount concern was with identifying crops suitable for specific agro-ecological regions and within those regions breeding a rice plant with architecture and characteristics suited to specific growing conditions. In the judgement of one IRRI scientist, 'Sri Lanka had the most sophisticated research structure of any country', including the United States. Whether that statement is true or not, the important point was that Sri Lankan agricultural research was being structured to fit the agro-ecological regions within the country. Underlying this structure was the idea that man's agriculture had to be adapted to the environment if it was to flourish over the long term. No other national agricultural research was so firmly grounded on an analysis of the variation in the cultivating environment and long-term need to adapt to that environment.

In contrast IRRI was concerned with breeding a few high yielding varieties of rice that would increase the wet-rice harvests throughout monsoon Asia. Where Sri Lanka started from the land and the environment, IRRI started with the rice plant. Where Sri Lanka built on local varieties, IRRI sought a universal solution. Where Sri Lanka examined the potential of many different crops, IRRI concentrated on irrigated rice. Where Sri Lanka built slowly, out of necessity, and planned for the long term, IRRI, at least in its early years, sought dramatic 'break throughs' and 'quick-fixes'. Where Sri Lankan rice science training was built on the analysis of the needs of the country, IRRI's training was conceived in terms of a universal, abstract science that would solve the food-shortage problems of Asia and that was not concerned about local field conditions or the practices of local farmers. Thus the gap between the two definitions of agricultural science was large, and the feelings of mutual exasperation expressed both by Sri Lankans and IRRI scientists were understandable.

Aggravating the gap was the arrogance of many of the Americans associated with IRRI who assumed that their definition of agricultural science was the only rational one and that all other views were backward and unscientific. One IRRI scientist baldly asserted that there was only one competent rice breeder in the whole of Sri Lanka; ironically this IRRI representative was one of the three IRRI staff members who was judged to be incompetent by the Sri Lankans and was asked to leave.

Thus, in review, IRRI scientists, filled with enthusiasm for their mission to relieve the rice shortages of Asia and confident that American rice sciences could duplicate in Asia the successes of American wheat sciences in Mexico, came as scientific missionaries to Sri Lanka. There the IRRI scientists met a long-established independent agricultural research establishment with a recent record of success in rice-breeding and a sophisticated understanding of the role of agricultural research as a means of adapting to diverse cultivating environments. It was scarcely surprising that such different views of agriculture should produce misunderstandings and limit the co-operation between IRRI and Sri Lankan scientists. It was scarcely surprising that many IRRI projects should be regarded as naïve and treated with scepticism by the Sri Lankan agricultural scientists.

6.6 New Research Strategies Emerge from Earlier Experience

Scientific research and plant-breeding were fundamental activities of the Department of Agriculture since its founding in 1921. Research activities which were originally located in Peradeniya Gardens were then centred in the Central Agricultural Research Institute at nearby Gannoruwa. Subsequently part of the research responsibility was shifted to the Agricultural Research Station, Maha Illuppallama, in 1950, and to the Central Rice Breeding Station at Batalagoda in 1952, and later to other smaller centres located in other agro-ecological regions. The staff were sent for advanced training to the UK, Canada, Australia, Japan, and increasingly the US. In at least the three major research centres, by the early 1960s, there were agricultural scientists with advanced-level training in botany, genetics, entomology, plant pathology, and soil sciences. These scientists with advanced training were supported by junior officers who had graduated from the Faculty of Agriculture of the University of Ceylon (now the University of Peradeniya). Their research findings and field observations were reported in the Department of Agriculture's own journal, *The Tropical Agriculturalist*, which had been founded around 1844 (new series begun in 1952) as well as in the *Journal of the National Agricultural Society of Ceylon* (begun in 1963). The researchers were few in number, but they had established the value of their work and were recognized internationally as a well-organized and competent group.

However, while a competent research tradition was being built, research in rice remained the poor cousin. From 1931 to 1981, far greater resources had been committed to developing or restoring large-scale irrigation in the dry zone. It was believed that the extension of area under irrigated rice would both meet the staple food needs of the nation and provide employment for the landless Sinhalese from the wet zone, in spite of the wartime Rice Committee's views to the contrary. Also, resources had been committed towards improving the supporting infrastructure for the small cultivator, particularly in the Kandyan areas. Construction of roads, small dams, and electrical distribution systems had been financed by the annual appropriations for the Peasantry Rehabilitation Department. While these measures were not specifically targeted at improving the rice production in the country, they were part of a broad-fronted agricultural development policy within which the more specific rice research strategy was located. It was in this context that the shift from pure-line seed selection towards systematic breeding of the rice plant occurred in the 1950s. However, at no time did the government leadership or the Department of Agriculture abandon the land development policy or the infrastructural improvement policy in favour of an exclusive seed-driven agricultural development strategy.

The alternative and competing research strategies of the plant scientists moved from pure-line seed selection through to breeding a succession of plants which were designed to meet the specific growing conditions of the different agro-ecological regions. The increasingly detailed identification of desirable plant characteristics which were to be realized by crossing was made possible by the greatly expanded pool of genetic materials housed both in Gannoruwa and at Los Banos upon which the plant breeders could draw. Within this general change in the sophistication and technical capability of the scientists, there had emerged two positions. Both attempted to explain why the increases in yields and total rice production had plateaued.

The first position, that of the plant breeder, argued that the existing varieties did not have the appropriate characteristics to bear plentiful harvests in different physical environments. The existing improved varieties lacked specific adaptability to the soils, water, and pest conditions of the twenty-four major agro-ecological zones and numerous sublocalities. The answer of the plant breeders was to call for more research and breeding programmes targeted at these deviant or 'unstable' areas. It was accepted that further research would provide the answers. To illustrate this view one senior plant physiologist argued that closer investigation of the requirements of the rice plant at each stage of its growth was necessary. Thus he argued that a four-month variety could be started off with a minimal amount of water, for before pannicalization the plant is very tolerant of dry conditions; but during the last forty days when about 75 per cent of the carbohydrates needed to fill the grain were synthesized the plant should have high levels of sunlight and should not be subjected to water stress. If these growth characteristics with the specific needs were clearly identified then with skilful timing of sowing and careful water management the plant could be brought to high yield levels. He summed up his views by saying that 'yields may ultimately be a function of sowing date'.[37] Comparable statements were made by other breeders and plant pathologists who suggested that limitations on adoption and yields could be overcome by more careful analysis of local soil types where the latesols of the wet zone, which were subject to recurrent wetting and drying, had released many iron salts in the soil. In extreme cases this manifests itself in a 'bronzing' of the plant leaf with a reduction of its capacity to photosynthesize and so fill out the grain. They argued that even iron levels that fell short of producing visible 'bronzing' were probably inhibiting the growth and yield of the plant.[38] More detailed research would reveal the extent of this limiting condition, the plant pathologists argued, as well as more information on viral and fungus-related diseases.

The plant breeders extended their analysis of the reasons for the falling off in the rate of improvement in yields to a concern for the supply of needed inputs and structural changes of the cultivating conditions that would make it possible for the farmer to realize the potential of new plants. The breeders spoke bitterly of their past achievement in hybridizing a wide range of high yielding varieties only to have the potential of the new plants wasted because the necessary fertilizer was not available to the cultivator.[39] One of the favourite targets for complaint was the Ceylon Fertilizer Corporation, which was judged as being everything from merely erratic in its supplies to being a total disaster. Closely following the supplier of fertilizer as a villain was the Irrigation Department's poorly designed, poorly maintained, or inadequately regulated irrigation works. It is doubtful if more than a fraction of either of these complaints were well founded, but they were clearly favourite whipping boys of those persons explaining the shortfall in production as being the weakness of other divisions or departments of government which were undercutting the contribution which the plant breeders had made to meeting Sri Lanka's rice shortage.

The second position was that which began with the cultivating environment rather than with the rice plant. This view was best articulated by C. R. Panabokke, Deputy Director of Research from 1974 to 1980. He argued that agriculture must be seen as a total system built on a foundation of soil types and hydrology which influenced not only the types of crops that could be grown but also the specific improved varieties of paddy that were optimum in each specific setting. These crop types and, more narrowly, specific paddy varieties required certain cultivating practices if they were to be grown successfully in a specific setting. It was less a matter of the organization of society affecting the agronomic choices and more the biological necessities of the crop requiring certain practices and making more advantageous some forms of agricultural organization. Each localized agro-ecological setting limited the crop choices but both factors together imposed patterns of activity on the cultivator, given his available technology and resources. From this perspective it was possible to modify the system and push it towards higher levels of productivity by offering, on the one hand, a wider range of high yielding plant types (not simply rice but a broader spectrum of plants), while introducing a multiple and varied cropping system, and, on the other hand, by enlarging the cultivation technology and resource base of the cultivator. It was a two-pronged strategy derived from a holistic perception of the relations of soil, plants, and man.

Associated with this cultivating–environment-based view of research were two other positions on agricultural development: a more general interpretation of conditions in the villages which led to the conclusion that the caste- and class-based control of land and capital inhibited the rapid modernization of production; and a policy position which gave priority to the extension work, arguing that research should be related to the problems identified by the cultivators and reported by the extension service.

The general interpretation of the conditions in the village came in part from a historical political view of colonialism that had been articulated by the Trotskyite party, the LSSP, and had become part of the intellectual baggage of the more progressive members of the educated élite even though they were not members or supporters of the LSSP.[40] It was their belief that major changes in the landholding and operating system as well as in the participation of local cultivators in development were essential if major long-term improvements in productivity were to be obtained. Some who held this position went beyond talking of increased production, saving foreign exchange, and attaining self-sufficiency in food grains, and spoke in terms of long-term economic growth and social justice. This interpretation of conditions in the villages was dramatically verified by the insurrection of 1971. Judging from the frequency with which such persons as Mahinda Silva, Permanent Secretary, Ministry of Agriculture and in 1962–5 Director of the Department of Agriculture, or Ernest Abeyratne, Director of the Department of Agriculture, 1974–7, referred to the insurrection, it is clear that this frightening event brought home to them the urgency of improving social relations in rural Sri Lanka. Reducing the effect of class- and caste-based control of agricultural resources was also seen by Hector Kobbekaduwa, Minister of Agriculture, 1971–7, as a means of reducing tension and heading off further revolutionary activity. If the countryside was troubled by intermittent sabotage, local jacqueries, or prolonged revolutionary unrest, agricultural production would decline no matter what new varieties of rice were available, no matter how efficiently fertilizer, water, and other inputs were supplied. On the positive side, it was expected that a more equitable organization of agriculture would lead to the growth in yields and greater national production of rice and other food grains.

The policy of supporting the work of the Extension Division meant that there would have to be a two-way flow of information, not simply distributing new varieties of seeds and teaching the farmers appropriate cultivating techniques, but also listening to the farmers' statements of their problems and learning from their practical experience of possible solutions. Recognition of the real importance of a two-way flow of information was common among the extension staff and at the senior administrative levels in the department, though not among the plant breeders and researchers. It was a policy position held by those who viewed research as environmentally based rather than as plant-centred.

Unfortunately little could be done to test the validity of this environment-based view of agriculture with its associated view of the importance of the social relations of production. Any implementation of a more holistic view of agriculture ran into the fragmented administrative responsibilities of the agriculture-related departments of government. There was an immense gap between the integrated, systematic conception and the articulation of the organizations to carry through coherent policies for improvement. The nearest approach to a policy guided by this conception was the further decentralization of research and plant-breeding to eight stations distributed through some of the major agro-ecological zones, the initiation of watershed management and cropping systems studies, and the bare beginnings of an integrated rural development programme in some selected districts.

NOTES

1. Barrie M. Morrison, 'Alternate Routes to Increasing Rice Yields and the Implications for Research in Sri Lanka', in Robert S. Anderson, *et al.* (eds.), *Science, Politics and the Agricultural Revolution in Asia*, Boulder, Colo.: Westview Press, 1982, table 2.
2. Ibid.
3. *Judgement of the Criminal Justice Commission (Insurgency)*, Inquiry No. 1, Colombo: Government Printing Office, 1976.
4. Interview, E. F. L. Abeyratne, 27 May 1978.
5. L. A. Wickremeratne, 'Planning and Economic Development', in K. M. de Silva (ed.), *Sri Lanka: A Survey*, Honolulu: University Press of Hawaii, 1977, pp. 164 ff.
6. *Five Years of ARTI: A Review of Activities of the Institute from 1972 February to 1977 February*, n.p., n.d.
7. Interview, Earl Jayasekera, Acting Director Agriculture, formerly Deputy Director Extension, 9 May 1978.
8. See for example Barrie M. Morrison, 'Meegama: Seeking Livelihoods in a Kandyan Village', and K. Tudor Silva, 'Welivita: The Demise of Kandyan Feudalism', both in Barrie M. Morrison et al. (eds.), *The Disintegrating Village: Social Change in Rural Sri Lanka*, Colombo: Lake House Investments, 1979. Also *The Agrarian Situation Relating to Paddy Cultivation in Five Selected Districts of Sri Lanka, Kandy District*, Colombo: Agrarian Research and Training Institute, 1974.
9. Interviews, H. E. Fernando, Director, and Dr Irwin Gunawardena, Plant Biologist, Central Agricultural Research Institute, Gannoruwa, 28 February 1977.
10. Interview, A. G. Kularatne, District Agricultural Extension Officer, Kandy, 16 June 1978.
11. The agricultural extension programme in theory and in organization follows the recommendations of Daniel Benor and James Q. Harrison, *Agricultural Extension: The Training and Visit System*, Washington, DC: World Bank, 1977.
12. Interview, Earl Jayasekera, Deputy Director of Agriculture (Extension), 9 May 1978.
13. Interview, D. D. Liyanage, AI Kandy District, 16 June 1978.
14. Interviews, P. Ganeshan, Geneticist; S. Ponnuthurai, Research Officer, and A. D. Somapala, Research Officer, Maha Illuppallama Research Station, 8 March 1977. Interviews, G. W. E. Fernando, Director, and M. H. J. P. Fernando, Entomologist, Maha Illuppallama Research Station, 13–15 May 1978.
15. Interviews, A. Piyadasa, AI Palagaswewa, 13 May 1978; B. Jinendradasa, District Agricultural Extension Officer, Anuradhapura, 15 May 1978; R. M. Karunaratne, Additional District Agricultural Extension Officer, Anuradhapura, 15 May 1978; Mr Jayasinghe, AI Mihintale, 15 May 1978; Mr Wijesuriya, District Agricultural Extension Officer, Kurunegala, 16 May 1978; J. A. T. P. Gunawardena, Additional District Agricultural Extension

Officer, Kurunegala, 16 May 1978; U. Tennekoon, AI Wariyapola, 17 May 1978; S. A. L. Senanayake, AI Nikaweratiya, 17 May 1978; and D. Senadhira, Officer in Charge, Batalagoda Rice Research Station, 7 March 1977 and 16 May 1978.
16. E. R. Leach, *Pul Eliya: A Village in Ceylon*, Cambridge: Cambridge University Press, 1961, pp. 61–4 and 289–95.
17. Interview, Mahindra Raja, Project Economist, Maha Illuppallama, 15 May 1978.
18. *Feasibility Study for a Sri Lanka–Canada Dry Zone Agricultural Research and Development Project*, Ottawa: Canadian International Development Agency, 1976.
19. Interviews, D. Senadhira, Officer in Charge, Batalagoda Rice Research Station, 7 March 1977 and 16 May 1978; V. A. D. Sumanasinghe, Research Officer, Batalagoda, 18 May 1978; Mr Wijesuriya, District Agricultural Extension Officer, Kurunegala, 16 May 1978; J. A. T. P. Gunawardena, Additional District Agricultural Extension Officer, Kurunegala, 16 May 1978; U. Tennekoon, AI Wariyapola, 17 May 1978; S. A. L. Senanayake, AI Nikaweratiya, 17 May 1978; B. M. K. Balasuriya, AI Karandagolla, 18 May 1978.
20. Interviews with two rice farmers, and with S. A. L. Senanayake, AI Nikaweratiya, 17 May 1978 and J. A. T. P. Gunewardena, Additional DAEO, Kurunegala, 16 May 1978.
21. The specific evidence about the smallholders under minor irrigation schemes and rain-fed agriculture was collected from interviews with agriculture instructors and other Extension Division staff. Understandably their emphasis is on those problems which fall within the administrative jurisdiction of their division rather than considering the agricultural system as a whole.
22. Interview, U. Tennekoon, Agricultural Instructor, Wariyapola, 17 May 1978, and field observations.
23. See for comparison averages reported from Polonnaruwa of Rs 568 per ha and from Hambantota of Rs 482 per ha, in A. S. Ranatunga and W. A. T. Abeysekera, *Profitability and Resource Characteristics of Paddy Farming*, Colombo: ARTI, 1977, Table 4.
24. Interview, B. M. K. Balasuriya, AI Karandagolla, 18 May 1978.
25. Unpublished report on Katupota by V. A. D. Sumanasinghe, Research Officer, Central Rice Breeding Station Batalagoda, n.d. (*c.* April 1978).
26. The relevance of this project is limited because of the prevalence of coconut which grows well as a predominant crop throughout this area, except for the wetter bottom lands which, during heavy rains and run-off, go under water and drain slowly. In terms of establishing improved land use it may be a better strategy to capitalize on the existing practices and explore the potential for undercropping the coconut plantations with a rice that could be grown in conditions of shading and dryness or testing various legumes as an underplanting. Whatever future direction the multiple-cropping systems study at Katupota might take, it does bring Batalagoda into closer contact with the cultivating conditions and the realities which shape the cultivator's choices. In this way it marks a much more intimate involvement with local

agricultural and socio-economic systems than existed earlier.

27. M. P. Moore and G. Wickramasinghe, *Agriculture and Society in the Low Country (Sri Lanka)*, Colombo: Agrarian Research and Training Institute, 1980; U. L. Jayantha Perera, 'Nigaruppe: Settlement and Conflict in the Coconut-Triangle', in Morrison *et al.*, *The Disintegrating Village*, pp. 243–66; J. Selvadurai, 'Palannoruwa Village', in N. D. Abdul Hameed (ed.), *Rice Revolution in Sri Lanka*, Geneva: United Nations Research Institute for Social Development, 1977, pp. 167–207; and *The Agrarian Situation Relating to Paddy Cultivation in Five Selected Districts of Sri Lanka*, Part 5. 'Colombo District', Colombo: Agrarian Research and Training Institute, 1975.
28. *The Agrarian Situation Relating to Paddy Cultivation in Five Selected Districts of Sri Lanka*, Part 5. 'Colombo District', p. 4.
29. Ibid. 68.
30. Interviews, A. S. Vivekanandar, Research Officer, Paul Pieris, Plant Breeder, Bombuwela Rice Research Station, 10 March 1977; and with Jerry Jayawardena, Research Officer and Head, Bombuwela, 23 May 1978.
31. Interviews, K. Wijaysena, AI Nagoda, 23 May 1978; H. D. H. Dayaratna, AI Panadura, 24 May 1978; G. P. A. DeSilva, AI Bandaragama, 24 May 1978.
32. Interview, Dr Ikehashi, IRRI, Los Banos, 27 April 1978.
33. Interview, J. W. L. Peries, 21 May 1978.
34. Interview, Hubert Zandstra, Department of Multiple Cropping, IRRI, Los Banos, 28 April 1978.
35. Interview, M. H. J. P. Fernando, Entomologist, Maha Illapallama, 13 May 1978.
36. Interview, Robert W. Herdt, Department of Agricultural Economics, IRRI, 28 April 1978.
37. Interview, M. F. Chandraratne, Director, Department of Agriculture, 1956–9, 31 May 1978.
38. Interview, J. W. L. Peries, 21 May 1978.
39. Interviews, J. W. L. Peries, 21 May 1978, and others.
40. This view was not stated by C. R. Panabokke, but by others who shared his environment-based position.

7

Bangladesh: A Research Tradition Becomes Absorbed in the Green Revolution

> You can no longer be satisfied with a variety that gives two or three maunds more yield than another. You are looking for new varieties that will give you 100% or 200% increase in yield under high level management.
>
> Robert F. Chandler (director of IRRI) to the Secretary of Agriculture, East Pakistan (1966)

SITUATED in the huge Bengal delta, the state of Bangladesh is one of the world's great rice growing areas, ranking third behind China and India in the area planted to rice and in annual production. The confluence of three great rivers—the Ganga, Brahmaputra, and Meghna—has built a delta of great fertility. The semitropical climate and generous monsoon rains, together with the constantly replenished alluvial silt deposits, have given the central part of the delta an unusually favourable natural endowment for growing rice. The fertility of the land has been exploited and its food grains have been shipped to deficit areas for thousands of years. Yet by 1980 Bangladesh still imported 809,000 tonnes of rice and over 2 million tonnes of wheat with a total value of US $623 million.

Research on rice was pioneered in Bengal—research stations were opened at Dhaka (Dacca) and Calcutta in 1911—but although a conception of research evolved oriented to the agro-ecological diversity of rice environments which held steadily into the 1960s, few full-time scientists were employed in the early days. Thus the organization and funding of research in East Bengal did not match the conception, and with the partition of Bengal into West Bengal and East Pakistan this under-supported network of rice research stations was broken. Attempts to regain links to international rice science continued into the early 1960s, but national attention was focused on other objectives. At the same time, officials from IRRI and the Ford Foundation began to act in the early 1960s as brokers to improve the status of rice research and to build a prestigious new institute. When these efforts finally succeeded

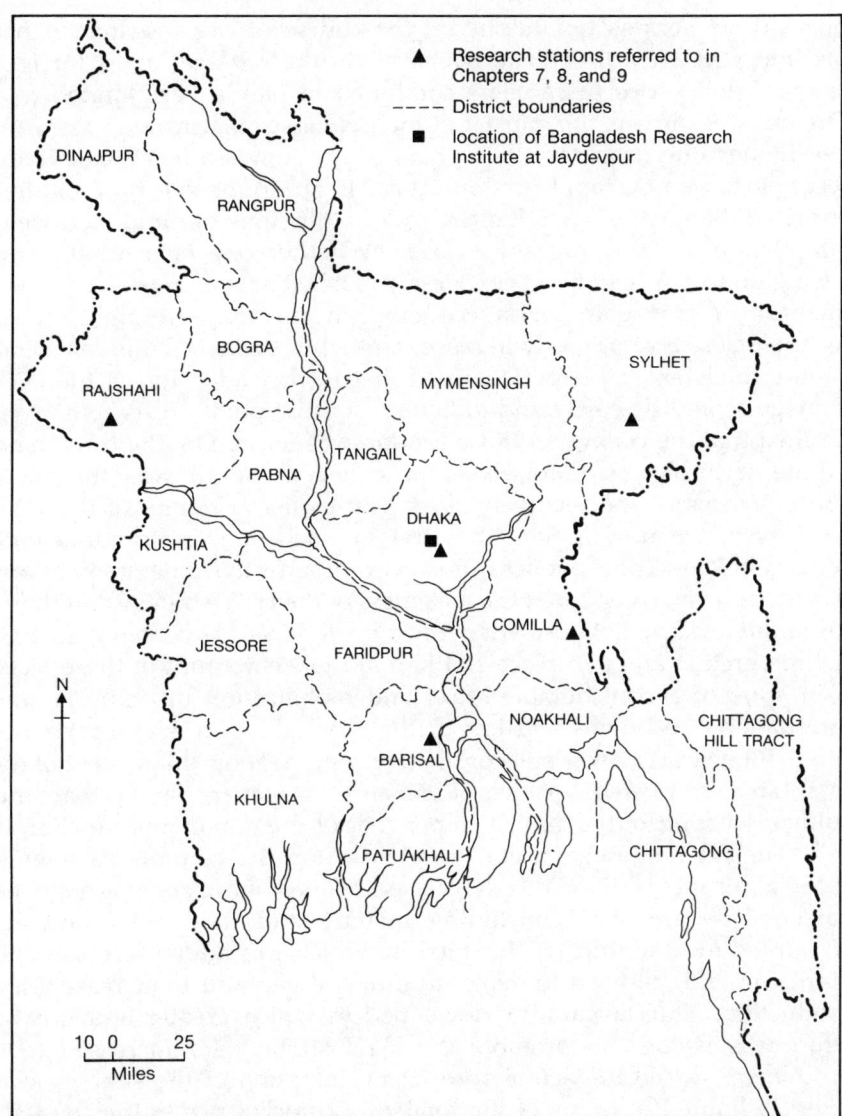

MAP 2. *Bangladesh: districts*

about 1970, with the inauguration of the rice research institute, the strategy of the Green Revolution had also been firmly planted in the country's outlook and massive investments already made in its technical requirements. Into this wider framework was set the intimate relation between IRRI and rice research in Bangladesh. By 1980 this dependency on IRRI set serious limitations on the course of rice research. It had become a pattern difficult to overturn. In the national quest for food self-sufficiency rice researchers continued to play a very limited role. The quest itself was the subject of increasing scepticism.

Whether one considers the spread of the new rice HYV cultivation technologies to be rapid or slow, there is no doubt that by 1980 their spread in Bangladesh was limited and that the time of rapid increase in adoption over large areas was over. Whether one believes that the ceiling on this spread had been largely reached or that there would be a quantum jump again when economic incentives were right, when technologies were properly in place, and when farmers' confidence had been established, all experts agreed that further adoption of the HYV strategy would be costly and difficult from that point on. Whether one claimed that the ceiling on this adoption was formed by the constraints in the technologies themselves, or believed that it was the social consequences of the relatively successful earlier adoption of the HYV strategy which now limited the strategy itself, it was clear that both constraints and consequences, however imperfectly understood at any particular level, constituted the weakness of the HYV strategy and therefore inhibited the government's quest for food self-sufficiency. In 1980 the research strategy in place still had not taken account of these facts.

In spite of a considerable effort and mobilization of domestic and international resources and knowledge, the adoption of the HYV rice production strategy was still confined by 1980 to about 15 per cent of the total area cultivated to rice. Estimates, however, show that this cultivation contributed almost 30 per cent of the national production of rice. The HYV strategy was dependent upon the continuous flow of external resources. Bangladesh received billions of dollars in foreign aid beteen 1950 and 1980, including military and food aid. Food aid accounted for one-third of this total, to which was added fertilizer and petroleum aid, all used to make up the food gap and to increase grain production. Thus a paradox developed in which greater and greater effort was needed to promote the HYV strategy for increased food production, while the actual spread and adoption of the strategy was severely limited in terms of the total cultivation of rice in the country. Absolute production increases had been achieved; yet the limits to the rapid adoption of the strategy were clearly visible by 1980. At the same time, in association with the HYV strategy, there was established a set of dependencies on imported food and imported technologies. While these dependencies were unpopular in many quarters, the country nevertheless lived with them and through them from year to year because food deficits had become chronic: varying in volume, imports remain necessary in political, economic, and social terms. As the authors of the 1938 *Report of the Bengal Paddy and Rice Enquiry Committee* noted, rice was and is still both a cash and a subsistence crop. In some areas it was a principal cash crop, and still is, while in neighbouring areas thousands of households seldom obtained enough rice for their needs from their own land (if they had any). This dual nature of rice means that while market considerations may be paramount in terms of capacity and incentives to produce more, non-market considerations are paramount in terms of the social stability of the country and fulfilment of its subsistence requirements.

Chapters 7, 8, and 9 trace the relation between scientific research, the objective of food self-sufficiency, and the HYV strategy for grain production. They deal with the origins of the HYV strategy before the 1960s, and how HYVs were introduced and promoted in Bangladesh, by following the relationships between scientific researchers, the HYV technologies and agrarian change. The evidence offered allows us to see beneath the surface of the national averages and national policies. Revealed are sharp differences between districts in terms of access to, and use of, each of the technological requirements of the HYV strategy and differences in terms of the agrarian and productive conditions which exist in each district. It is thus possible to compare these districts in terms of the extent to which they are self-sufficient in the production of grain. In particular it is possible to trace the weakening relation of Bangladesh research institutions with the historic values of crop improvement due to the increasing national commitment to the international HYV strategy. This approach so affected the research strategy that the result was a kind of entrapment, in which so much had been invested that alternative paths were not seriously considered for a long time. These relationships are the context of this story: its central figures are scientific researchers and planners in their struggles to address the fundamental agricultural diversity characteristic of rice cultivation.

7.1 Agricultural Diversity, Social Complexity

In the period up to 1980 Bangladesh was (and still is) characterized by profound diversity in agriculture and complexity in the composition of agrarian society. This variation in physical parameters of agriculture (rainfall, soil, temperature regime), and variation in population dynamics is to be understood against a background of relatively static factors. Such continuities as competition for land were under a process of intensification, but taken together present a situation of internal differentiation and variation which resists generalization. Having understood the trends which changed the area from rice surplus to rice deficit, and having grasped this diversity and complexity in agricultural environments, we can see how and why any research strategy which did not account for this diversity would not succeed. The claim to be established in this chapter is that conceptions of agriculture and of the role of rice research promoted officially from the early 1960s tried to ignore this diversity and complexity, and thus it was years until the necessary grasp of the situation was evident in decision-making circles. In some ways this grasp required a return to an earlier conception of how a research strategy should address complexity and diversity. But obviously the processes of intensification, the scale of food deficits in various districts, and the constant enlargement of the portion of the population which could not or did not subsist on the land meant that, by 1980, researchers could not simply resurrect long-neglected institutions.

Agroclimatic and Agro-Ecological Zones and the Districts

Bangladesh is a small country of 142,500 square km., half the size of Uttar Pradesh state in India or Arizona state in the USA. It had a population in 1980 of about 90 million people, or 1,600 per square km. in average density in the 1981 Census. Its most striking feature is the variation in physical and climatic conditions, which is the basis for the sociological and agricultural variation so important for the outcomes of the HYV strategy for food self-sufficiency. By using only two variables—rainfall and temperature—Rashid has arranged the country into seven climatic subzones ranging from 'A—Heavy Rainfall, Small Range of Mean Temperatures' in the south-eastern coastal districts to 'E—Very Hot Summer, Relatively Low Rainfall' in the western districts, and to 'G—Mild Summer, Fairly Heavy Rainfall' in the south-central districts, including Mymensingh, Dhaka, and Comilla.[1]

Manalo of IRRI has made possible a systematic comparison of the nineteen districts in terms of a greater range of climatic variables oriented to correspond closely to the cropping system's potential in rain-fed areas of Bangladesh.[2] He proposed six climatic zones showing a very wide range of climatic influences on cultivating conditions: the water balance regime, length of time of recharging ground water, evapo-transpiration levels, relative humidity, hours of bright sunshine, probability of cyclonic disturbance, wind speeds, cloudiness, early or late arrival of the annual monsoon, flooding probability which affects the rate of soil leaching and nutrient deposition, range of cold and heat extremes, and drought possibilities. Rainfall is heaviest in the north-eastern district of Sylhet (5,000 mm) and lightest at the same latitude across the country in the district of Rajshahi (1,200 mm). The district with the greatest areas planted to HYV rice, Chittagong, received a mean annual rainfall much lower than Sylhet, that is, 3,000 mm. Another district across the Padma River, Faridpur, which had very little area planted to HYV rice had half the rainfall received by Chittagong, that is, 1,500 mm—closer to the minimum in Rajshahi. When discussing rainfall, drought, and the monsoon, Manalo says, 'nowhere in Bangladesh can we find a district with a gradual start or finish of the rainy period. Monsoon activity is clearly delineated.'[3] The districts are grouped by Manalo in terms of probability of 'climatic stress' due to imbalance between the vectors of water recharge and water depletion. Stress is lower in the districts of Chittagong, Noakhali, Barisal, Faridpur, and Sylhet and higher in the rest of the districts. Because the concept of stress has not yet been systematically developed for Bangladesh, Manalo presents it in a preliminary state,[4] emphasizing the distinctive forms of climatic stress differences faced by cultivators and rice plants in the nineteen districts.

Soils of Bangladesh Districts

The interaction between climatic forces and the earth more or less determines the suitability of soil for rice agriculture. Rashid lists twenty-four physiographic subregions in the country—flood plains, valleys, basins, ranges, hills, etc. These are distinct from, but sometimes coincide with, the twenty soil tracts in the country, ranging from the widespread calcareous dark grey and brown flood-plain soils west of the Padma River, to clays, piedmont, acid basin, and non-calcareous flood-plain soils east of the Padma and in other regions to the north and south.[5] Farmers differentiate very clearly between the qualities and capacities of soil in their own holdings and in their cultivating regions, and there is an inflation of words for soil types and cultivating potentials similar to the inflation of Inuit words for snow types. In particular, there is a classification in terms of whether soils will support double- and triple-cropping, or only single crops, or should be left on long fallow or grazing. Of equal importance is the distinction between high ground (above normal flood levels) and the low land that is given over to rice. While it is true that these physiographic regions and soil tracts do not coincide with the political boundaries of the twenty administrative districts, these underlying differences serve further to make the districts very different regimes in which to cultivate rice.

Related to the variations in the physical setting are the man-made changes to the environment and the associated infrastructure which, when taken together, we have labelled the macrostructure of agriculture. One of the most visible historic modifications of the environment can be seen on the geographic feature called the Tippera surface which covers much of Comilla District. There the dendritic drainage systems running into the Gumti River or into the confluence of the Ganga, Brahmaputra, and Meghna have been so modified by the hands of men and women over the centuries as to create a rectilinear drainage grid.[6]

The channels of the small rivers and their tributaries have been straightened and squared off. The hazards of flooding have accordingly been reduced, and more land is available for cultivation. Comparable changes in containing the smaller rivers and improving the drainage have been carried out in parts of the Rajshahi and Dhaka districts.

Changes in Population Density in Districts

The density of the population varies widely in the 1974 census, as it did in the 1961 and 1981 censuses. Density varies from 850 per square kilometre in Comilla to 350 per square kilometre in the northern districts of Dinajpur and Sylhet, to lower densities of 300 per square kilometre in Patuakhali, 270 per square kilometre in Khulna, and 40 per square kilometre in Chittagong Hill Tracts. The range of the increases, as a percentage of the 1961 census, was from 61.5 per cent in Kushtia to 25.5 per cent in Patuakhali. The rural population of all districts increased between 1961 and 1974, including 56 per cent in Kushtia, 50 per cent in Dinajpur—both of them north-western border districts with no large urban centres. This population increase occurred in spite of the fact that by 1978 less than 20 per cent of the area of those districts was planted to HYV rice. Districts differed widely in their population increases, just as they did in their agricultural productivity; the mean population increase for all districts was 40 per cent between 1961 and 1974. There were also districts with markedly lower density changes between 1961 and 1974. For example, Faridpur was third in density in 1961, and dropped to sixth; neighbouring Barisal was sixth in density and dropped to eleventh; these are two districts which had high Hindu populations who remained refugees in India, large migrant worker and labourer populations, and riverine activity which eliminated villages and parts of villages, and which were prone to cyclonic activity.

As the population density increased the pressures on rice cultivation were becoming extreme, by world standards. The HYV strategy was first tested and strongly promoted in Comilla. Comilla ranked second in density in the censuses of both 1961 and 1974; in 1974 it had a population density of 1,000 per square kilometre, if water surface is excluded. In 1981 it was Bangladesh's most densely populated district. In 1961 the density (excluding water) was almost 690 per square kilometre. By 1978 over 40 per cent of the land cultivated in rice was planted to HYV rices, and in some thanas (like counties within a district) over 60 per cent of the area was under HYVs. Another district, Faridpur, separated by river but at the same latitude and of the same physical size (6,200 square kilometres), had a population density in 1961 of 500 per square kilometres (excluding water) and in 1974 of 630. Faridpur was ranked third after Comilla in density and had very little of its area cultivated in HYV rice.[7]

But the remarkable differences between regions was also evident in growth of towns and cities. In the period between 1961 and 1974, the unprecedented growth of cities and towns of Bangladesh had created new market opportunities and improvements in the communications surrounding agricultural areas. Farmers living in proximity to Khulna or Dhaka–Narayanganj during the advent of the Green Revolution thus had much greater opportunities than existed fifteen years earlier to market vegetables and other high-value crops. These large-scale changes in population and city size directly affected the adjacent agricultural areas. New crops became profitable, smaller scale operating units were economically viable, demand for labour increased, and so on. Rice in these peri-urban agricultural areas thus usually became less significant.

Complexity in Agrarian Conditions

Given these differences in rates of population growth in various districts and the significant climatic differences, it is clear that at any time in the 1950s or 1960s planners could have acknowledged the variation between districts in terms of land productivity (as well as the entire system of agrarian relations of production). For example, the area planted to the major rice crop, district by district, conveys some of the variety of growing conditions that influence rice plantings. The districts of Noakhali and Comilla in the east, with their ample rainfall, shallow flooding, and relatively good draining, plant a large fraction of the land to some variety of rice during the aman season, July to December. Khulna and Barisal, while subject to deeper flooding and poorer draining, also planted about 80 per cent of their crop land to an aman rice crop. The variation in cropping patterns suggested by the aman cropping figures is brought out more fully by looking at the range of the major crops sown throughout the year, such as jute, sugar-cane, tobacco, fruit, tea, potatoes, and wheat. Differences in cropping intensity in various districts was always related to the value of the other crops grown as well as rice. There was wide variation: for example in 1971 the value of output per unit of gross cropped area and of net sown area was more than twice as high in Chittagong (with high adoption rates of HYVs and Pajam rice) as in Faridpur District.[8] But overall there was a gradual decline in cropping intensity in many districts, in spite of the increasing population densities.

The differences in density of population of districts, while a very crude indicator of the pressure on the agricultural land, does underline the probability that there are substantial social and economic differences district by district. The proportion of the agricultural labour force (aged twelve or more) who are owner–cultivators or who are landless labourers provides an indication of the social organization of agriculture. In 1971 owners ranged from 50 per cent of the agricultural labour force in

Dhaka District to 25 per cent in Chittagong District. Labourers varied from under 10 per cent in Comilla to 30 per cent in Bakerganj. Alerted by these socio-economic variations, one is not surprised to discover that the average size of landholding was quite different from district to district—ranging from a low of 1 up to 2.1 hectares. Districts with more fertile soil and adequate moisture support denser populations with smaller-sized farms which are also associated with a higher proportion of owner–cultivators and fewer labourers. In Comilla a 1-hectare holding could be large enough to provide a subsistence living for a family, but in Bogra District, where the soil is less fertile, a household required 3–4 hectares to produce a surplus beyond subsistence needs.[9] Nevertheless, both in 1938 and 1978 Bogra was a leading surplus rice-producing district. Given the variations in the physical environment, in the infrastructure, and in the social organization of agriculture, there were also variations in the adoption rates of the HYV of rice or, more fundamentally, even of the area planted to rice relative to other crops, as Maps 3–5 showing HYV rice cultivation demonstrate. These distinctions are of long standing. For example one of the roots of the process by which people became landless is indebtedness. In his seminal studies which compared Barisal and Faridpur at the beginning of the century, J. C. Jack found that levels of debt were three times as high in Faridpur as in Barisal.[10] Indebtedness, Jack proposed, would inexorably lead to loss of land by cultivators.

The dynamics of landlessness has been studied by Brammer and colleagues, in ten districts.[11] The number of households without land to cultivate was in excess of 30 per cent in all ten districts, 'and higher in the northern and western districts where there are still many large land holdings'. There were thanas in five of the ten districts in which the number of landless households exceeded 50 per cent, and neighbouring thanas where the landless number was minimal, as can be seen in Table 7.1. It is of course from the landless population which does not migrate out of the region that the bulk of the daily wage-labour in agriculture is

TABLE 7.1. *Landless Bangladesh agricultural households, 1976–1977 (%)*

District	Mean	Max. thana	Min. thana
Rajshahi	30	61	6
Noakhali	23	58	11
Sylhet	31	53	7
Kushtia	36	52	23
Khulna	29	50	15

Source: Brammer *et al.*, and *Incidence of Landlessness*.

drawn. As this population expands the HYV strategy will have to consider the absorptive or retentive capacity of the rural economy.

The same study of ten districts found that in six districts more than 10 per cent of the rural households farmed more than 2 hectares of land; in the definition of that study they were big farmers, who lived mainly by renting out land to other farmers and by farming part of their land with labour. The district with the lowest mean of big farmer households was Noakhali, with 5.4 per cent. The six higher districts, with over 10 per cent, are shown in Table 7.2. Many of these farms were very much larger than 2 hectares, and some larger than the legal ceiling of 13 hectares; it will be recalled that the bulk of procurement by government for the public grain system took place from the marketed surplus from these northern districts, presumably from this category of big farmers. In the districts of Mymensingh and Sylhet the adoption of the HYV rices had been greater and more rapid than in adjacent north-western districts.

Intensification of these processes—population growth, peaks in cropping intensity, landlessness, share-cropping, etc.—have had other effects, such as conflict over land. In the nineteenth century Faridpur led the districts of Bengal in court cases over land disputes, while in other districts violence and conflict were uncommon. Zaman provides an assessment of forty major conflicts on new land or *char* (river islands) in nine districts, most of them resulting in injury or death.[12] These conflicts generally occurred between groups competing for the opportunity to plant rice in land arising from changes in the course of rivers; in some cases hundreds of people were involved in the violence, many dying as a result. Even within specific thanas some areas have been well known for conflict over land, while others were remarkably settled and stable. All of these processes have led to polarization, one of the results of which has been the creation of greater and greater numbers of landless people.

There are numerous other differences between districts, such as the relative involvement of women in the agricultural economy or in various

TABLE 7.2. *Big-farmer agricultural households, 1976–1977 (%)*

District	Mean	Max. thana	Min. thana
Dinajpur	14.9	33.6	7.5
Rajshahi	13.3	27.5	3.0
Kushtia	13.3	25.7	6.0
Mymensingh	11.5	31.9	4.0
Sylhet	10.7	20.7	4.6
Khulna	10.6	14.3	4.6

Source: Brammer *et al.*, *Incidence of Landlessness*.

stages of the production or processing of rice.[13] When asked about the origins of these differences, Bangladeshis usually point out that the secondary settled quality of life in a region, in which long-distance communication was difficult, encouraged localized cultural development. They also point out that each district had a very different kind of contact through trade with other districts, or with the world economy—Sylhet and tea, Faridpur and jute, Chittagong and trade with the Middle East, etc. The characteristics mentioned by Bangladeshis, of 'orthodoxy' or 'tolerance', of 'fierceness' or 'peacefulness', are usually employed in order to explain the relative 'backwardness' or resistance to social changes which the speaker considers desirable, such as 'literacy' or 'education'. One rarely heard anyone explain the adoption or resistance to the HYV strategy in a district, or even the next thana, in terms of orthodoxy or tolerance in religious matters, or in terms of the history of relations of a district to the world economy. But to the extent that these differences have played a role in the politics and economics of rice cultivation, as in Chittagong, they help to account for the pattern in which HYV cultivation has spread or been limited: that is, not 'backwardness' or 'orthodoxy' alone, but the peoples' perception of their resources, skills, advantages, and limitations which also reflect the agro-ecological conditions in their district, conditions which vary considerably within districts, as studies of power structures and economy in various villages have shown.[14]

7.2 From Historic Surplus to Rice Deficits

The complexities of the situation of surplus and deficit in Bangladesh are exemplified in the following example. There was a record rice crop in 1977–8; 13.396 million tonnes of rice were produced, up 13 per cent from the previous year. Never had production been so great in history, and seldom had such an increase occurred over the previous year. Some districts had record surpluses, yet the total 'required' in the country was calculated officially to be 14 million tonnes, and a minimum 10 per cent of the total production had to be subtracted for seed and wastage. Thus the total grain available was 12.056 million tonnes, leaving an official gap of 1.944 million tonnes at least. Thus in this year of record grain production over 1.7 million tonnes of grain was imported, as much as had been imported in previous years when such a record harvest had not been achieved. Only 88 per cent of the national 'requirement' could be met in 1977–8; Bangladesh was dependent on external sources of supply for the rest. Food production increased following that year, but still five years later, 1.5 million tons of wheat were imported to Bangladesh from the US, Canada, European Economic Community, and

Australia. The proportions of production and imports varied, but each year since independence in 1971–2 the quest for self-sufficiency had to be politically reaffirmed in the light of these facts. Sufficiency was an elusive goal; although it had been a preoccupation since the 1940s, in 1980 it still appeared to be a long way off.

In 1943 Bengal was shaken by famine. By creating the buffer war zone between Burma and Calcutta, military and government authorities inhibited the movement of rice and confiscated (and in some cases destroyed) transport boats. Rice imports from Burma ceased. Cultivators in Bengal also withheld rice from the market against difficulties they anticipated in feeding their own households. Demands in towns and cities increased simultaneously. Foreign troops were stationed throughout the region and needed food. A few influential speculators also entered the market to buy up stocks which were then ready for market. The system of 'entitlement' which conveyed essential food to the poor people in periods of stress broke down absolutely, with tragic results.[15] Estimates of deaths attributable to this famine vary from the official 1.5 million to the unofficial 3 to 4 million persons. The famine was a *cause célèbre* at the first international agricultural conference held in Washington to establish the Food and Agriculture Organization (FAO) of the new United Nations and was used as proof of the urgency of its negotiations.[16]

This crisis forced some small changes in the official approach to agriculture. The 'Grow More Food Campaign' was launched in 1943, busying most of Bengal's district-level agricultural officers in promoting increased production. Later evaluations showed little result from this effort. (In Bengal, Islam finds an increase in area planted to rice from 1942–3 onward; this 'coincides', he says, with the campaign, but it is probable that area was expanded by reducing the customary underestimate of the rice crop which was reported by the Department of Agriculture.[17]) While officers, particularly extension workers, were busy with the campaign, they were further burdened by the instruction that they were now to operate the district research demonstration farms under the aegis of the District Boards. To promote the 'self-reliance' of these farms and to save money, trials of the new rice introductions were suspended, essential staff were moved to Dhaka (Dacca) or Calcutta, and research work (such as it was) wound down.[18] Extension workers did not have the training or motivation to maintain these demonstration farms because it was simply a task added to their other tasks, and most research farms thus withered away through neglect.

At the same time the governments of Bengal and India approved the Damodar Valley Project in western Bengal, which included both irrigation and hydroelectric systems, as well as a fertilizer factory. The aim of the project, modelled on the Tennessee Valley Authority (TVA), was to stop the Damodar River from flooding (which it had done in 1943), to supply power to the industrial belt and coal mines in central Bengal and to Calcutta, and to modernize agriculture. The Chief Planning Officer for the TVA was appointed to the Central Technical Power Board in 1943–4 in order to advise the government of India on how to integrate industrial and agricultural objectives in this project. Finally, in this same post-famine context, the government of India proposed a cess on milled rice in 1944 in order to fund a new central rice research institute. The government of Bengal proposed that it should be located in Bengal. Although both the cess and the location were turned down, there was now no doubt about the intention to establish a Central Rice Research Institute (CRRI) somewhere in eastern India. It was a tangible response to the famine, and another expression of the idea that there should be a separate national institution to study every problem in India (nuclear physics, fisheries, statistics, etc.). By the time CRRI was established in 1947, India was dividing into two countries. But the underlying imbalance between food supplies and population growth remained, and the new government in East Bengal would have to do much more about the consequences of the imbalance since it would be cut off from the research establishments of both India and West Pakistan.

Because all crops in Bengal experienced 'the lowest rate of expansion' in the whole Indian subcontinent, according to Islam, 'the disparity between crop production and growth of population became very marked' in the period preceding independence. By 1920 Bengal's population was already more dense than that in any other region of the subcontinent, and during the period 1920 to 1946 'food production remained stagnant as against an annual growth of 0.8 per cent in population'.[19] Area planted to food crops accounted for 80 per cent of the total cultivated area. Though there was an annual increase in area planted to food crops of 0.2 per cent, this increase was probably neutralized by a decline in yields on the overall area. Food crop output per hectare was 9 per cent lower in the period 1940 to 1944 than it was twenty years earlier in 1920 to 1924; 'the reduction,' says Islam, 'was probably due to declines in output of both winter and autumn rice.'[20] In keeping with the decline in food output over this period, the per capita availability of food grains declined through the whole of British India between 1911 and 1941 at 0.83 per cent per annum. Recent studies for Bangladesh have also shown that the decline in per capita availability of food grain continued in the period 1950 to 1963.[21] No satisfactory explanation of the static or declining yields has been offered. The most careful study concludes that the decline was due to the effect of the transfer of better lands from winter rice to higher-value non-food crops and the parallel reallocation of resources and labour to the higher-value crops and away from rice.[22]

Nevertheless there were marked differences between districts in East Bengal during the period before 1947, differences in population density, farm size, land tenure arrangements, secondary crops or non-food crops, and the district's position in the rice market. The 1938 Paddy and Rice Enquiry Report gives a breakdown of East Bengal districts on the question of their self-sufficiency, as shown in Table 7.3.

Forty years later, in 1978, the same rice surplus districts of Dinajpur, Rangpur, and Bogra remained the sources of most of the procurement of the government of Bangladesh of rice for the state rationing system. In addition to interdistrict trade in rice before independence, Bengal depended upon imports of rice from the early 1930s. Though in the 1927–32 period exports of rice exceeded imports, and these exports accounted for 2 per cent of the total production (not a great amount in contrast to other exporters in the period), from that year (1932) imports exceeded exports.[23] In the period 1933 to 1942 *net* imports of rice amounted to 3–4 per cent of total domestic supply. The sources of this rice were Burma, Assam, Orissa, and (occasionally) the United Provinces. In the years 1942–5 there was a 'sudden drop' in net imports, down to 1 per cent of total domestic supply due to the 1943 famine and its aftermath, and the concurrent reduction in sea shipments of rice from Burma because it was controlled by Japan. Until then rice, most of it coarse rice, from Burma had entered Bengal with no barrier. The coarse grains were going to feed the urban population and the plantation workers while the higher-quality rice was shipped out. The price differentials made it an attractive market decision. Rice from Burma, Thailand, and Indo-China was now successfully penetrating markets which had been Bengal's. Fall in external demand (for example in Madras) depressed prices. The middle and top qualities of the rice export trade from Burma were controlled by people from Bengal or Madras, who were extremely knowledgeable about the market in

TABLE 7.3. *Self-sufficiency in rice of eastern Bengal districts, 1938*

Self-sufficient	Surplus	Deficient
Rangpur	Chittagong	Mymensingh
Bogra	Bakerganj	Pabna
Jessore	Dinajpur	Rajshahi
		Khulna
		Noakhali
		Faridpur
		Dhaka

Source: Government of Bengal, *Report of the Bengal Paddy and Rice Enquiry Committee*, Alipore, 1940, i. 23.

Bengal. Rice prices in Bengal, however, were deliberately kept lower than jute prices so as not to draw areas away from jute production. The ceiling on jute prices, in turn, was set by the interests and needs of foreign jute mills in the UK and Europe and jute mills in Calcutta which had to make their purchases in East Bengal. This depressed pricing structure and availability of imports are reasons why rice agriculture did not expand.

There was an expansion of jute area between 1920 and 1946 at the expense of the aman rice crop, though jute accounted for only 9 per cent of cultivated area in the period. Though in some districts other non-food crops played a role in land use (cotton, sugar, tea, tobacco, oil seeds, etc.), only jute seems to have been in continuous competition with rice for land in the same season. By 1930, 50 per cent of the jute area was sown to improved seed in Bengal, compared to 6 per cent of the rice area.[24] Cultivators tended to transfer their more productive land to jute, to use improved varieties where possible, to take better care of this crop (which requires a lot of labour during cultivation and processing), and also to reduce the amount of rotation and fallow land. This affected rice cultivation on both marginal and productive lands.

Within the territorial bounds of modern Bangladesh, the population has more than trebled since the turn of the century. In 1901 the population was reported as 28.9 million whereas the estimate for 1980 was 90 million. For the first forty years of the century the growth rate was around 1 per cent a year. In the second forty years, 1941–81, the rate has been in excess of 3 per cent a year. This enormous and, in the last decades, rapid increase in population has put great strains on the food-production and distribution system. Complicating the supply of food to the growing numbers of people were two other developments: the growth of an urban population and the increasing withdrawal of land and labour force from growing food to cultivating industrial crops. In addition, subregional patterns of agricultural specialization began to emerge. The hills and valleys around Sylhet in the north and more recently the hills to the east have been developed as tea plantations. Dhaka District, with its urban concentration and area committed to jute-growing, was deficient in rice, as were Chittagong, Faridpur, and the old, large Comilla districts. Yet in spite of the growth in population of the cities, the numbers of plantation workers, as well as the general growth in population, the larger Bengal delta still exported rice to parts of India and overseas until 1931.

With the centuries of rice surplus followed by a period of cheap rice imports from Burma, it is not surprising that neither the British government of India nor the Bengal provincial government had seriously commited resources to improving the yields and production of rice in Bengal. The earliest agricultural research centre was at the Royal

Botanic Gardens Department establishment in Calcutta, which was largely preoccupied with the improvement of export crops of jute, sugar, tea and, in its substation in the mountains near Darjeeling, the production of chinchona bark, source of an antimalarial drug.[25]

7.3 Rice Research and Agricultural Development in Bengal before 1947

Rice research began in Bengal when the Central Farm was opened at Dhaka (Dacca) in 1911. At first only an 'economic botanist' was assigned to rice, but in 1917 an agricultural chemist was also given partial responsibilities. Still, by the time the School of Agriculture was opened in Dhaka University in 1930, fewer than half a dozen scientists had responsibility for rice research. This school was to train potential extension and demonstration workers for district level positions but not to conduct research. It took in twenty students annually and was separate from the rest of the university. In 1934, partly as a result of the 1928 Royal Commission on Agriculture, two rice research substations were opened, one in Bankura (West Bengal) for rice in laterite soils and the other in Habibganj (East Bengal) for deep-water rice. These stations opened a few more positions for rice researchers. The Home Rule Act of 1936 assigned most of the positions in the Ministry of Agriculture to Indians, thus opening more scientific posts to Bengalis. In the 1940s, partly as preparation for extension of irrigated agriculture associated with the Damodar Valley Project, an intensified agricultural course was started at Dhaka, and some people went for training in other parts of India where there were private agricultural colleges.[26] When it became clear that the subcontinent would be partitioned, more emphasis was put on training East Bengali Muslims for agricultural work. But still at the time of partition in 1947 about 80 per cent of the researchers in Bengal were Hindu, and most of them chose to work in India, leaving East Pakistan with very few research resources except the Central Farm, the Habibganj substation, and a number of small district farms.

Research positions before 1947 were less prestigious and lower paid than administrative positions, a pattern that continued for research in East Pakistan and Bangladesh. There were certainly fewer positions for research than for administration. The administrative post of District Agricultural Officer was the equivalent in salary and status to the District Engineer or the District Civil Surgeon, and the Agricultural Officer did no research and was not trained to do so.[27] Research was not the path of upward mobility in the civil service, and rice research was less attractive than research on jute, cotton, sugar, or tea; these crops had autonomous research institutions which were jointly funded by government and by a cess on production and trade. These were, after all, international plantation crops which offered to researchers an opportunity for international recognition and travel or work which working on rice did not. Promotion was based on publishing in foreign journals or Indian journals patterned after them. 'Our rice agriculture was not taken seriously. I don't like to say it but we Bengalis pursued typical middle-class fantasies, emulating and copying the research of other countries and disregarding our own skills.'[28]

Ideas about the role of research in rice cultivation seem to have followed the dominant theme of the development of science and technology in Bengal. The common argument was that a backward and poor country would be improved only if it was civilized. Civilization occurred through the application of science to the problems which obstructed 'progress'. The backwardnesss of smallholder agriculture was to be overcome by the application of science and the displacement of the old 'backward practices'. The argument also continued that when science had produced a good idea, it was necessary to diffuse it throughout Bengal, relying on the 'progressive classes'. This diffusion could be ensured by creating an agricultural extension service which would demonstrate the benefits of these new techniques or seeds, thus convincing the progressive classes.

In the simplistic conception of solutions, furthermore, specialists were clearly segregated—botanists, for example, need not know about soils—and people were trained separately. In addition, some problems in agriculture did not enter the curriculum because, like land tenure or irrigation systems, they were considered external to the academic discipline of agriculture. It was recognized that there were institutional obstructions to progress which could reinforce 'backward practices', such as rack-renting and extortionist moneylending practices. While it could not have escaped scientists' notice that the landlords of Bengal had invested their capital in the further purchase of property, in moneylending, in the trade of commodities, and not in improving the productive quality of agriculture from which they drew their income, whether or how these practices were to be changed or removed was deemed a question fit only for their private opinions.

In contrast to this rather simplistic approach there was an analytic and empirical approach to the 'problems which obstructed progress', an approach seen in newspapers, journals, books, or the Royal Commission on Agriculture. This grasp of problems was often expressed in the Indian Association for the Cultivation of Science from 1890, the Indian Science Congress from 1911, the *Modern Review*, the *Calcutta Review*, and the *Statesman* throughout the 1920s and 1930s.[29]

The most important activity of the rice scientists before 1947 was systematic seed selection. The selection of superior strains of rice and their propagation in pure lines, as well as the introduction of varieties from other countries, were part of the general effort to spread improved varieties through Bengal. By 1930 only 6 per cent of rice area was sown to improved varieties, in contrast to 50 per cent of jute area. Before the major effects of the improved varieties began to show in the 1930s, there was evidence that scientifically selected varieties could yield more in farmer's fields than local varieties; in 1928 the selected variety Surjamukhi yielded 3.3 tonnes per hectare on fields in Sylhet, in contrast to yields of 1.8 tonnes per hectare of local varieties in the same area. In 1930, the government's economic botanist at Dhaka, G. P. Hector, published a major study of classification of rice types, pointing out which varieties would meet the strong commercial demand for a medium to fine rice. By 1930 there were 1,200 pure-line rice varieties being studied at Dhaka; while 200 were rejected on the basis of yield or quality, 1,000 lines were maintained.

Also researched was rice which would respond to difficult or stress conditions; a saline-resistant variety Patnai 23 was developed on the district fram at Goshaba in Twenty-Four Parganas. This variety was widely adopted in saline areas. Also in the 1930s varieties from other countries—the US Bluestick and Latishail from Assam, both fine rices—were successfully introduced and widely adopted in 1934. Three years later six rices from the Philippines and three from Guyana were tested and introduced. And in 1941 the now famous Nigershail was brought from Nigeria. This was a rice developed in Madras and transported about 1930 to Nigeria for testing; ten years later, in a routine exchange of genetic material between colonial administrations, it arrived in Bengal and was named after Nigeria. Released in 1944-5, it was widely adopted by farmers over the next thirty years. All these rices were initially intended for the dominant aman crop or the secondary aus crop.

In a brief to the Paddy and Rice Enquiry Committee in 1938 the economic botanist at the Dhaka Farm defended the two techniques he had used: selection and breeding. Yield characteristics had to be weighed against resistance and quality characteristics. Testing was to be done on district farms. Good varieties must be superseded by new improved varieties in a continuous never-ending process of selection. In this way, twenty 'improved pedigree varieties' for upland aus and transplanted aman cultivation suited to 'different climate and other environmental conditions of the different districts' had been identifed or bred by 1938.[30] This was the way in which both difficult conditions and favourable conditions were being addressed by research.

An important subset of the selection and breeding programme was the attention paid to varieties that would grow and yield in the flooded conditions of many parts of the central delta. From 1911, it was recognized that deep-water rice was important in Bengal. In 1917 the agricultural chemist began manure trials with deep water rice at Dhaka, though the results were not successful. The traditional deep-water rices seem to get most of their nutrients from the water itself after the initial growth stages. By 1922 the deep-water rice collection included 170 separate strains, and the deep-water rice substation was staffed by eleven people: one trained botanist, two clerks, a field assistant, and four field workers, in addition to a farm manager and his two staff.[31] Much work was done on listing and classifying deep-water rice between 1934 and 1944, when the substation was transferred to the responsibility of the government of Assam. From the beginning, researchers were conscious of the deep-water problem as it was perceived by farmers: 'Unfortunately typical deep water paddies are almost universally coarse grained, red kernelled and awned and thus they fetch a lower price.... Consequently cultivators very often cannot resist the temptation of growing a medium grained variety even in the low lying areas with the result that a flood of moderate nature causes damage to the crop.'[32] Study of deep-water rice was carried on fitfully until it was more fully revived in the 1950s, when eight new deep-water varieties were released.

The other crop which had received little attention until the 1940s was boro rice, planted in the cold season of December, transplanted in January, and harvested before the monsoon in late April or early May. Most of the work on boro rice was also carried out at the Habibganj Substation in the 1940s, following a classification of these rices in terms of groups, Bengali names, and botanical names. Rice was grown in the boro season at the edges of swamps (*bil* or *haor* in Bengali) and wherever irrigation existed, since some of the seed used in warmer seasons failed to germinate in the winter planting, and there was a fear of lodging damage from the early storms and cyclones of March and April when the grain was at its height. The work at Habibganj was mostly on investigating cultural practices of the boro season and conducting trials of varieties grown in other seasons. There were studies of double transplanting, application of manure, mixed cropping of varieties, drought tolerance, time of irrigation, seed rates in seed-beds, broadcasting versus transplanting, use of water hyacinth as fertilizer, and so on. Between 1937 and 1945 yield results fluctuated, depending on the trial, with top experimental boro yields of 2.3 to 2.5 tonnes per hectare and low yields of 1.1 to 1.26 tonnes per hectare. There were reports of boro yields in farmer's fields of 2.9 tonnes per hectare in the 1940s, and it appears the potential of the boro season was understood as well as the risks. But the official promotion of rice-growing in the boro season did not occur until the 1950s.

Aside from the seed selection and breeding programmes for aman, aus, boro, and deep-water rice, some attention was given to experimenting with improved cultivation practices. At the Central Farm in

Dhaka between 1912 and 1916 the Japanese method of selecting healthy seed by floating seed in salt water to allow the specific gravity of healthy seed to float it on the surface was tested; the test was declared 'very successful', but we do not know, for lack of evidence, whether the result was propagated or widely adopted at the time.[33] In 1917 and almost each year thereafter, the effect of manure on yield was studied, with the conclusion that depending on soil and type of manure, the results were very good for yields. Unfortunately competing use of animal manure for fuel and cultural avoidance of human manure (unlike, say, in China) kept manure uses to a minimum. Also in the 1900s, there were studies of the effect of early or late seed-sowing on yield. In research begun on the boro rice in the 1930s, cultural practices were carefully examined.

After the decades of work on selecting and improving the varieties of rice and with some attention paid to the cultivating conditions, it was understandable that the Ministry of Agriculture decided to gauge more accurately the actual yield in the farmers' fields. Remarkably, there was no evidence available to the government from which to calculate the 'true position regarding yield per acre' in Bengal at this time.[34] Galvanized by the experience of the famine in 1943 and of the plot-by-plot survey of land use in Bengal in 1944–5 (ordered after the famine), a sample crop-cut to establish yields on fields was carried out between 1943 and 1947 both by ministry officials and independently by the Indian Statistical Institute (ISI). The official estimates were higher in these years for both winter and autumn rice than were the ISI sample estimates, between 11 and 21 per cent higher.

What was striking to researchers, and to others no doubt, was that average experimental yields were so much higher than these yields on farmers' fields. For example, the average yield of Nigarshail was reported to be 3.06 tonnes per hectare, and good indigenous pure-line varieties gave average yields of 1.8 to 2.7 tonnes per hectare. It is understandable that researchers were troubled by the yield gap, even though they had no responsibility to intervene and promote cultural practices which would result in higher yields.

Thus at the time of independence, rice research was conducted in relatively isolated and understaffed institutions on a far smaller scale than other agricultural research. Not tied to any politically powerful interest group of 'rice farmers', the research effort was supported by government officials concerned about the condition of the Bengali cultivator and the supply of food grains in India. By 1946 the food crop output was 9 per cent lower than it was in 1920–4. Distinguished government commissions enquired into paddy and rice prices and into the famine of 1943. The 'Grow More Food' campaign and the Damodar Valley Project irrigation developments were expressions of official concern. But the land-tenure situation remained unchanged: minute fragmentation at the bottom and remarkable concentration among landlords at the top. Rice prices remained softer than jute prices except during and immediately after the famine. There was a climate of great uncertainty and instability in Eastern Bengal until the abolition of zemindars (the local revenue collectors who had become entrenched landlords). While coping with the vicissitudes of natural disaster and communal violence, farmers worked with the materials which they knew and which were predictable. Research and development ideas took a back seat, and expedient governments ruled without long-term planning.

7.4 The Continued Neglect of Rice Research 1948–1960

By the acts of partition and creation of Pakistan, East Bengal was changed from its status of a province of British India to become half of a new nation joined to another half lying sixteen hundred km. away to the west. Without the help of the imperial research institutions, such as the Central Rice Research Institute or the Indian Agricultural Research Institute, the Pakistan government now had to sort out the rice research needs of the Bengal delta. Greatly complicating the task was the exodus of agricultural researchers from East Bengal, mostly Hindus who chose to live in India, although most of the seed collections and records remained in Dhaka at the Central Farm. It took the new government eight years, until 1955, to draw up a plan for a rice research institute.

Accompanying these massive and violent exchanges of population was a profound structural change in the agrarian system. The outflow of Hindu capital accelerated when it became clear that the landlord system (zemindari) would be abolished. The new East Bengal State Acquisition and Tenancy Act of 1950 contained a ceiling on legal holdings clause which allowed ownership of up to 13 hectares and no more. Some exceptions were made for co-operative farms and for commodities like cotton. There were some major Muslim landlords in East Bengal; of the 2,237 largest estates, 358 (about 15 per cent) were owned by Muslims.[35] Some of the Hindu and Muslim families who owned a lot of land fought hard in the courts to preserve part or all of their estates. But two processes were under way. The first was that these estates, lands, and buildings were often taken over by the local agents, *talukdars* (landlords beneath the zemindars), and others of the former landlords, zemindars. Even those lower down the feudal land hierarchy struggled for years to gain more and more control of these properties, both by legal fiction and by illegal means. It was clearly understood that these concentrations of property were the keys to wealth and influence in the countryside, and some of these families, such as the Lal Miah/Mohan Miah family of Faridpur, used these estates to enter national politics.

The other process was that the government could not and did not enforce the 13-hectare ceiling. Much of the former zemindari land was unregistered, and the process of registration took years. Meanwhile its cultivation was profitable. In 1954 when the United Front government came to power in East Pakistan there was no agreement on enforcing the ceiling on landholdings. When Ayub Khan came to power by coup in 1958 he waited three years before increasing the land ceiling to 50 hectares. Only 27,000 hectares of the former zemindari land was finally placed in the government's Khas category as a result of the 1950 Act, and it was not distributed among landless or land-poor families.[36] Abolition of the landlord system did not destroy all concentrations of property, but simply created a vacuum into which new groups and forces moved. Taxes still had to be collected from small farmers. Tenants, the majority of farmers, still had to pay a share in kind for rent of the land from the owner in addition to tax. Rice still had to be grown in small plots. There was still no answer to the question of the imbalanced population and food supply equation. Islam reviewed the records and decided that:

The argument that the change of tenurial arrangements does not automatically solve the problem of low productivity is clearly borne out by the experience of Bangladesh. Here superior landlordism was abolished in 1950, but until the middle of the 1960s productivity remained as low as it was before the abolition of the Permanent Settlement. The proximate economic effect of the abolition of the Permanent Settlement remained confined to an enhancement of government revenue.[37]

In the 1950s the major efforts of the new government to improve the condition of the Bengali cultivators were directed to community development and irrigation, with rice research lagging far behind. Specifically the Village Agricultural and Industrial (AID) scheme, with its offspring, the Comilla Project, and the Ganges–Kobadak Irrigation project received unprecedented amounts of foreign aid while rice research was relatively neglected.

In 1953 the Village-AID scheme began as a joint programme between the Technical Cooperative Mission of the US, the Ford Foundation, and the government of Pakistan. It was based on the sociological and communication assumptions which underlay the role of the US extension agents and county agents in US agriculture. Large numbers of people from Pakistan were sent for a year's training in the US where Michigan State University was responsible for much of the training, using grants from the Ford Foundation. In 1956, out of the need for informed and trained extension workers, came the idea of building local training and research academies in both East and West Pakistan. Prior to establishing the Academy for Rural Development at Comilla, the future principal Akhter Hameed Khan and his staff spent a year at Michigan State University preparing for their own training role.[38] Khan's iconoclastic approach to rural mobilization through small-scale village co-operatives run by, and accountable to, small cultivators and the landless eventually attracted the support of the new President, Ayub Khan. Though Village-AID was terminated in 1961 when Ayub Khan presented his new constitution, the Comilla Project, with its associated Academy of Rural Development, carried on the integrated approach to economic betterment with its associated concerns of basic education, health, and community institution building. The project was given full government support, which was supplemented by funding from the International Cooperative Administration of the US State Department and the Ford Foundation. The Comilla Project area later became the test site for new rice varieties in the 1960s.

The Ganges–Kobadak Irrigation Project was begun in 1954 when the government of East Pakistan saw the benefits of large-scale irrigation in West Pakistan from projects completed by British engineers before 1947. This became the first of fifty-one proposed water projects to be implemented by the Water and Power Development Authority (WAPDA) formed in 1959. Such projects and bureaucracies were advised by a UN technical mission in 1957, led by the former head of the US Army Corps of Engineers. Subsequently the American firm of International Engineers, Inc., began preparing the master plan for irrigation, which proposed expenditure of $3.2 billion over the next twenty years. After expenditure of $550 million by 1980, much of it from the World Bank, these projects remained classic examples of large-scale engineering, absorbing vast sums of local funds and foreign exchange but of marginal use to most farmers.[39] For example, $132 million had been spent by 1970 on the Ganges–Kobadak irrigation and flood-control project. After sixteen years of construction, redesign, and reconstruction, it failed to perform at even 50 per cent of original design capacity. Millions had therefore to be spent in efforts intended to 'rehabilitate' this scheme. Only 7 per cent of Bangladesh's cultivable area had been irrigated by these means by 1980, yet there was the well-founded perception that water control and management was crucial to rice production in the country. However, small-scale irrigation technologies might have had a major impact from the beginning.

The earliest known analysis of the contribution that rice research might make to the improvement of production and to the condition of the cultivator repeated the fashionable wisdom of the day. A former economic botanist for rice, S. Hedayatullah, was the Director of Agriculture for the whole province of East Pakistan. When he presented the government's rice research and development philosophy at the International Rice Commission meetings in Indonesia in 1952, he spoke from experience beginning in 1935.[40] The first priority, he believed, was to improve rice varieties through breeding rather than through pure-line seed selection. There had to be more attention to selection and breeding of rice for special regimes, namely flood, drought, salinity, and for upland conditions where irrigation was impossible. Pest resistance and disease resistance had to be re-emphasized, he said. New seeds would have to be multiplied and distributed. Then, Hedayatullah asserted that fertilization was essential to increase yields. In 1951 fertilizer research at Dhaka showed that 54 kg per hectare of nitrogen gave 67 per cent higher yields with local varieties over the same varieties which received no treatment. Rice researchers used human manure from the Dhaka sewer and found that yields could be increased almost as much, at least on experimental plots. Irrigation water was the next priority. While Hedayatullah agreed that small implements for farmers had to be improved and in some cases invented, the main need was for water-lifting machinery. He acknowledged the use of the traditional *doon* (water lift) and Egyptian water wheel already in East Pakistan, 'but the most effective implement is the diesel oil irrigation pump'.[41] He was referring to the boro season in which yields had been high if water was available. He noted the importance of the development of standard quality grades of rice and marketing control which along with a rise in rice prices would provide the only economic incentive to the farmer to produce more food. Finally, he said that when this reserach was ready he had almost 500 extension workers in contact with 411 small demonstration farms throughout the country; they would spread the message.

While the analysis of the potential contribution of a rice research establishment was clear and plausible there was little progress in creating a modern research establishment. The apparent opportunity to create such a modern establishment depended on foreign support. Hedayatullah met US agriculturalists who visited Pakistan for various development programmes in the early 1950s. Among these visitors were delegations from the Rockefeller Foundation, sounding out reaction to the idea of a rice research centre in Asia. Hedayatullah announced the 'Rice Institute Scheme' in Dhaka in 1955, with FAO and International Rice Commission support. However this separate scheme for a rice research institute, to be financed by the Pakistan Agricultural Research Council and by the provincial government, remained dormant until 1959, the same year that the decision was taken in New York to establish the International Rice Research Institute (IRRI) in the Philippines. In 1959 Hedayatullah was reappointed from retirement to 'recast the Rice Research Institute Scheme in the light of various suggestions made by various authorities'.[42] There was some feeling, even at this late date, that perhaps the Rockefeller–Ford institute might be located in Pakistan. One official historian, Zaman, later recalled that the location of IRRI in the Philippines 'disappointed many people in East Pakistan'.

In the absence of a modern and independent research centre, the Department of Agriculture continued with its fertilizer trials and testing and releasing new rice varieties. There were more introductions of imported rice varieties, more experiments with the irrigated boro rice crop, and an increase in study of the deep-water rice. Researchers hoped to circumvent the long ten-year process of breeding a new variety by releasing foreign varieties. Imported US varieties first outyielded the local variety (Badamkali), but imports experienced a steady deterioration in yield after the third season.[43] One researcher recalled the expectations surrounding these rice introductions: 'there was no question whether it actually went into our fields—our job was just to test it and release it—the rest was up to the extension department, if they cared'.[44] In 1952 Pakistan decided to participate in the *indica–japonica* crossing programme mounted by the FAO, and sent four varieties to CRRI at Cuttack; Nigershail (of Indian and Nigerian origin), Latishail (of Assam origin), Patnai-23 (saline-resistant), and Indrashail. Crosses between Norin rice of Japan and both Latishail and Indrashail failed to produce any seed, but the other twenty-two crosses were bred to the F2 generation. Between 1955 and 1958, the Central Farm participated in the International Cooperative Varietal Trials co-ordinated by the International Rice Commission (IRC) of the FAO. A number of scientists visited Dhaka during the 1950s; as one senior researcher recalled, 'a lot of good people came here in those days but we weren't in a position to benefit from their presence'.[45] Although this project was deemed unsuccessful owing to the production of a number of sterile crosses, the fact is that much was learned about the *indica* rices, and regional scientific co-operation began.

In spite of the work of the researchers on the boro crop from the late 1930s, it was the judgement of the official historians in 1962 that 'no attempt had been made to improve the type and quality of boro paddy before Independence though the researches and investigations were undertaken on Aus and T. Aman right from the beginning'.[46] This failure seems to have been remedied quickly. Between 1952 and 1962, 113 varieties from twenty-three countries were 'tested for adaptability in the boro season'. In 1954 tests of Latishail resulted in yields of 5,274 kg/ha in the boro season compared to 2,250 kg/ha in the aman season in which it was commonly grown. With these clear yield differences a big opportunity presented itself to rice researchers. Through their participation in the IRC, East Pakistani rice researchers learned of Japanese methods which could be used in the boro crop. These required denser seeding, more fertilizer, more weeding, line sowing, and greater spacing. These tests, conducted (in 1956) at the same time as Indian rice researchers were testing and deciding against Japanese methods, showed 'the superiority of Japanese methods for boro' in terms of yield, though more weeds were also produced than under the local boro methods.[47] What was crucial was that the Japanese methods could not be used without artificial irrigation. Nevertheless, they were promoted by the Department of Agriculture, concurrent with its new 'Grow More Food' campaign. By 1958, the demand for diesel pumping machines for irrigation of boro rice was very great, and by 1961 the 'tremendous' demand for these machines could not be satisfied, according to government officials. Already in the 1960–1 season 44,500 hectares were irrigated for the boro season although half of this work was done by traditional hand-irrigation methods. Boro rice from Assam and Sylhet, particularly Latishail, was being grown, though some farmers also experimented with foreign rices introduced by the Department of Agriculture. Average yields on farmers' fields were higher in the boro season than in the aman. Because the average yield in the main transplant aman crop remained under 900 kg/ha, official opinion during the 1950s appears to have supported research investment in the boro rice crop as the path of quickest return. By 1960 there were more than 405,000 hectares under rice in this season, a tenth of it mechanically irrigated. The boro season favoured mechanical irrigation, and richer farmers could afford mechanical irrigation. This season thus seems to have been ripe for a major promotion and the application of new production technology, both of which occurred with the importation of ten tons of IR 8 seed in 1966.

Research on deep-water rice was stimulated by the return, after border changes in 1944, of the Habibganj substation to East Bengal from Assam. Selection of finer-quality rice and higher yielding rice for deep-water continued; in fact during the 1950s there were eight new deep-water varieties selected and released, the fruition of work that had gone on since 1935, under extremely difficult conditions.[48] But the work stopped about 1960 and the programme was closed for a period; it is true that most of the emphasis was then being placed on the row crop concept for rice, and that international advisers and supporters knew little or nothing about deep-water rice cultivation. But it seems, based on a number of private judgements, that the reason for terminating the deep-water research programme in 1960 was internal politics in the Ministry of Agriculture, including some disagreement between central officials of Pakistan and provincial officials of East Bengal. After the change in international opinion in 1973 regarding the potential of deep-water rice, the work at Habibganj was started again. Informed estimates place deep-water rice area in 1980 at about 20 per cent of all area cultivated to rice in Bangladesh.

7.5 Forces for the Green Revolution

The beginning of the first serious attempt to establish a separate rice research institute in Dhaka occurred in 1960. The decade of the 1960s also begins with the founding of IRRI in the Philippines and ends with the establishment of the East Pakistan Rice Research Institute in 1970. Most important, it was during this decade that the initiatives called 'the Green Revolution' were made, particularly the importing and propagation of the 'packages of new technology' for rice cultivation. Because there was a bewildering series of proposals and initiatives, the progress of the Green Revolution during the 1960s has been placed in a chronology at the end of this chapter. In order to understand why the Green Revolution proceeded as it did, and to understand why rice researchers did not establish a strong autonomous institution based on their earlier and independent conception of rice environments, we must analyse external forces in three separate domains: (1) national politics and food supply, (2) the growth of official institutions and policies for agricultural development, and (3) why and how American development institutions had a special role in agricultural and rural development in East Pakistan. The fact is, of course, that government efforts spent in promoting the Green Revolution did not stop another kind of revolution in East Pakistan, red in blood at least, or the establishment of the new country of Bangladesh.

National Politics and Food Supply

Under the reign of President Field Marshal Ayub Khan, Pakistan experienced a steady military expansion based on foreign aid and on domestic allocation of revenues. This expansion was accelerated following the war with India in 1965 and involved between 40 and 50 per cent of the state budget during the decade. Closer relations with China after the 1962 Indo-Chinese war were both natural and provocative. Pakistan was a full member of the SEATO and CENTO military alliances and of the Regional Cooperation for Development alliance with Turkey and Iran. During this decade industrialization in West Pakistan was paid partly from foreign loans and partly from foreign exchange earnings from exports, half of them agricultural exports such as jute and tea from East Pakistan.

We will argue here that the causes of the uneasy relation between the two wings of the country, and of the events which led to its dissolution, affected the whole attitude to agricultural and rural development and to new rice technologies in particular. With military power concentrated in the Punjab, with the political capitals in Rawalpindi and Islamabad, and with industrial investment in Karachi, Lahore, and other towns of the

West, East Pakistanis felt they had little influence over national decisions. After the turbulent 1952 Language Movement both sides were aware of a latent unease in their relationship. An opposition coalition of Fazlul Huq (Krishak Praja party), Hussein Suhrawady, and Moulana Bhashani (National Awami party) formed a United Front and swept the elections of 1954, leaving the Muslim League with only ten seats in Bengal. While both Huq and Bhashani spoke about rural problems and agrarian interests, they spoke for a rural and small-town middle class. Sharp criticism of the West's indifference was voiced in the Constitutent Assembly in Dhaka from 1954 onward. This unease occasionally turned to specific opposition and agitation. Ayub Khan's new constitution in 1962 spoke of the need for an 'honourable partnership' between the two wings, but when this failed, Ayub admitted later that he expected dissolution of the country to occur even earlier than it did in 1971. Bhutto, although not unbiased, spoke in 1972 of Ayub's 'intense prejudice' against East Pakistan.[49] Feldman concludes that all signs pointed by February 1968 to the likely breakup of the country. The tendentious Six Points were already announced by Sheikh Mujibur Rahman and the Awami League in 1965, calling for maximum autonomy between the two wings. This unease and opposition increased nearer the end of the decade as more Bengalis became aware of the unequal flow of resources between East and West. This flow was in addition to a kind of internal brain drain of able and talented Bengalis who lived and worked in the West. President Ayub and his advisers apparently awoke to their vulnerability after the war with India in 1965 and the Awami League's declaration of the Six Points.

In this year also the decision to import IRRI rice technology was taken. Agricultural earnings from jute and other commodities were still the most important source of income to the nation as a whole. Rice production and distribution questions were still more important to subsistence producers—the majority of East Pakistan's population. Groups and classes in command chose to emphasize industrial, urban, and military development, and their international supporters largely acceded to this emphasis: the consequences of the emphasis culminated in the civil war of 1971.

In West Pakistan the foreign policy issues were Kashmir, relations with India, Iran, and Turkey, and selling consumer goods in the world market. At this time East Pakistan was the major market of goods made in West Pakistan. In February and March 1969 there were antigovernment demonstrations throughout Pakistan. Ayub Khan was overthrown in a coup by Yahya Khan. The military was used to suppress Bengali nationalist activities in East Pakistan. Within eighteen months elections were held in which the Awami League of East Pakistan gained the majority of seats on a national basis (Bhutto's political party had never been able to build a link to the political sections of the middle class in East Pakistan). The chaos caused by the devastating cyclone and the extraordinary consequences of the election in late 1970 are well known. Four months later in March 1971 there was a military crackdown; a government in exile gathered around Sheikh Mujib; war raged within the country, in which India joined in December 1971—all of which resulted in the foundation of the independent country of Bangladesh.

These were the conditions under which a decision was taken to pursue the Green Revolution in roughly the same fashion as it had been pursued in the 1960s. During the 1950s a few northern districts of East Pakistan were self-sufficient in food production, and deficit districts like Faridpur were usually supplied from the surplus of districts to the north. Poor harvests of 1958 and 1960, however, caused a shortfall in production, to which there was 'a mild reaction'.[50] Food aid from the USA and cash purchases in West Pakistan offered security; while the actual imports were small, there was an official belief, based on foreign assurances, that there was more food where that came from. This belief, and the small amount of the shortfall, may explain the mild reaction. Moreover, the problem lay in the East, where there had always been a deficit, and there were doubts in the West whether anything could or should be done about it. In addition no senior officers had any interests in East Pakistan, though Ayub Khan had been stationed there for a number of years. Pakistan had an autocratic military regime in which senior officers had only indirect interests in agriculture. Thus for the next five years, during the Second Five Year Plan (1960–5), 'between 24% and 49% of agriculture's annual income [depending on the year] slipped through subsidies into the manufacturing sector'.[51] This was consistent with the pattern before 1965 in which East Pakistan earned 60 per cent of the country's foreign exchange and received 33 per cent of its use. In addition one should remember that in West Pakistan food production doubled in this decade and that much of the increase was based on new wheat seeds used in very large irrigation schemes and grown on large farms. At this time very little wheat was grown in East Pakistan.

Production of rice in East Pakistan in the 1960s, as before, was oriented to subsistence. Official surveys in 1963 and 1968 estimated marketed surplus at 10 per cent of the total production. Ahmed calculated that the average annual crop for this decade was 10.8 million tonnes, so the average amount traded was just under 2 million tonnes. Government distribution through rations accounted for about 25 per cent of the total trade in most years. This ration rice was either procured in the local market or was imported—usually as US food aid. Imports were always less than the marketed surplus. Rice was procured from farmers on a voluntary basis in eight years of this decade and annually did not exceed 26,000 tonnes. In the two years in which there was compulsory procurement the amounts were a more significant portion of the marketed surplus: in 1964, 130,000 tonnes and in 1966, 91,500 tonnes. This compulsory procurement was in years of total production of over 10 million tonnes.[52]

Prices remained fairly steady throughout the decade, and showed a remarkable consistency from district to district. The price 'ceiling' favoured the non-producing consumer, particulary urban consumers, civil servants, military personnel, and industrial workers. These groups, not rural people, were seen as potential troublemakers. It was common in the early years of Bangladesh to hear people of these groups say that 'at least under Pakistan rice was cheap'. The cost of imported rice was close to retail prices of local rice. The government of Pakistan did not wish to allow rice prices to rise too high or too rapidly. That this was an important political decision is demonstrated by the later public consensus that Mujibur Rahman had allowed the price of rice to rise, and when it doubled he was unable to to anything about it even though he tried to. This was considered one of the causes of his downfall in 1975. Retail prices were remarkably consistent from district to district, rising in the short months of April, May, and June when the small boro crop was harvested but the larger aus crop was not ready. This consistency held for the 25 per cent difference between the growers' price and retail price. But there was a supply and price effect from the irregular natural disasters and from the 1965 war between Pakistan and India. Certainly in 1964 and 1966 the government found it necessary to make much larger compulsory procurements of rice in order to enhance the public grain-rationing system. But we are left with a sense of relative official disregard about the question of food self-sufficiency until the late 1960s.

Institutions and Policies for the Green Revolution

The planning on which the second and third five-year plans (1960–5, 1965–70) were based was presented to President Ayub by American-trained economists and guided by the Development Advisory Service of Harvard University. This planning activity was funded by the World Bank or Ford Foundation directly or by the government of Pakistan (which in turn obtained earmarked funds from foreign grants and loans). These plans were characterized by a deliberately overvalued rupee which favoured imports, the transfer of agricultural incomes through subsidies into the manufacturing sector, and centralized decision-making. The trickle-down philosophy was further supported by the view that the cities and industry should be rewarded and encouraged, even if the rural and agricultural sector had to be neglected or ignored.

The Agricultural Commission, set up by the government in 1959, made its report in 1960. Enquiring into every aspect of agriculture, it concluded that there were five top priorities: a fertilizer industry, better seeds, irrigation, agricultural credit, and agricultural extension. In East Pakistan the commission's recommendations did lead to strengthening of the extension services, partly through an appeal to the new USAID; the American advisers in the early 1960s were trying to continue the sort of work begun in the Village-AID programme but stressed the use of new agricultural technologies. Agricultural credit was more difficult: abolition of the zemindari system also removed many of the largest moneylenders and their capital. Replacements were incomplete. Private banks of Pakistan remained reluctant to make loans outside the urban, commercial, and industrial community, except for a limited number of irrigation and tubewell investments. Banks believed, like the planners, that industry was the leading sector, and so power grids or textile mills took their attention and received precedence. Thus rural credit was seldom contracted at rates lower than the exorbitantly high (50 per cent) levels of the moneylenders. For these reasons the Agricultural Development Bank of Pakistan was established in 1961, though it did not make loans directly to small cultivators and so did not disturb the exploitive rural credit system.

Irrigation, the commission's third recommendation, was unquestionably the biggest form of state and foreign investment in agriculture in East Pakistan. Modelled on big irrigation projects in West Pakistan, fifty-one large schemes to irrigate thousands of hectares were drawn up. In addition there were to be embankments and canals, partly for irrigation and partly for flood control. Funds for construction came from the massive Rural Works Programme or Test Relief accounts, and US food aid was used in partial payment for labour. Completion of these massive projects was the responsibility of the Water and Power Development Authority (WAPDA), a huge bureaucracy surrounded by foreign consulting firms. These projects took more than ten years to complete, eventually operating at half capacity (or less), in the estimate of Bangladesh and World Bank experts. WAPDA was more interested in project design and construction completion than in project implementation and utilization.

'Better seeds' are at the heart of our story. Since the Agricultural Commission did not specify the development of indigenous seed varieties, the result was the introduction and propagation of foreign high yielding seeds; thus East Pakistan became the world's largest importer of HYV rice seeds. The recommendation also led to construction of a 'seed store' in every Union (unit below the thana) in East Pakistan in 1964–5. Though largely an expensive gift to building contractors, and paid for by US aid, these concrete buildings were intended to store seeds, pesticides, and fertilizer before distribution by the Union Agricultural Assistant. His status, it was correctly judged by the government, was low in farmers' eyes, yet he would be on the front line of agricultural change. A new building under his jurisdiction would be enough to communicate the appropriate message to his clients.[53] While there is no evidence that the status of the Union Agricultural Assistant was raised as a result of this building, it was certainly made clear that seeds were supposed to be part of an important new policy.

As the response to the first recommendation of the Agricultural Commission, a fertilizer factory was begun in 1962, with two more planned. The first was ready for production at the time of the war for independence in 1971. It was funded by a foreign loan and was largely a 'turnkey' project. Meanwhile, during construction, massive amounts of fertilizer were imported. To manage the importation and distribution of these new inputs, and to give coherence to new projects, the government relied on the Pakistan Agricultural Development Corporation to conduct buying and selling arrangements entirely separately from the Ministry of Agriculture.

These recommendations of the Food and Agriculture Commission in 1960 reshaped the hierarchy of institutional interest in agriculture, and led to the creation of the Agricultural Development Bank (1961), the Agricultural Development Corporation (1962), the fertilizer and pesticide factories (1963), and the Water and Irrigation Master Plan (1964). Credit and extension questions did not receive nearly as much attention as the procurement and supply of technical inputs, into which the fertilizer- and irrigation-responsive rice seeds were designed to fit. The assumption was that capital-intensive technologies developed for the 'efficiencies' of labour-scarce economics would in themselves produce the desired social change.

Setting aside for the moment the disagreements between the government of East Pakistan and the central government, what factors led the government to bless, promote, and pay for the marriage of the boro season and the new HYVs and to approve an accelerated rice research programme in 1966? Clearly Ayub Khan needed a set of links between a technocratic civil service and his Basic Democrats in the rural agrarian society.[54] This link could reduce the probability of rural instability. If Basic Democrats were big farmers and received a new technology which was profitable, they would then play an essential role in reducing rural dissatisfaction, a role which could not be played by the police or the army. The government received advice from the very highest sources that to foster HYVs grown in the boro season was the optimum use of its resources. The transfer of technology to an underdeveloped season seemed natural, particularly in the light of demand by farmers for irrigation in this season.

Even those who would not believe that instability in East Pakistan could lead to Pakistan's dissolution seemed to have agreed that promoting a Green Revolution for rice could do no harm. The strategy had the chance to do some good; food shortages in East Pakistan were mounting. When import purchases originated in the wheat fields of West Pakistan that was useful to the national government. The government of Pakistan was also underwriting a major change in the wheat economy of Punjab, the heart of the military and industrial complex which ruled the nation. Imports which were assured from America, although not stimulating to the economy, were always accompanied by a great deal of assistance of other kinds, including military. Increased food production in East Pakistan was a solution to a problem in which few Pakistani government officials were much interested unless they had specific responsibilities in that province. There was therefore no need to study the possible effects or consequences very thoroughly, particularly when reputable economists were so firmly in favour of the strategy, nor to choose carefully between one cultivation season and another. Tractorization, fertilization, and new seeds were being embraced in West Pakistan, and some government officials and military officers had roots in the farms which opted for this new technology. Furthermore, it was well known that this technology was also being promoted in India, Pakistan's adversary in the 1965 war.

The Special Role of American Development Institutions

Above all, the strategy appeared to have a low immediate cost because it was to be financed or guided by a whole spectrum of American institutions—the USAID, the Ford Foundation, the Rockefeller Foundation, prestigious universities like Harvard—and the World Bank. All these foreign institutions were already actively engaged in adjusting and reforming the structure of economic planning, industrial development, rural mobilization, water management, military modernization, etc. The government of Pakistan could not afford *not* to buy the new HYV rice seed and have it planted in the boro season. By 1966 it was perceived that an autonomous research institute for rice was also overdue. The government accepted the idea more to make a symbolic display of acceptance of a 'scientific' research base for a 'modernized' agriculture than with a clear conception of the role that such an institute should play.

American strategic interests were strong in Pakistan and, by extension, in East Pakistan. Asia was the number one 'regional' priority for US strategic attention during the 1960s. Already supplying food aid, it was beginning to plan, train, and build institutions in East Pakistan. US support to Ayub Khan would seem justified if it promoted a strategy for both the conquest of hunger and the production of new forms of rural wealth. Moreover US institutions and corporations would be a natural source of supply for some inputs (except perhaps the seed) and some services under the usual terms of tied-aid contracts. Moreover, individuals in these US institutions came from their world based on cheap energy and cheap food, and these assumptions were naturally built into almost everything they thought and did. Neither energy nor food was cheap for people in East Pakistan, so such assumptions were alien there. These individuals would be returning to that world when their time came. Meanwhile they would participate as a community in the gradual construction and reform of the infrastructure of a modern state in East Pakistan. Few foreigners believed any more than government officials did that changes to rural social organization could be substituted for certain kinds of investment in technological change. They believed in the technology of a rice plant, a rice fertilizer factory, a grain storage elevator, new ports, etc.; this was either *all* that was necessary, or was the *only* thing which could 'get traditional agriculture moving'. New rice technology would cause economic growth, and would thus probably lead to desired social changes, or so they thought. If these changes occurred then all the major US investments in other institutions in East Pakistan, and in countries like it, would be justified. Such growth would have a salutary effect on the dignity and budgets of the development institutions for which they worked and would enhance the viability of external intervention and their own careers.

The relatively modest level of US official development assistance for new agricultural technology programmes should be put in perspective here. USAID calculated that between 1947 and 1971 the total expenditure by the government of Pakistan on agriculture and water development was $1.3 billion. In addition, approximately $200 million was spent during this period for rural works and related relief-oriented programmes. The Pakistan government's total expenditure was thus $1.5 billion towards agricultural development before 1971. Between 1955 and 1971 US official assistance to East Pakistan was $30 million for agricultural development, excluding food aid and commodity loans; the US contribution was thus 2 per cent of total development expenditures. Of this $30 million, $4 million was in technical assistance, grants, and loans between 1955 and 1959 and $7 million between 1960 and 1965.[55] These official US expenditures do not include sources like the Ford Foundation or US voluntary agencies like CARE. In comparison to $7 million for 1960–5 (the second Five Year Plan) from USAID, financial assistance from UNDP, the World Bank, Australia, and the Federal Republic of Germany totalled $50 million for the same period.

The USAID office in Dhaka explained this 'relatively modest grant and loan level' in 1975 by saying that it was limited by the fact that the government programmes were all initiated and run by the central government, not the provincial government. Only in the 1965–70 plan period, said USAID, did the the central government emphasize work in East Pakistan and therefore the direct US assistance increased.[56] In fact US official support went overwhelmingly to West Pakistan, and much more for military and industrial sectors than for agriculture. What USAID does not say is that the real allocation of US assistance was in response to decisions by the central government made on the basis of earlier economic planning and development theory done by Pakistani and American economists. Also, even 2 per cent of total investment in agriculture, if added to the sums spent by the Ford Foundation and the World Bank and if spent as foreign exchange for specific projects, had a profound impact on the institutions and the thinking which guided agricultural development in East Pakistan.

7.6 Researchers' Values and Green Revolution Practices

Before the dissolution of Pakistan in 1971 a Green Revolution was widely expected as the consequence of three processes having been completed, in turn. The first process was identification of the winter (boro) season and the promotion of the boro rice crop as the greatest potential for increased productivity; the second was the choice of the new HYV rices developed by IRRI (and investment in their associated technical requirements) as the 'appropriate technology'; and the third was the establishment of a breeding-oriented rice research institute at Dhaka which was intimately related to IRRI. What follows is a closer examination of the link between the second and third processes as seen from the perspectives of East Pakistan's research and of IRRI.

The role that East Pakistan researchers were supposed to play in the Green Revolution was certainly ambiguous: marginalized by events in the early 1960s, they found no significant allies within the government and thus began to rely on the lobbying power of foreigners. Until 1968 there was no perception among these lobbyists or the government that fundamental studies of rice-cultivating environments might be necessary because by 1964–5 the idea of convenient technology-transfer had taken firm hold. Meanwhile, during the 1960s, an entire conception of research which was adapted to the different cultivating zones and cultivators had been neglected and ignored. Indeed that conception and its declining institutions and substations were absorbed in the simplified research strategy of the Green Revolution. Alternative conceptions of research and its role were in an incipient state at IRRI; indeed IRRI's staff active in East Pakistan were largely ignorant of the alternative conceptions, as well as earlier strategies for research in Bengal. In this way the process of increased dependence on IRRI had a stifling effect. It was, therefore, a shock in official circles when it was announced in 1970 that more fundamental research was required, including the breeding and adapting of new rice varieties. By that time, of course, under the wartime conditions, most normal activities of the government had come to a halt.

Rice Researchers in East Pakistan: Their Institute and the Green Revolution

In 1960 the Dhaka rice research scheme was given final approval by the government of Pakistan. Construction began in 1961, personnel were appointed, and equipment ordered from abroad. But within five months the project was cancelled, and rice research was merged with another scheme for an Agricultural Research Institute in Dhaka. The official rice research history states that 'this tragic end' was entirely 'due to the action of an individual'. The reason for cancellation was never explained (beyond 'bureaucratic conflict'), but two individuals were considered by East Pakistan researchers to be directly responsible: a member of the Planning Commission of Pakistan (Shafi Newaz) and East Pakistan's Director of Agriculture (Abdur Rahim).[57] It is likely they were carrying out instructions of much higher authorities. The cancellation meant rice research would not have autonomous status in the administrative system of agriculture. In fact it was ten years (1970) before rice research became a significant centre of activity in Dhaka.

Researchers had been struck a blow by the 1961 termination of the planning for an autonomous institute for their work; rice research would again be submerged amid work on other crops, and researchers would lose the chance to distinguish themselves from others. There is no doubt that agreement to promote the HYV rices in the boro season was seen as a means to join forces with an international movement backed by IRRI, thus possibly improving their status within their own ministry and directorate. They knew that higher yields could be obtained on research plots by the same variety in two different seasons: for example, the Assamese variety Latishail in 1954 test plots yielded 5,274 kg per hectare in the boro season compared to 2,250 kg per hectare in the aman season. Though until the mid-1950s these researchers had been breeding rices which did not require fertilizer, they had always experimented with fertilizer and knew that many varieties responded to artificial fertilizer or manure. They knew of the effects on yield of irrigation and that the demand for low-lift pumps in 1961 could not be met by the government. They were also aware of the very large irrigation schemes either under construction in the 1960s or called for in the master plan, schemes which were underwritten by the government and American development institutions.

Researchers also thought that the transfer of technology might be successful if it went beyond simply introducing new varieties, machines, or chemicals as had been done in the early 1950s. They certainly could understand the concept of the genetics and architecture of the rice plant itself being a new form of technology. The official history by Alim, Zaman, and colleagues in 1962 reports on the performance of Japanese and Egyptian varieties in tests between 1953 and 1957 in comparison with local varieties:

The japonica varieties have got short stiff straw combined with high non-shattering and non-lodging character are extremely essential for the successful cultivation of the Boro paddy as the season is stormy and frequent hail storm, heavy showers always damage the local Boro varieties to a considerable extent [sic].[58]

The introduction of exotic new varieties was one of their basic stocks in trade. A Japanese team had begun working at the Comilla Academy in 1959, focusing on changing cultivating techniques in order to raise yields in both aman and boro seasons.[59] By 1962 they had selected the most productive of the local varieties, tested them against Japanese varieties, and started to popularize the best local ones. On the whole the highest yields obtained were equivalent to average yields in Japan in 1962. Limits on yields were noted: cyclones spoiled whole fields, sparrows consumed the grain, and insects such as caterpillar worm, leaf hopper, stem borer, mealy bug, and hispa were more voracious when there were no neighbouring fields to attack. In addition, random-method transplanting was preferred locally because it required less labour than rows, but it also led to lower yields and did not allow the use of suitable machines. In fact the use made of project machinery was judged unskilled and wasteful by the Japanese. Nevertheless by 1962 the team had demonstrated the use of chemicals and machinery on the Academy farm, persuaded trainees to do manual labour personally in every phase of rice cultivation, and increased yields on one hundred neighbouring 'model' farms.

Along with local researchers, the Japanese became committed to the view that the boro rice crop held the greatest potential for the country. Local researchers were generally responsive. In 1962–3 Japanese agronomists were planting demonstration fields of Taichung Native One (TN 1—a parent of IR 8) and the Egyptian variety Zaibani in the Comilla Project area, using low-lift pumps and rototillers. The Japanese straight-row planting method with denser seeding had already been judged by researchers in East Pakistan to give better yields in the boro season than local methods.[60]

In addition to these factors, rice researchers saw clearly what was going on in the big irrigated areas of West Pakistan, not only with rice but with wheat. In 1962 Borlaug gave some improved seeds to two trainees from Pakistan at the research centre in Mexico; 205 kg of seed were planted at Lyallpur in 1963. Borlaug visited Pakistan in 1964 and 'secured governmental and foundation support for the varieties'.[61] Pakistan purchased 350 tons of Mexican wheat seed for planting in the 1965–1966 season, and 42,000 tons for the 1967–1968 season. Rice researchers knew about the tractorization of the larger farms in the West, and that there was government and foreign support for it because of the wide promotion of the HYV wheat seeds.

Finally, these researchers were inclined to take some risks because they knew they could then bargain for the autonomy for rice which they felt they deserved. No doubt some of them, then the servants of the government of East Pakistan (theirs was a provincial service appointment, not the Civil Service of Pakistan), felt that rice in their province should receive at least a portion of the attention received by wheat in West Pakistan. In 1966–7 there were 1,524 tonnes of rice seed imported to the East compared to 42,700 tonnes of wheat imported to the West. It is suggested that the West withheld HYV rice seed from the East, but levels of investment were clearly higher there. It is possible, though only hinted in interviews, that some researchers perceived new rice technology and its possible benefits as a path to 'getting even' with West Pakistan. Certainly when IRRI officials began to lobby in East and West Pakistan for an accelerated rice research project at Dhaka, local hopes were raised even higher.[62]

Thus when research organization was at its lowest point in 1964–5, East Pakistani researchers were listening to the IRRI plan for an autonomous institute in the context of a planned agricultural revolution. At this stage, when there was no government commitment to a home for the Agricultural Research Institute (ARI) or to the expansion of rice research, the Director of Agriculture for East Pakistan, S. H. Hazarika, proposed a new research centre at Jamalpur near the new Mymensingh Agricultural University. Hoping to hurry the process, he was willing to locate the national rice research facility far out in the countryside on immediately available land, far from the capital. But IRRI scientists, among others, objected to the location and the proposal was dropped.[63] The decision was eventually taken in 1965 to locate researchers at a temporary site at the dairy farm at Savar, thirty kilometres from Dhaka. A year later the big new Agricultural Research Institute scheme was approved for 350 hectares near Dhaka. No special provision was made for rice, but rice researchers kept their idea alive by relying on the lobbying power of IRRI and the Ford Foundation.

Their reliance entailed a certain amount of time away from research: senior people were involved in so much organizational politics in this period that it is difficult to see how work was done in the fields. In one sense the period was politically tumultuous, but in another sense changes in research strategy were occurring at a glacial pace. An enormous amount of valuable time was wasted by bureaucratic delays. Frustration was thus a chronic condition. More attention and money were invested in other aspects of agricultural development, to irrigation, chemicals, mechanization, credit, etc.—things which seemed to have bigger and faster pay-offs than plants. Indeed, the money spent on research on other crops during 1964–5 was much greater than on rice, as Table 7.4 shows.

Rice researchers and officials were in an uncomfortable bind. Party to the co-operative research with IRRI, testing its new varieties, going abroad for training and tours, they were seen to be promoting the idea that self-sufficiency could soon be achieved, as some foreigners claimed. Although they would be delighted if sufficiency were achieved, obviously it would be difficult to show that sufficiency was largely the result of their own work when so few resources were given them for it. IRRI loomed very large on the horizon. But privately, like some foreigners, East Pakistani researchers were doubtful about how soon or even how sufficiency might be realized, and when the early enthusiastic predictions were withdrawn, these very researchers were implicated in the disappointment. In their December 1967 report to the Director of Agriculture on rice research, government scientists did little more than repeat the predictions made by IRRI scientists. 'By introducing IRRI rice selections and new rice production techniques from IRRI, East Pakistan may save 15 to 25 years in its accelerated rice research.'[64] They also reiterated that despite the fact that flooding renders short rice unsuitable for the aman season and despite the bacterial and virus diseases affecting these new rice varieties, they thought the food gap could be eliminated by 1970 if 1.2 million hectares of IRRI rice were grown with the proper management and inputs. They confirmed that the boro season strategy (which clearly favoured richer farmers), was the solution to the problem. While HYV varieties were ready for the boro season,

TABLE 7.4. *Agricultural research budgets compared, Pakistan, 1964–1965* (Rs '000)

Budget	Crop
2,900	cotton (all Pakistan)
2,000	jute (E. Pakistan)
1,200	wheat (W. Pakistan)
500	rice (E. Pakistan)

Source: Department of Agriculture, Government of East Pakistan, 1965; and Rodney Tyers, 'Optimal Resource Allocation in Transitional Agriculture', unpublished Ph.D. thesis, Harvard University, 1978.

East Pakistani researchers already knew what the IRRI researchers had yet to learn, that in a country which gave the appearance of agronomic homogeneity there was great regional difference in rice-cultivating conditions right down to the village level. They noted also in their 1967 report that results from different test sites of new varieties were contradictory. Scientists later recalled being doubtful about grand predictions in the 1960s, but we can assume that at the time their position was a mixture of public caution, private scepticism, and secret hope. They knew big issues were at stake, and they were hoping for their own Institute.

During all the organizational politics the community of rice researchers experienced a curious paradox: when the need was greatest for an expansion of their work force, the greatest number of scientists were selected out for advanced training at IRRI or American universities such as Texas A & M. In effect much of this scientific community was out of the country on foreign financial support. In the sociology of provincial government employment, of which agricultural research formed a tiny part, foreign training for a degree constituted a chief mode of career advancement. If such foreign training was a part of the earliest stage of employment, the job was even more attractive. The Secretary and the Director of Agriculture are both reported to have wanted staff to have quick and practical training, not long periods abroad doing doctorates. But everyone else knew that it was the degree or the certificate which counted in the system, for with it one could get out of the slow-moving provincial research stream into better paid positions elsewhere both inside and outside the government. Until 1969 they saw a small programme with a small budget, without a permanent research centre: most positions at that centre would eventually be filled by senior researchers so recruits calculated that they might be posted to the isolation of the research substations. These were the conditions under which they began work. So loyalty to rice research was not universal among young recruits; it was often just a path to advancement in the bureaucracy.

Meanwhile the main research activity was breeding adaptive varieties and running test plots at substations. Because of problems with IR 8, researchers were now remembering the relative importance to the economy of the aman and aus crops; thus they adapted advanced (F3) selections from IRRI specifically for these seasons and released them during 1969 and 1970. With these releases, which had symbolic importance, and with the achievement of formal autonomy of the East Pakistan Accelerated Rice Research Institute (EPARRI) within the Ministry of Agriculture early in 1970, the morale of researchers is reported to have greatly improved. Their autonomy permitted them to receive the higher salaries for which IRRI and the Ford Foundation had been agitating, to follow different rules governing career progress and evaluation, to operate the institute without interference from the bureaucrats in the secretariat and to block uses of its funds by the ministry for purposes other than rice research. A special board of governors was created for the institute, of which the Secretary of Agriculture was a member; but day-to-day autonomy was a great achievement in formal terms, although as we shall see its practical realization was still some years away. All of these efforts were intensely centralizing; consequently the development of the substations was neglected; this neglect had important effects during the 1970s, as we shall see. The decade of the 1960s was strange for researchers because many important and far-reaching decisions were taken in agriculture without reference to their work, yet their importance was trumpeted so long that they finally got the status and resources which they had been so loudly denied. This strange condition would follow them throughout the 1970s as well.

Dealing with the High Expectations of Researchers at IRRI

Their autonomy, though, was already mediated by special dependence upon IRRI. The first visits by IRRI scientists to East Pakistan occurred in 1962, although IRRI was known to researchers there through its publicity and invitations to conferences. Disappointed by government rejection of a 1960 plan for an autonomous rice research institute, East Pakistan's researchers hoped for help from IRRI officials to lobby for a new organization for their work. When IRRI visits began, the farm land previously used for rice research in Dhaka was being appropriated for a new 'Second Capital' project, designed to give East Pakistan a symbol of higher government status within Pakistan. No irony could be greater: the indifference of the government to the need to apply science and resources to the most pressing problem in the province, reinforced by (interminable) construction of a symbolic parliamentary edifice in a country governed by martial law. Years later the edifice remained incomplete and the farm land had disappeared.

As the accompanying chronology shows, research organization declined over the next ten years, the same years in which American institutions became more and more involved in agricultural development. For example, the Ford Foundation was not only heavily involved in IRRI, but also in rural development at Comilla and in economic planning in the central government. Together the foundation, East Pakistan researchers, and IRRI sought an autonomous institute for rice. Unable to decide on a permanent site for research, the government was persuaded to adopt a temporary scheme in 1965, which continued unchanged until 1969. This scheme, 'The Accelerated Rice Research

Program', had the Ford Foundation channelling money designated for research in East Pakistan through IRRI (see chronology). Lobbying for an autonomous institute ultimately succeeded in 1969–70, and from the land allocated to the huge new Agricultural Research Institute was carved a 36-hectare research area for EPARRI, with its own buildings and services. IRRI remained the financial agent for grants from the Ford Foundation to EPARRI—the sole source of foreign exchange except for small USAID training grants. This struggle to organize and to gain autonomy drained much of the energy of East Pakistan's researchers, making them more likely to believe in or accede to a kind of technical fix for the country's food problems. They were getting stronger domestically while greatly increasing their dependence upon IRRI.

The link between IRRI and East Pakistan was probably deeper than with any Asian country except the Philippines. Regular visits begun in 1962 by Chandler, Wortman, and McClung kept IRRI's staff and board informed of developments in the province, and most IRRI specialists met regularly with their counterparts there. By 1965, when IRRI's Assistant Director reported that rice research activities in East Pakistan had reached their lowest point in half a century, nine top officials had already visited IRRI, scientists had attended conferences, and young researchers had been sent for training at IRRI. Architects and planners from the Philippines appointed by IRRI designed the new facilities for EPARRI (modelled directly on IRRI's building) and ordered and purchased equipment or supplies. The Vice-Chancellor of Dhaka University was made a member of IRRI's board in 1965, and IRRI's director eventually became an ex-officio member on EPARRI's board. IRRI was not only financial agent, but through its liaison officer in Dhaka, had considerable influence on selecting and promoting individuals for the project. In addition to Lee Johnson (a Canadian geneticist), IRRI posted three other people (Walker, Golden, and Leopold), to East Pakistan for rice research in the 1960s. By the end of the decade the Ford grants through IRRI accounted for about 50 per cent of the total rice research budget of East Pakistan and all of its hard currency.

The central expectation among foreign planners and researchers throughout the 1960s was that a technical fix could be obtained for East Pakistan's growing food deficits. At the height of this expectation the term 'Green Revolution' was coined in 1968; its work was believed to be revolutionary, but not violent; indeed it might delay or stop other revolutions. Nevertheless by 1968–9 few observers (and particularly not well-informed foundation officers) could possibly have believed that violence in East Pakistan would be avoided. Looking for this elusive technical fix occupied a great amount of time and energy for Pakistanis and foreigners. But, as a result of this search for a fix, the role envisioned for researchers was limited and almost incidental.

A number of visitors carried IRRI rice seed from Los Banos to Dhaka and Comilla in 1964, including Pakistani scientists and the Ford Foundation chief representative in Pakistan, Haldor Hanson (Hanson later became director of CIMMYT in Mexico); these seeds were all grown successfully at Comilla in the 1964–5 boro season. In 1966 10.1 tonnes of IR 8 seed was imported in order to plant 490 hectares supervised by government officers at Comilla; this government cultivation was complemented by distributing free seed, fertilizer, and irrigation pumps to the demonstration farms and by supplying the services of IRRI extension scientist William Golden in Comilla.[65] Costs were borne by the Ford Foundation and USAID; in addition, there was indirect support from various programmes like Rural Works, the Peace Corps, etc. To augment the breeding activities already under way at Dhaka 303 lines of rice selected at IRRI were transported to East Pakistan. The result of the 1966–7 boro season was the achievement of yields of up to 5.4 tonnes per hectare at Comilla from the 10 tonnes of rice brought from IRRI, contrasting with typical yields of 1.8–2.7 tonnes per hectare of existing varieties.

For the next two years, until about 1968–9, the popular expectation in rice circles was roughly that a few varieties could lead East Pakistan to food self-sufficiency if planted properly on 1.2 million hectares, in the boro season. The expectation usually was that these varieties would be from IRRI and would need only to be adapted to local conditions. From 1965 onward predictions were being made of self-sufficiency in rice production within five years, but this immensely attractive idea continued to elude all efforts right to the end of the period under study (1980). Indeed, waves of enthusiasm and disappointment moved through East Pakistan depending on the results of the rice crop. As can be seen in the maps of the spread of HYV rice cultivation (Maps 3–5), the actual adoption of the new varieties was extremely uneven across the land.[66]

The high yields of IR 8 in the boro season led to 'wild enthusiasm' for the new crop in 1967: farmers were impressed with the yields and loved the distribution of free inputs and irrigation pumps. This enthusiasm led to the government's decision to import over 1,500 tonnes of IR 8 seed from the Philippines for the 1967–8 season. A finer IRRI rice, IR 20, and a more disease-resistant variety, IR 5, were also imported for testing that year. This large importation of seed formed the basis for the government's announced 'Programme for Food Self-Sufficiency in East Pakistan' to begin in 1969–70. One of the senior rice researchers in East Pakistan, the Chief Economic Botanist A. Alim, declared himself to be confident in the strategy and enthusiastic about the new HYVs from IRRI: they were, he said, like the work of 'creative genius'.[67] In 1968 Pakistan also imported an unknown quantity of a highly responsive semidwarf rice from China called 'Purbachi' in East Pakistan. Pakistan and China were pursuing closer diplomatic relations at the time. None of the public relations efforts given to the IRRI varieties were given to the Chinese variety. Yields of 4.3 tonnes per hectare were recorded for Purbachi. This rice remained in circulation and is popular to this day; it became known among farmers by the intriguing title 'China-IRRI'.

When IRRI officers visited East Pakistan they were talking about their ideas for rice research and development, studying the institutional arrangements in the government and countryside, and lobbying for a co-operative research programme between IRRI and East Pakistan. IRRI needed access to the research and demonstration potential of the country; it had a long-term interest in access to the rich genetic materials, both collected and still uncollected, in East Pakistan, which could be included in their own bank of genetic materials. IRRI officials moved in the same paths as other Americans in East Pakistan; they moved in a community of interests with Ford and Rockefeller Foundation officials, with representatives of the Agricultural Development Council, with the Harvard Development Advisory Service, with those on USAID contracts. They were part of a larger established US team, but had not yet made their mark in it. The success in Pakistan of wheat seed developed by their 'rival' international research centre CIMMYT in Mexico certainly urged them to promote their own products more vigorously.

Each of these foreigners or each group of foreigners had to work with and through government institutions such as ADC, WAPDA, and the Ministry of Agriculture. These relationships were necessary but not always smooth or productive from the point of view of either side. The delays and intrigue of the government infuriated the foreigners, and the ignorance or insensitivity of most foreigners infuriated the Bengalis. Nevertheless the geopolitical conditions of the 1960s ensured that these relationships survived personal storms, and departures or transfers. The relationship between IRRI and EPARRI was difficult and complicated by the actions of people who knew little about rice research, but it survived. The officers of the Ford Foundation monitored the relationship continually, developing it or restoring it when necessary: throughout this period they were its underwriters. This is why some of their views on rice research are included here, because their views were decisive in questions of finance and remained so throughout the 1970s, as we shall show.

When IRRI scientists visited Pakistan before 1966, they did not believe that much basic research could be done there. Therefore they communicated the message that East Pakistanis could take advantage of what had been done elsewhere and do adaptive and applied research in order to 'finely tune' new rice technologies which would increase the yield per hectare. In May 1965 the assistant director of IRRI (McClung) toured the province and reported that rice research activities had reached their lowest point in half a century: he observed that, 'No experiment station...claims to be getting rice yields equal to the best farmers.'[68] This view also circulated among USAID agricultural specialists, who concluded that there was little need for further basic research in East Pakistan because farmers were smart enough to capitalize on research already done elsewhere, to take what the Japanese were demonstrating at Comilla, and to increase productivity immediately. The experimental stations could be disregarded. The current goals and organization of researchers in government, said one seasoned USAID official, bore little resemblance to the current need of farmers, which was to acquire effective new information about rice quickly, and the means to implement it.

As IRRI's efforts in East Pakistan progressed, its influence on the organization of research increased: 'follow the guidelines shown for IRRI,' said McClung, and concentrate 'on problems for which solutions are...expected in the shortest possible time'.[69] In 1966 IRRI's director (Chandler) reiterated the advice to break up research disciplines in the ministry and to organize multidisciplinary mission-oriented research teams for rice. But this advice was not fully heeded for some time, even when the Ford Foundation was willing to pay for these teams. Getting specialists to cross disciplinary boundaries (entomology, agronomy, etc.) inside the institute took many years and was a subject of international review mission reports during the late 1970s. Not only were foreign scientists planning changes in organization, they were trying to establish new working goals for researchers. 'The rice breeder in [East] Pakistan must now seek a radically different plant type, if he hopes for a major break-through in yields,' said McClung in 1965.[70] This breakthrough, comparable to that in the Philippines, could come in five years, he thought. By December that year IRRI scientists expected HYVs from IRRI to give 50 per cent higher yields per hectare than any variety then grown by farmers in East Pakistan. The following year IRRI's Director informed the Secretary of Agriculture that average yields can be doubled or tripled if water, diseases, and insects can be controlled.[71] 'Your best farmers,' said Chandler in 1966, in the best weather, now seldom reach one-half the yield which IRRI scientists achieved with the new rice varieties. What is more, said Chandler, 'virtually' no rice-breeding was being conducted in East Pakistan. The implication was that given these possibilities, and given the previous major investments in the requirements of the HYVs, there was little reason to take any basic research, including rice-breeding, seriously in East Pakistan. That was the position of IRRI in East Pakistan.

How did this position translate into advice? 'You can no longer be satisfied with a variety that gives two or three maunds [a maund weighs about 37 kg] more yield than another,' Chandler lectured the Secretary in 1966; 'You are looking for new varieties that will give you 100% or 200% increase in yield under high level management.'[72] Listing the eight characteristics of the varieties with their new architecture which IRRI had adopted (short sturdy stems, etc), he added

> Those are objective characteristics. They can be measured. They permit the plant breeder little room for bias about the origin of the variety, or the protection of earlier breeding achievements which are now outmoded. All Bengali and imported materials would be held up to the same criteria. Superior economic characteristics alone would dictate the selection. ...it is not likely that the basic characteristics of the new high yielding plant will be found among them [among the 1,000 pure lines of Bengali rice varieties maintained since 1912] because we have already screened most of these plants at IRRI for that purpose.[73]

Chandler zealously argued that the government could decide to reduce the number of pure lines preserved at Dhaka 'without the loss of any economic characteristics required for future breeding purposes'. While his views had already generated some sensitive reactions (according to documents and interviews), he had the confidence of believing he was both scientifically and morally correct. This righteousness arose from the conviction held at IRRI that East Pakistan could achieve rice self-sufficiency quickly if it would stop wasting time and do what it was told. By looking at the annual food deficits, IRRI scientists knew the available rice had to be increased only between 10 and 15 per cent annually. If yields per hectare were doubled this increase could be achieved easily. In 1965 McClung predicted a breakthrough by 1970. The following year Chandler told the government that East Pakistan could reach and surpass sufficiency in rice within five years, not only solving an important food problem, but perhaps triggering an agricultural revolution that would be widely observed by other countries in Asia.[74] A few months after the Indo-Pakistan war, he pointed out that the same HYV rice was spreading rapidly in India, and that country was one crop-season ahead of Pakistan. This argument, too, must have appealed to the competition and fear between the two countries.

Following Chandler's first 1966 visit, the IRRI liaison scientist Lee Johnson, a geneticist, announced his calculations, done with Gulam Mohammed, that sufficiency could be reached within two years if seed multiplication was accelerated. A foundation account of a 1966 meeting with government and foreign officials shows that Johnson acknowledged and accepted the risks of his prediction of sufficiency by 1968.

> As a scientist, he [Johnson] warns that moving this quickly with a new variety in an untested area involves certain risks and violates sound plant-breeding principles. The danger is that the new variety will fail, thus bringing discredit to

the entire program and creating suspicion of any future new varieties. Unlike Canada, however, where the difference between a new variety and the best existing varieties is small, the potential increment of IR 8–288 is so great over existing Pakistan varieties that Johnson believes the risk is worth taking.... Johnson predicts that the accelerated multiplication program can have either of two results, both favourable: (1) IR 8–288 will be a big success, and will have a tremendous psychological impact, even as a stopgap variety until the new Pakistani-developed varieties are ready. (2) If terminated before the 7-million-acre stage, the program will still provide a good 'dress rehearsal' for the multiplication of later hybrids.[75]

With these high expectations, scientists became preoccupied with bottlenecks. Convinced they had the right technology, IRRI scientists looked into seed multiplication, irrigation, extension training, chemicals, etc. Speed was everything. They sought to reduce the growing period of HYV rice from 130–55 days to 100 days. Chandler advised the government to ignore the slow-adopting farmers or non-adopting farmers and to concentrate entirely on fast adopters. This was the pattern repeated among researchers in the US since the nineteenth century. When seed multiplication was slow, East Pakistan became the world's largest importer of HYV rice seed. In fact Chandler had argued in 1966 that the success or failure of the entire work of the Green Revolution depended on the degree to which the government accepted its responsibilities and how effectively it transferred existing technology and conducted an extension programme with adopting farmers.

Although not himself a conformist among IRRI scientists, Richard Bradfield came from IRRI to review the progress of research in 1967. 'My research at IRRI,' Bradfield wrote to the Secretary of Agriculture, 'convinces me that [in] East Pakistan, agriculture has the potential to become as productive of food—per acre, per year, and per day—as the agriculture of Taiwan, which is presently the most progressive tropical agricultural nation in the world.'[76] Nevertheless, Bradfield hit the system hard: it had to convert 'an attitude of passive acceptance' to an 'active determination'; it had only to double the cropping intensity, and food production 'would be increased four times'; it had to shift marginal rice land to other crops—soya, sorghum, sweet corn, vegetables, potatoes; and it had to adopt 'a modest amount of mechanization'.[77] Ten years earlier, Bradfield had made the same analysis of agriculture in the Philippines. Despite this advice few of his expectations were realized by 1980.

But evidence was slowly appearing to support doubts which had lain outside the field of IRRI's vision in East Pakistan. By mid-1967 Ford Foundation officials reported that IR 8 was not living up to early expectations, and by mid-1968 it was clear that IR 5 would not do so either.[78] Disease and insect problems were found to be severe, and a great deal of adaptive research seemed to them to be required for both varieties. The negative impact of this news had already been reported by government scientists, and Johnson's worst fear of disappointment had been realized. Alim, Zaman and their colleagues had months earlier advised the government to be careful of the 'miracle rice' slogan, 'because the ups and downs of performance in different seasons were not anticipated and caused excessive enthusiasm, followed by excessive pessimism in the Government'.[79]

IRRI scientists learned that 'IR 8 was not the answer to East Pakistan's needs' and that it had little long-term impact on commercial cultivation where cold injury, virus diseases, and insect pests were major production problems, according to Athwal.[80] So in order to ensure the wider adoption of IRRI varieties, a short training course, covering the production of IR 8, IR 5, and IR 20, was conducted by a Philippines extension specialist from IRRI in preparation for the 1968–9 season. The efforts did not stop there. In preparation for the 1970–1 boro season, the 'largest single consignment of improved rice seed imported by any country', 1,800 tonnes of rice seed, was brought from the Philippines.[81] The Pakistan Agricultural Development Corporation was responsible for importation and sale or distribution of all these inputs, and the Integrated Rural Development Program (based on the Comilla model and adopted in 1970 for East Pakistan as a whole) was responsible for the actual operation of the 'IRRI blocks' in which rice grew around an irrigation pump in the villages. But it is not clear what kind of responsibility the East Pakistan Accelerated Rice Research Institute (EPARRI) was expected to assume beyond the testing of imported varieties and the slow indigenous breeding of HYV rice varieties. EPARRI's importance seems by this stage to have already been symbolic rather than economic, given the level of imports. Nevertheless EPARRI released three HYV varieties at the time of its establishment at Joydevpur. Each was selected from IRRI advanced generations. Irrisail, released in 1969, was intended for the transplanted aman crop; Chandina, released in 1970, was intended for the aus and transplant aman crop; and Mala, released in 1970, was intended for the aus and boro crop. (These names appeared to have been assigned some years after release.) It is clear that researchers already felt that the HYV strategy would have to diversify beyond the boro season.

Still committed to East Pakistan's self-sufficiency (in 1969 it was predicted for 1972 by the IRRI liaison scientist), IRRI was forced by these setbacks with IR 8 to view its relation with EPARRI differently. The IRRI associate director concluded in 1969 that 'since the diseases and insect problems of the area are extremely complex, some rather basic work will be required'.[82] Six months later in 1970 he advised that 'purely adaptive research will not be enough'.[83] McClung acknowledged there had been

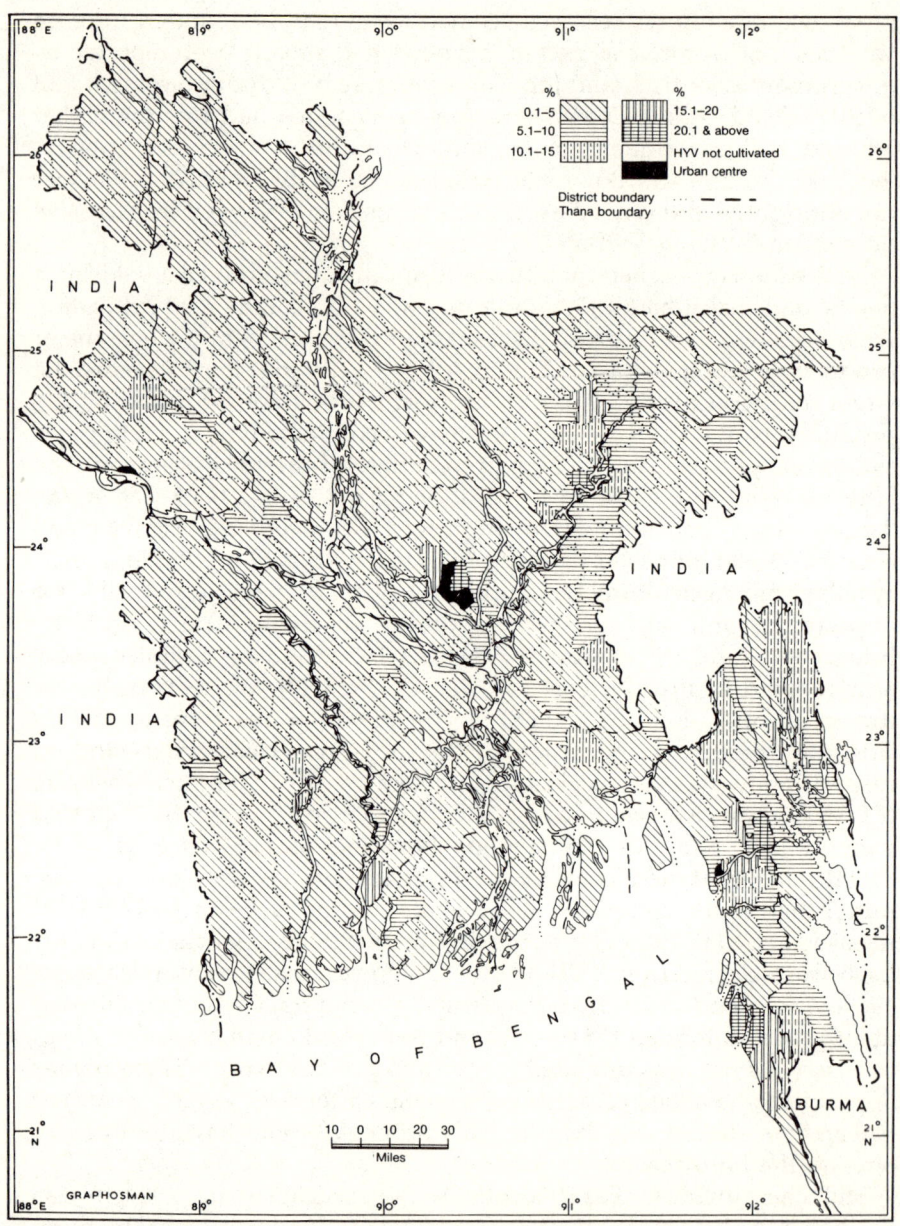

MAP 3. *Bangladesh: HYV rice cultivation as percentage of total rice area, 1969–1970, by thana*

a loss of government confidence because the promised technology had not been viable: this disappointment in outside circles stood in contrast with the higher morale among local researchers due to the establishment of an autonomous EPARRI. In 1969 the IRRI liaison scientist was reported reluctant to release IR 5 'because it is inferior to rice varieties on the way; the Government, however, feels under pressure to pass out something new'.[84] The government, in 1970, continued the habit of chasing after the year's most likely rice variety, so it decided to import 1,800 tonnes of IR 20, although even IRRI did not promise it to be a strong candidate for success. The cautious side of IRRI was now showing.

When McClung described the recent change in rice production in West Pakistan, however, his values seem to have resonated happily with the situation there. A few small research stations had adapted certain fine high yielding rices, and progressive seed companies and bigger commercial farmers growing for export had managed to more than double the crop output. Foreign exchange earned in the Middle East had encouraged other farmers, and the results in the West were spectacular. It was closer to the North American agribusiness experience and closer to IRRI's values, in which a few high yielding varieties would replace the plethora of local varieties and non-adopting subsistence farmers could be ignored by researchers. National problems could be solved without engaging the whole agricultural economy.

At the end of the decade IRRI scientists found they had almost as many problems in East Pakistan as when they began in 1962, and although EPARRI had been established it appeared to be just a beginning, not an end. Wanting a productive research centre, like IRRI, they had encouraged the centralized model which also fitted naturally with the ideas of government bureaucrats. Yet from 1965 onward Chandler and McClung mentioned the need to have strong substations in different regions. Wanting to have an army of trained extension agents, they held IRRI courses in East Pakistan and cut the few months' training to a few weeks; they also sent many people abroad for training. But all this had 'not resulted in a viable self-sustaining program', admitted McClung in 1970.[85] People trained abroad or at IRRI were, on return, eventually reconciled with the old disciplinary order in the institute, or they left. Wanting to break up this order, IRRI posted more foreign scientists: by 1970 four permanent scientists and a succession of consultants were planned. Their work crossed divisional and disciplinary boundaries, and they were told to communicate enthusiasm for research. Finally, IRRI scientists agitated with senior government officials to enshrine research as a key element in the solution. They struggled to get the attention, and then the commitment, of distracted officials, urging that the best leaders should be appointed in EPARRI to give it direction. These very officials and directors, however, were subject to transfer, especially in the chaos of the 1969–71 period. So IRRI and foundation officials had continually to do new homework which had little to do with research. Meanwhile they do not appear privately to have had much confidence in the entire undertaking.[86]

Inadvertently, in 1970, IRRI's associate director summed up the entire situation in a phrase: 'some of these objectives have conflicting elements'.[87] Indeed. How can one attempt to confer bureaucratic autonomy and yet avoid excessive independence? How can one ask for imitation and get independence? Can one engage in instant technology-transfer and simultaneously demonstrate the virtue of knowledge which takes years to acquire? How can people trained rapidly be told to stay put where they started? Can one promise the moon and avoid disappointment, or avoid the charge that one does not know where the moon is? These 'conflicting elements' had profound consequences throughout the 1970s.

S. M. H. Zaman later said that it is an extremely easy task to criticize the HYV strategy, but many internationally reputable experts found it the best choice at the time. 'Those who advocated the Green Revolution did so honestly,' he said.[88] Chandler told Zaman and others 'as early as 1965' that IR 8 was just a modest beginning, and Zaman recalled that authorities and rice breeders alike understood and accepted responsibility to develop suitable varieties. In this case, as in others, recollections of what was said, to whom it was said, and when it was said, seem to have varied. If there is criticism of the HYV strategy here it is not because people were not honest but because they failed to understand the consequences of closing off alternative strategies.

By 1970–1 the official plan for the Green Revolution appeared totally confused. Moreover there was a devastating cyclone with heavy crop damage in 1970, followed by a routine and bureaucratic response from the government. Voters in East Pakistan, in the election held soon after the cyclone, favoured a Bengali nationalist party, the Awami League: thus it, and its leader Mujibur Rahman, captured a majority of seats in the Pakistan legislature; the contradictory relation between the two wings of the country finally came to a crisis. The food deficit was 12 per cent, as it was in 1967: government efforts for four years had not changed the overall food picture. It is quite remarkable, during the chaos and military violence of 1971, that any rice was planted at all, yet cultivation and harvesting continued because people needed to eat. Even rice research, in its small way, continued amidst the massive importing of foreign HYV seed. During this time of crisis, knowledge about rice among rice researchers was disregarded and took a back seat to other kinds of planning intelligence. The consequence of this back seat was that in the national decisions taken in 1972 and 1973 regarding self-sufficiency (taken with IRRI's guidance) it is clear that the Bangladesh Rice Research Institute (BRRI) still took a back seat; it chose a narrow conception of its work, and this limitation allowed others to ignore it in the crisis, as we shall see.

7.7 Chronology of the Green Revolution in East Pakistan

1960
- major rice research institute planned by East Pakistani researchers
- Pakistan's Agricultural Commission completes report
- Japanese team of experts experimenting at Comilla

1961
- plan for rice institute blocked by government officials
- 6.3% food deficit in East Pakistan

1962
- rice research lands expropriated for Second Capital project
- Comilla experiment yields encouraging results
- first visit by IRRI officials

1963
- last rice research plots planted at Dhaka
- 13.8% food deficit in East Pakistan

1964
- last rice research plots planted at substations
- Agricultural Research Institute has no home and no director
- research budget for agriculture (food-crops) Rs 600,000

1965
- temporary rice research site approved for 'adaptive' testing
- permanent IRRI representative arrives
- East Pakistani appointed to IRRI Board
- East Pakistan population 65 million, at 3% annual growth
- 7.5% food deficit in East Pakistan

1966
- 303 rice varieties transferred to East Pakistani rice researchers from IRRI
- 10.1 tonnes of IR 8 seed transferred

- IRRI tells government 'self-sufficiency' possible by 1970 or earlier
- joint plan approved between East Pakistan, Ford Foundation, and IRRI
- Ford Foundation allocates $316,000 to IRRI for East Pakistan research
- government approves permanent Agricultural Research Institute site

1967
- over 1,500 tonnes of IR 8 imported from the Philippines
- government-supervised cultivation gets high yields at Comilla
- government researchers say self-sufficiency possible by 1970
- second permanent IRRI representative arrives
- Ford Foundation allocates $400,000 to IRRI for East Pakistan research
- work begins on rice research centre near Dhaka
- 12.0% food deficit in East Pakistan

1968
- doubts about IR 8 yields circulating publically
- self-sufficiency targets questioned
- Ford Foundation allocates $280,000 to IRRI for East Pakistan research
- work going full speed at rice research centre near Dhaka
- little progress in securing autonomy for rice research within Ministry
- HYV rice grown on 121,000 ha, mostly IR 8

1969
- doubts about IR 5 and self-sufficiency predictions circulated
- IRRI promoting new IR 140, moderately tall rice
- need for basic research in East Pakistan acknowledged at IRRI
- adapted IRRI varieties released by East Pakistani researchers
- Ayub Khan's government removed by coup

1970
- EPARRI made autonomous institute in law, given 36.4 ha land
- IRRI proposes four foreign experts plus consultants for EPARRI

- 1,800 tonnes IR 20 seed imported from the Philippines
- 12.0% food deficit in East Pakistan
- cyclone and election set the stage for 1971 confrontation

1971 • East Pakistan becomes Bangladesh

1972 • EPARRI becomes BRRI

NOTES

1. Haroun Er Rashid, *Geography of Bangladesh*, Boulder, Colo.: Westview Press, 1978, p. 90.
2. Engenio B. Manalo, *Agro-Climatic Survey of Bangladesh*, Los Banos: IRRI, n.d., c.1977.
3. Ibid. 14.
4. Ibid. 22.
5. Rashid, *Geography of Bangladesh*, p. 119.
6. J. P. Morgan and W. G. McIntire, 'Quartenary Geology of the Bengal Basin, East Pakistan and India', *Bulletin of the Geological Society of America*, 70 (1959), 319–42.
7. For the best study of this subject see Alia Ahmad, *Agricultural Stagnation under Population Pressure: The Case of Bangladesh*, New Delhi: Vikas Publishing, 1984.
8. James K. Boyce, *Agrarian Impasse in Bengal*, New York: Oxford University Press, 1987, p. 145, table 5.6.
9. Kirsten Westergaard, *State and Rural Society in Bangladesh: A Study in Relationship*, London: Curzon Press, 1985, p. 107.
10. J. C. Jack, *Final Report on the Survey and Settlement Operations in Faridpur District*, Calcutta: Government of Bengal Press, 1916.
11. H. Brammer, E. J. Clay, and N. I. Bhuiyan (for Directorate of Agriculture Extension and Management), *Incidence of Landlessness and Major Landholding and Cultivation Groups in Rural Bangladesh*, Dakha: Bangladesh Agricultural Research Council, 1978.
12. M. Q. Zaman, 'Endemic Land Conflict and Violence in Char Villages of Bangladesh', in V. S. Pendakur, O. P. Divedi (eds.), *South Asian Horizons*, v. Ottawa: Canadian Asian Studies Association, 1987.
13. Note that there is a marked difference between households in the same village in terms of women's role in the rice economy, depending on the season. See Ben Wallace, Rosie Ahsan, Shahnaz Huq Hussain, and Ekramul Ahsan, *The Invisible Resource: Women and Work in Rural Bangladesh*, Boulder, Colo.: Westview Press, 1987. There is no trace of the gender question in the documents reviewed or interviews conducted with respect to rice research planning in the 1950s and 1960s. This became an issue only in the mid-1970s when women in Bangladesh began to press the subject into public view. The

Ford Foundation in Dhaka pioneered in giving support at that time to empirical research about the role of women in the agricultural economy, so that by 1980 this was an established field of enquiry and action. See also Latifa Akanda and Roushan Jahan (eds.), *Women for Women: Collected Articles 1983*, Dhaka: Women for Women, 1983.

14. Representative of these studies are those of. Peter J. Bertocci, 'Elusive Villages: Social Structure and Community Organization in Rural East Pakistan,' Michigan State University Ph.D., 1970; A. K. M. A. Islam, *A Bangladesh Village: Conflict and Cohesion*, Cambridge, Mass.: Schenkman, 1974; M. Ameerul Huq (ed.), *Exploitation and the Rural Poor*, Comilla: Bangladesh Academy for Rural Development, 1976; Willem van Schendel, *Peasant Mobility: The Odds of Life in Rural Bangladesh*, Assen: Van Gorcum, 1981; Betsy Hartmann and James K. Boyce, *A Quiet Violence: View from a Bangladesh Village*, London: Zed Press, 1983; Bangladesh Rural Advancement Committee, *Who Gets What and Why*, Dhaka: BRAC Publications, 1983.
15. Amartya Sen, *Poverty and Famines: An Essay on Entitlement and Deprivation*, Oxford: Clarendon Press, 1981; Paul R. Greenough, *Prosperity and Misery in Modern Bengal: The Famine of 1943–1944*, New York: Oxford University Press, 1982.
16. M. S. Venkatramani, *Bengal Famine, 1943: The American Response*, Delhi: Vikas Publishing, 1973.
17. M. M. Islam, *Bengal Agriculture, 1920–1946: A Quantitative Study*, Cambridge: Cambridge University Press, 1978, p. 35.
18. A. Alim et al., *Review of Half a Century of Rice Research in East Pakistan*, Dhaka: East Pakistan Government Press, 1962, p. 13.
19. Islam, *Bengal Agriculture*, p. 50.
20. Ibid. 55.
21. M. M. Alamgir and L. J. J. B. Berlage, 'Foodgrain (Rice and Wheat) Demand, Import and Price Policy for Bangladesh', *Bangladesh Economic Review*, 1, 1 (1973), table I.
22. Islam, *Bengal Agriculture*, p. 82.
23. Ibid. 57 ff.
24. Ibid. 86.
25. Lucile H. Brockway, *Science and Colonial Expansion: The Role of the British Royal Botanic Gardens*, New York: Academic Press, 1979, p. 122.
26. Discussion, S. M. H. Zaman, Dhaka, January 1978. Zaman was an author of the *Review of a Half a Century of Rice Research in East Pakistan* (see n. 18).
27. Ibid
28. Ibid.
29. This judgement is based on extensive review of these primary sources for the history of scientific institutions from the mid-nineteenth century. For analysis of the ramification of these opinions in the industrial development under private initiative, and, to a lesser extent, in commercialized agriculture (e.g. sugar), see A. K. Bagchi, *Private Investment in India, 1900–1939*, Cambridge: Cambridge University Press, 1972. We are also grateful for a lengthy discussion of this subject with Professor Bagchi in 1978. See also Shiv Visvanathan, *Organizing for Science*, Delhi: Oxford University Press, 1984.

30. Ibid. 77–9.
31. Ibid. 90 ff.
32. Ibid. 99.
33. Alim *et al.*, *Review*, p. 23.
34. Islam, *Bengal Agriculture*, p. 34.
35. J. F. Stepanek, *Bangladesh: Equitable Growth?*, New York: Pergamon Press, 1979, p. 95.
36. Ibid. 95.
37. Islam, *Bengal Agriculture*, p. 198.
38. A. H. Khan, 'The Comilla Experience in Bangladesh—My Lessons in Communication', in W. Schramm and Lerner (eds.), *Communication and Change*, Honolulu: University Press of Hawaii, 1976.
39. Stepanek, *Bangladesh*, ch. 8, pp. 118–33.
40. FAO (United Nations), 'Statements Submitted by Dr. S. Hedayatullah, Director of Agriculture East Bengal (Pakistan),' 3rd Session, Bandung, Indonesia: International Rice Commission, 1952 (IRC/52/85p).
41. Ibid. 3.
42. Alim *et al.*, *Review*, p. i. Also S. M. H. Zaman, personal communication, 30 August 1979.
43. Alim *et al.*, *Review*, pp. 44–51.
44. Discussions, BRRI scientist, January 1978.
45. Ibid.
46. Alim *et al.*, *Review*, p. 68.
47. Ibid.
48. See BRRI, *Review of Research in Deep Water Rice in Bangladesh (1917–1960)*, Dhaka, 1974; also BRRI, *Proceedings of the International Seminar on Deep Water Rice*, Dhaka, 1974.
49. Quoted in Herbert Feldman, *The End and the Beginning: Pakistan, 1969–1971*, London: Oxford University Press, 1976, p. 2.
50. Raisuddin Ahmed, *Food Production in Bangladesh: An Analysis of Growth, Its Sources, and Related Policies*, Dhaka: Bangladesh Agricultural Research Council, 1977.
51. Premen Addy and Ibne Azad, 'Politics and Society in Bengal', in Robin Blackburn (ed.), *Explosion in a Subcontinent*, Harmondsworth: Penguin Books, 1975, p. 264.
52. Ahmed, *Food Production*, p. 95.
53. Although over a thousand of them were built in the country, 90 per cent of the 100 seed stores seen between 1972 and 1976 were not used for the intended purposes. Sometimes meetings were held there, or refugees squatted there, or private business was conducted there, etc. Occasionally, at the boro season, fertilizer or pesticide was stored for a few days. A number of the buildings were unsafe because the contractors had mixed more sand than cement in the roof and walls. In general the seeds and other inputs were in such short supply that the farmers had to go begging at the thana level or higher up even to get a few bags. Little was ever stored at the Union level, and rice seed was always stored at home.
54. Government of Pakistan, 'The Basic Democracies Order, 1959', Karachi: Ministry of Law, (Manager of Publications), 1962, pp. 122–74.

55. USAID, *Development Assistance Program FY 1975: Bangladesh*, Washington, DC: Department of State, December 1974, unclassified, pp. 85–6.
56. Ibid. 86.
57. Alim, *et al.*, *Review*, p. ii. Also Zaman, personal communication, 30 August 1979.
58. Alim *et al.*, *Review*, p. 13.
59. Tsao Matsuda *et al.*, *Annual Report of the Japanese Experts 1961–1962*, Comilla: Pakistan Academy for Rural Development, 1962.
60. Alim *et al.*, *Review*, p. 13.
61. Dana Dalrymple, *Development and Spread of High Yielding Varieties of Wheat and Rice in the Less Developed Nations*, Washington, DC: USDA Foreign Agricultural Economic Report 95 (6th edn.), 1978, p. 40.
62. Discussions, various rice researchers and officials, Dhaka, January 1978.
63. Haldor Hanson to L. P. V. Johnson, 5 December 1965, p. 8, FFA PA 66–125.
64. A. Alim *et al.*, *Annual Progress Report on Accelerated Rice Research Program of East Pakistan*, Dhaka Agricultural Rice Research Institute, December 1967, p. 5, FFA PA 66–125.
65. S. A. Quadir, 'Institutions and Linkages in Rural Development in Bangladesh', in G. C. Chu, *Communication for Group Transformation in Development*, Honolulu: East–West Communication Institute, 1976.
66. These maps are based on reports of HYV cultivation: see Chapter 9 regarding historic exaggeration in such reports.
67. A. Alim, 'Creative Genius and the IRRI Rice', *Agriculture of Pakistan*, 1968.
68. A. Colin McClung, 'Accelerated Rice Research Program for East Pakistan in Cooperation with IRRI,' Dhaka, May 1965, pp. 16 and 27, FFA report 001616 from PA 66–125.
69. Ibid. 18
70. McClung, 'Accelerated Rice', p. 4.
71. Robert F. Chandler, 'Progress Report on the Accelerated Rice...with IRRI,' Karachi: 12 February, 1966, p. 3. FFA report 001676 from PA 66–125.
72. Ibid. 21.
73. Ibid. 20, 24.
74. Ibid. 36.
75. W. J. Hertz to H. Hanson, 28 May 1966, FFA PA 66–125.
76. Richard Bradfield to Karem Iqbal, 12 September 1967, FFA PA 66–125.
77. Richard Bradfield, 'Increasing and Diversifying Food Production in East Pakistan', December 1967, FFA PA 66–125.
78. 'Activities in East Pakistan under Ford Foundation Grant No. 66–125 during the period March 1, 1966 to February 28, 1968,' FFA PA 66–125, p. 5.
79. Alim *et. al.*, Annual Progress Report p. 12.
80. D. S. Athwal, 'Semidwarf Rice and Wheat in Global Food Needs', *Quarterly Review of Biology*, 46, 1 (March 1971), p. 22.
81. Ibid. 22.
82. A. C. McClung to Secretary of Agriculture, East Pakistan, 13 October 1969, FFA PA 66–125.
83. A. C. McClung, 'Report on Rice Research Activities in Pakistan...', Islamabad: Ford Foundation, 13 March 1970, FFA PA 66–125.
84. B. Edwards to R. Smuckler, 12 April 1969, FFA PA 66–125.

85. McClung, 'Report on Rice Research', p. 14.
86. Based on the tone of internal correspondence at the time, on interviews, and Oasa, 1981 (see above, Ch. 3 n. 12).
87. Ibid. 15.
88. Personal communication, January 1981.

8
Bangladesh: Independent Nation, Dependent Research

> The goal of raising average yields of HYV rice by 22% and of local rice varieties by 20% is 'a rather ambitious program'.
> Robert Chandler, *Report of the Special Agricultural Mission*, Dhaka: United Nations Relief Office, 1972.

FOLLOWING the surprising results of the extraordinary election in 1970, in which the Awami League won the majority of seats, and following the devastating cyclone in December 1970, in which the government did little to relieve or rehabilitate the population, political events moved swiftly. Rice research was far from anyone's mind when, in March 1971, the army initiated a rapid reign of terror, complete with public executions and secret assassinations. Eight months of guerrilla warfare later, the country, on the eve of its independence from Pakistan, was beginning to address its bankruptcy. High on the list of things people needed was food. And here lay the political tension between the consequences of continuously receiving "free" food aid and the efforts of pursuing food self-sufficiency. This tension continued through the decade, profoundly affecting BRRI. In this chapter we examine the marginal role of BRRI in the pursuit of self-sufficiency, as well as how the government and its supporters planned and pursued self-sufficiency while simultaneously relying on international food supplies.

At the end of the 1960s, despite some doubts, the planners of the HYV strategy thought that a few varieties would be grown universally (as, for example, in the American rice industry) and that the cultivation of inefficient varieties would be reduced if not eliminated, thus driving out subsistence farming and promoting commercial agriculture. With massive importation of the HYV seed and its requirements there would be a dramatic jump in average yields and total production. But because most farmers resisted or did not conform to this plan, and because food deficits remained constant, a shift in the strategy seemed indicated by the mid-1970s. Still convinced that the revolution lay in the seed, planners in the 1970s admitted that the expansion of summer harvest (boro) rice cultivation might have reached its limit because it was

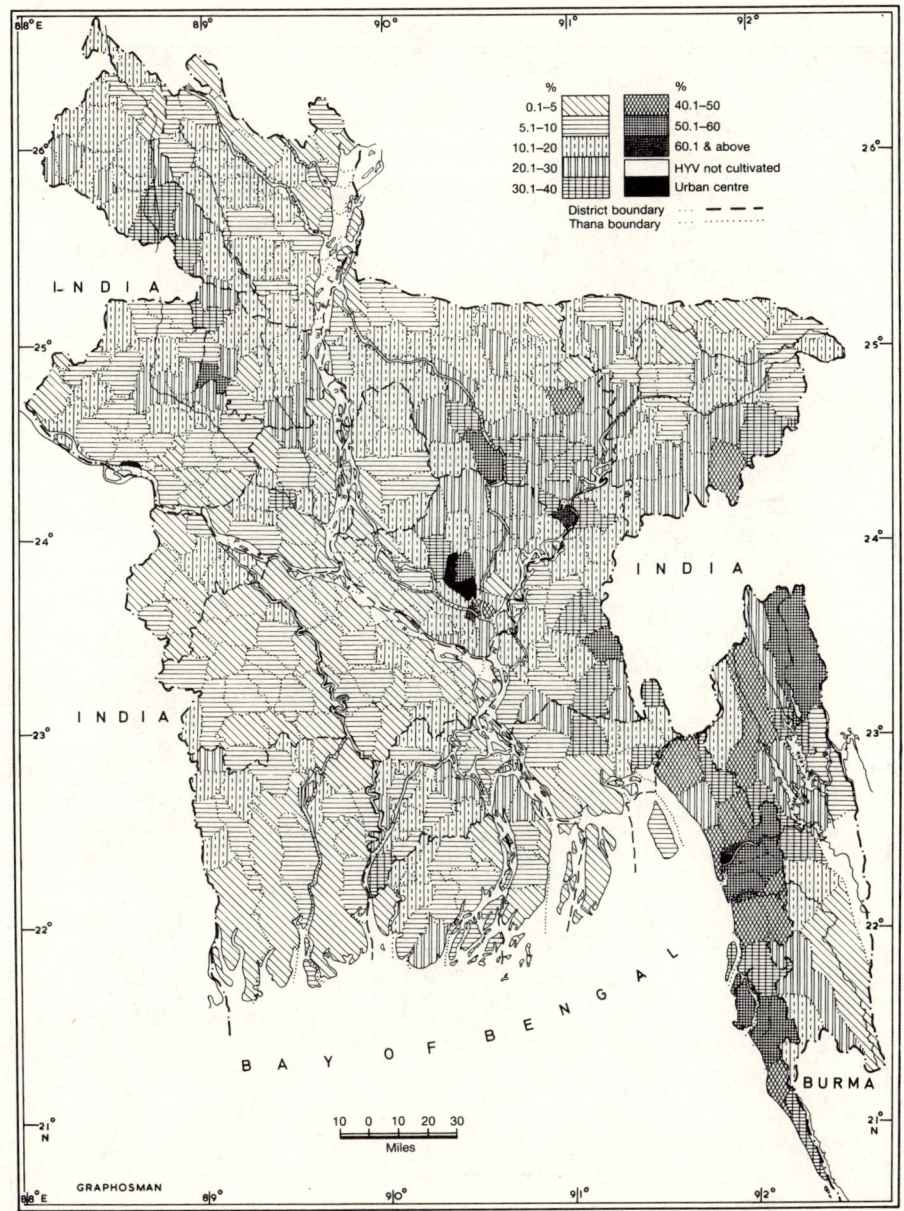

MAP 4. *Bangladesh: HYV rice cultivation as percentage of total rice area, 1973–1974, by thana*

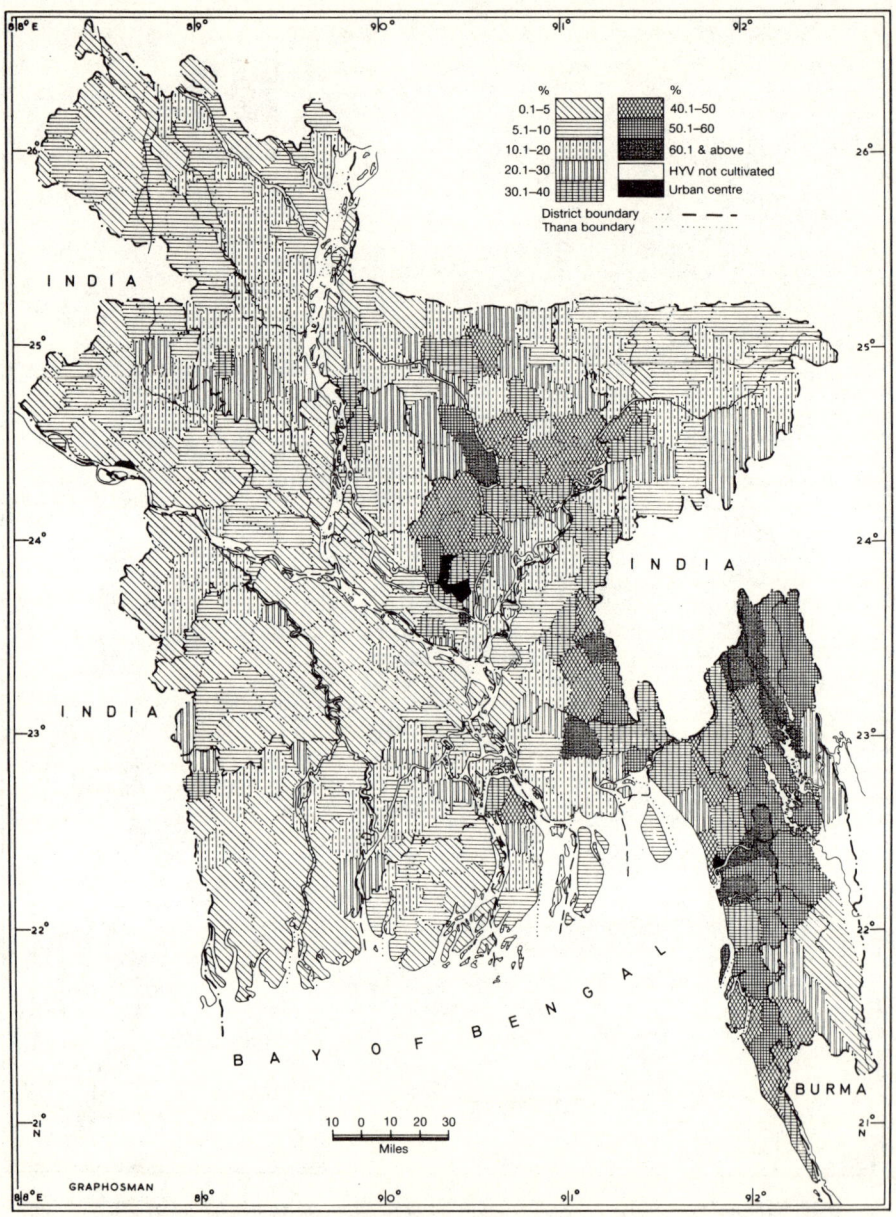

MAP 5. *Bangladesh: HYV rice cultivation as percentage of total rice area, 1977–1978, by thana*

necessarily tied to the slow increase in irrigated area largely under the control of large farmers. Although BRRI was, by the mid-1970s, expected to breed varieties for other seasons and for adverse cultivating conditions, the importance of the summer season in the strategy continued undiminished because HYV varieties were not so widely grown in any other season.

Providing HYV rice for seasons other than summer was roughly equivalent to providing HYV rice for farmers without mechanical irrigation, who usually had small farms of 1 to 3 hectares. Smaller farmers practised the art of trapping and manipulating water because they also had little capital for HYV requirements such as pumps. The strategy shifted to address small farmers because they were so numerous; there were simply not enough big farmers to constitute a strong political base or source of production increase. Big farmers were also visible, vulnerable, and difficult for the government to protect. Of course the connection with them would not be neglected—it produced rice and political rewards—but the smaller farmers had to retain their land if they were to become a political force and mobilized in a political party such as the Awami League, particularly as a bulwark against the growing number of landless people. The HYV strategy thus might enable the small farmer to retain his land, especially if, as conceded by some researchers, the smaller farmer is a more efficient user of technology. This idea was also important in the thinking of the underwriters of the costs of the HYV strategy, the World Bank, the United Nations, bilateral donors like USAID and CIDA, and others interested in supporting the country's political establishment. The idea combined two virtues—political expediency and an enlargement of the scope, and the market, for new rice technology.

Because the strategy could no longer be simply that of only 'building on the best' (irrigated cultivation by big farmers in the summer season) there was naturally much less confidence among planners about big pay-offs and big jumps in yield. Scientists seemed also to have experienced a loss of confidence in these simple objectives. While politicians and planners spoke vaguely but grandly of their expectations of the strategy, researchers were obliged to specify procedures and methods. Cognizant of the political objective of self-sufficiency, charged with supplying the 'magic beans' for the HYV strategy, BRRI researchers had not reconciled their private doubts about the strategy with pronouncements by the government. In addition it was widely recognized in Bangladesh that thinking and talking about other cultivating seasons or about adverse rice environments was much easier than doing something creative about them. These environments were adverse to most cultivators and most researchers alike.

Throughout this decade BRRI scientists may have had their doubts about the official strategy, but what is striking is their marginalization: they were not involved in calculations of self-sufficiency, in the choice or control of the HYV's requirements, and they were no longer oriented to the agro-ecological zones of the country. Despite the fact that there were remnants of an old system of rice substations for specific zones and specific problems, the majority of BRRI's budget and staff in 1980 was concentrated at Dhaka. Feeling compelled to do something, isolated from the larger politics of self-sufficiency (and criticized for 'not delivering'), tugged in differing directions by IRRI and donors, BRRI researchers were entrapped, after almost twenty years of this pattern, in a course of action which had not succeeded! This research strategy called for greater commitment by BRRI, more knowledge, and more money. With so much invested, personally and collectively, could scientists really change the pattern? When much more powerful institutions and persons around them, including their financial benefactors, had reinforced the strategy, year after year (building fertilizer factories, importing seed), could they make their private scepticism the fulcrum of public research reorganization? And who was to know whether calls for the reorganization of the strategy and rethinking of researchers (evident by 1980) would sustain scientists' narrower administrative interests, the autonomy of their institute, and safety of their careers? Entrapped, they had given increased commitment despite the absence of evidence that this commitment was leading to a successful outcome.[1] But who cared about what they were really thinking? BRRI researchers were evidently not taken seriously and were simply left to produce a stream of new varieties.

8.1 The Politics of Self-Sufficiency and the Weakness of Rice Research

Amidst reports of extraordinary drought and predictions of a 1 million-tonne drop in food grain production and the need for massive food imports, President Ziaur Rahman announced in May 1979 that national food self-sufficiency would be achieved by 1984—within five years.[2] One recalls that seven years earlier, amidst the chaos of the post-war period and a drought which led to massive, continuous food aid and food imports in 1972–3, Prime Minister Mujibur Rahman proclaimed food self-sufficiency as the country's top priority; the Minister of Agriculture announced in 1973 that national self-sufficiency would be achieved by 1978—within five years.[3] Even earlier, in 1965, self-sufficiency was predicted by 1970.

The quest for self-sufficiency, alternatively called 'food autarky', received continuous support and praise by both Mujib's and Zia's administrations; Mujib's for forty-four months, until his assassination in

August 1975, and Zia's for forty-five months, until his assassination in 1981. Politicians and officials of these administrations did not define 'sufficiency' or 'autarky' with any greater precision than to calculate the per capita daily intake requirement of 425 grams for the whole population against the annual food grain production and to make projections of increased production in the future. The circumstances of an early calculation in 1972–4, in which IRRI, in the person of Robert Chandler, was intimately involved, are discussed below. By 1977 an eleven-volume study of self-sufficiency written by Bangladeshi officials and academics and by the Philippines Secretary of Agriculture and the United States Ambassador to the OECD was presented to the FAO and the government of Bangladesh. 'Sufficiency' in Bangladesh was therefore still a matter of the greatest continuing importance to international development agencies, as well as to the government of Bangladesh. This importance led to an extraordinary amount of planning.

Unfortunately, the plan initiated by Prime Minister Rahman did not succeed in 1978, and the plan initiated by President Zia did not succeed in 1984. Although mathematically calculable, sufficiency remained an elusive goal. The reason why a variety of governments since the mid-1960s endorsed the HYV rice variety was that they believed that it would lead quickly to self-sufficiency. We have already discussed the weakened status of rice research in the context of the Green Revolution in the 1960s, and this weakness became more pronounced in the early 1970s. During 1971, 1972, and 1973 rice research efforts again were interrupted: these were tumultuous times for farmers and for food supplies, and rice researchers played only a minor role in the plans for the HYV strategy for self-sufficiency. In this chapter the central political importance of self-sufficiency is established, and how the HYV strategy was judged as the only route to it. The strong politics of the food supply picture, however, is contrasted with the weak state of rice research.

The Political Importance of Food Self-Sufficiency in the 1970s

The political and social importance of the government's insistence on food sufficiency lies in the underspecification and ambiguity of the objective. An average month's news in Bangladesh in the 1970s contained at least one major speech or announcement of the absolute and irreducible goal of self-sufficiency and a vague description of how it was to be achieved using HYV seeds and modern methods of cultivation. Both Mujib's and Zia's governments offered the following reasons for this priority of autarky: it would reduce Bangladesh's dependence on foreign sources of supply and spending of scarce foreign exchange on the unpredictable world food market; it would mean that increases in production had caught up with increases in Bangladesh's population. The consequence of sufficiency would be that foreign sources of supply could be approached for other forms of aid and trade, and scarce foreign exchange could be spent on other purchases. Self-sufficiency in food production clearly offered a measure of political freedom for a government in the form of 'freedom from want'. Thus autarky was the chief mobilizing slogan for both governments; self-sufficiency would be a demonstration of any government's control or mastery of a difficult production situation in Bangladesh.

In all political administrations the additional reasoning behind the autarky objective was that the strategy towards increased production really lay in building stronger links with the dominant landed section of rural agrarian society. These linkages were both part of the economic instruments of production increase *and* the political income from the investments of the strategy itself. As both instrument and political income the HYV strategy could be maintained only by a continuous flow of modern inputs. The ambiguity of the meaning of food sufficiency was also reflected in the underspecification of the effects of the strategy to increase production. There was an unspoken official acceptance that the means to produce more food were unequally distributed, that access to food was unequal, and the distribution of both the food and the benefits from production increases would be unequal. Nevertheless, even when egalitarian concerns occasionally enjoyed political favour, the HYV strategy actually changed little: it had a natural affinity with the prevailing political trends.

Mujib's administration had an effective political party machine in the agrarian society. The party leaders attempted to restrict distribution of HYV requirements to party faithfuls, using the inputs as rewards for support. In fact one of the reasons Mujib abolished all other parties and constituted himself President in 1975 (for five years) of the monopoly BAKSAL (Bangladesh Krishak Shramik Awami League) party was to bring about the 'Second Revolution'. This move was supposed to effect an agrarian transformation based on new political and economic discipline at every level. His choice of strategies, his party's questionable conduct, and the sharp rise in rice prices brought about his downfall and virtual destruction of his party after the two coups.[4] Zia's martial law administration (until 1979) had no such party machine and relied heavily on the supply of modern agricultural inputs to forge a successful connection to rural electorates before Zia himself formed a political party. That is why sufficiency took on an additional importance for Zia, the military, and his appointed cabinet advisers.

Given that the two political formations of Mujib and Zia were so different, in style, in their relations with the powerful civil service and the dominant rural classes, and on the question of law and order, the continuity of the objective of food autarky or self-sufficiency throughout the 1970s is remarkable. One should not imagine, however, that because this objective was the number one *announced* priority that it received more of the budget. Agricultural expenditures were usually only third in priority, after urban and industrial infrastructure, and after the military establishment. While these latter had their symbolic value and strategic importance, 'importance' should not be weighed in terms of the budget alone. Food self-sufficiency concerned the greater number of people in the country and was closest to the question of political stability and the distribution of power. The stability of the state rested on a balance between government influence on the cultivation, availability, and price of food, particularly rice, and the private profit to be made from these activities. Instability in the 1970s can partly be explained by instability of these specific factors. The calculation of self-sufficiency on a per capita basis comes one step closer to the issue of national well-being. This very step was defined to be outside the work of the 1938 Bengal Paddy and Rice Enquiry:

What we are concerned with, in this context, is not the food requirements of this province, but the effective demand for the food crop by its rice-eating population. The former is a problem of nutrition, which raises far reaching issues of national well-being; the latter is a severely limited problem of economics, in which we are interested at the moment.[5]

And this exclusion of self-sufficiency and well-being was profoundly costly in the famine of 1942–3, adding greater force to the movement to end colonial rule.

How was national well-being connected to food sufficiency in the thinking of planners and politicians in the 1970s? On an abstract level self-sufficiency, through the HYV strategy, would result from the application of science and technology for agricultural improvement and this progress would, in turn, lead to growth. On a less abstract level, food sufficiency would lead to a surplus, which would allow for agricultural income or food savings. Savings would mean accumulation; and accumulation, so the thinking went, would break the vicious circle of poverty and malnutrition.

The sufficiency calculations were made in order to establish the official level of annual food imports. Even when made at the district level, these 'sufficiency quotients' show that some districts were not sufficient in grain production from year to year. Finally, it was thought that savings and accumulation would go proportionately to those households and classes which appropriated the surplus not only in its production but in its processing, marketing, and consumption. This dynamic circle of progress and growth would conquer, in value terms, the static and backward nature of traditional agriculture. That was the vision of planners.

Immediate increases in production were viewed as impossible unless the government followed through with the adoption of the HYV seeds and promotion of their associated technical changes. Technical change thus was the only framework through which to see the possibility of increases in production. Aggregate production increase, at any cost, wherever and however achieved, was to be the best outcome of these technical changes. There was a race with population and with dwindling world food supplies which cost more and more; the perceived need was for immediate aggregate increases. 'There are some people who want self-sufficiency in one year,' said the Director of BRRI in 1975.[6]

Other political and economic plans were usually concerned with distributive questions: revenue, income, nutrition, health care, law and order, the ration system, and so on. These choices ran headlong into the simpler strategy for self-sufficiency and were usually subordinated to it. There was a contradiction between distributive questions and the values which underlay the objective of aggregate increases in food production at any cost. This contradiction was usually sidestepped by experts and officials who said that benefits would flow from the aggregate increases in food production and trickle outward or downward to the rest of society. The dynamic, progressive, responsive agents of change in the agricultural sector would then be moving and would stimulate other parts of the system. The HYV seed package, with other technical changes, including rural electrification, roads and communication, etc., would be the locomotives to pull the train of agrarian development and distributive justice—the railway image was actually used.

The choice of immediate increases in production meant that the government and private sector should work with those producers who could and who would maximize their profits and their production. Only the lack of information or shortage of inputs and credit inhibited these individuals; otherwise they were perceived as very responsive to modern agriculture and scientific ideas and would be efficient users of public resources for their own profit. (The fact that rice cultivators operate as households and not individuals was ignored.) It was assumed that these farmers had risen to their present position of strength through competition and so had already proved their efficiency. 'Efficiency' was the fundamental value throughout the discussion of food self-sufficiency; officials were persuaded by economists that efficiency could be mathematically demonstrated. This is why advocates of immediate and maximum increases in food production were unhappy when other choices were made along distributive lines. Their unhappiness was expressed as opposition to the kinds of distortion which resulted from government policies, distortion of what would otherwise be a competitive situation in which the true efficiencies of new agricultural technologies would be revealed if only these efficiencies were left alone without interference. In their opinion, pursuing equity introduced inefficiency.

All these distributive questions were not formally seen by planners as rooted in political and cultural reality. But in fact equity concerns were as deep in Bangladesh society as efficiency concerns were in planning, and required continuous attention (or masking). There had to be an appearance of attention to these concerns. There was intense struggle within the Planning Commission over these questions. The other side of the Bangladesh economic planners' unhappiness with equity questions is that they felt publicly and politically vulnerable if they were seen to subscribe to the doctrinal correctness of inequity and injustice simply because it might lead to economic growth.[7] Some planners avoided this vulnerability by claiming that the efficiencies of the market would act to rapidly diffuse benefits from increased production towards those who were not endowed with advantages enjoyed by more responsive producers.

Given these government priorities and conflicting values, how was food self-sufficiency to be achieved? The strategy throughout the 1970s had three components: (1) immediate increases in gross aggregate production brought about by rapid adoption of HYV seeds and associated technical changes, (2) a secular decline in the consumption of rice by changing the national diet to wheat, and (3) continuous imports of cheap food (mostly wheat) and procurement of rice and wheat within the country—all to be sold relatively cheaply to urban populations through the public grain system. These populations were the civil service, the military and police, and the industrial workers whose goodwill the government was buying. It is ironic how little security governments actually obtained by this kind of use of these massive public funds and how little stability the country enjoyed as a result.

And what were the results? There was a per capita decline in the consumption of rice between 1962 and 1976, according to the Food and Nutrition Study of the Dhaka University Institute of Nutrition and Food Science, published in January 1977. The intake of rice and pulses (legumes) declined in the Institute's sample households in rural areas; a 2.4 per cent decline for rice and 15 per cent decline for pulses. Wheat intake in the same households increased by 61 per cent, that is up from 18 grams daily in 1962 to 29 grams in 1976. Wheat was cheaper in the market throughout this entire period. By 1980 it was commonplace to hear that wheat was eaten once a day. Although some men who did continuous heavy labour, on the docks, for example, told us they believed that only wheat would give them the strength required to perform their tasks, wheat was not a preferred food for either rich or poor people. To this day rice remains at the conservative dietary core of Bengali cultural values. For rural Bangladesh people the world is divided into rich and poor countries, Muslim and non-Muslim countries, distant and near countries, and rice- and wheat-eating countries. The dietary distinction is thus considered as fundamental as power.

The increase in the consumption of wheat and decrease in the rice intake was congruent with one of the three components of the strategy for food self-sufficiency throughout the 1970s. Shortly after the foundation of the country, the Minister of Finance, Tajuddin Ahmed, told a Commonwealth Finance Ministers Conference in London that 'now is the time for us to get used to eating wheat, and give up our dependence on rice, since rice is scarce'.[8] Again, in July 1977 while exhorting the people to greater production of food grains, President Zia said, 'People must strive to change the dietary habit. Rice is not the only food.... We must reduce pressure on rice which is short.'[9] To encourage this change the English and Bengali newspapers carried such front page slogans as 'Wheat is More Nutritious than Rice' and 'Wheat Contains More Calories and Proteins than Rice'.

Because of the increased production of wheat, but more because of the great volumes of wheat imports which were sold cheaply in the city to the government's main clients, the cheaper wheat was moving into the countryside as flour and the more expensive rice was moving into the city where there was greater effective demand. More wheat than rice was available as food aid from Canada, the US, Australia, and the EEC. Only the US tied the receipt of wheat to the receipt of rice and vice versa. And if grain is to be purchased the world price difference between wheat and rice usually ensures that more wheat can be bought for the same amount of scarce foreign exchange.

But the change was not simply from rice to wheat consumption. The 1977 Food and Nutrition Study noted that in 1962 about 10 per cent of households obtained 80 per cent or less of the then caloric requirement of 2,150 calories per person per day, but by 1976 the number of deficit-households had risen to 30 per cent of the total sample. The average caloric intake decreased by 7 per cent, and the average protein intake decreased at the same rate. Despite shifts from rice to wheat in some households, there was an increasing number of households where consumption of either cereal was simply inadequate, not to speak of protein deficiencies.

A Case Study of Food Self-Sufficiency Calculation, 1972–1973

When EPARRI was granted its charter for autonomy in 1970, the next two years could have been extremely significant for researchers, if they could have used the new resources to tackle the interesting problems which had emerged during the previous few years of the Green Revolution. But 1971 and 1972 set rice research back a long way: there was chaos in the government administration, and the only research conducted at Joydevpur was routine maintenance. After the cyclone and election of late 1970 everyone was watching for the confrontation between the country's east and west wings. In March 1971 the military staged a violent crackdown upon individuals and institutions in East Pakistan whom they expected to aid the nationalist movement. Guerrilla forces of the movement began to confront the military and police in order to retaliate for the brutality of the crackdown. Civil servants had to choose whether to appear for work, risk being branded as collaborators, or hide and perhaps be dismissed. Researchers working for the government had to select what to do, when, and where. Certain districts became known as liberated areas, under non-government control and management. Nine months later, following a swift Indian invasion, Pakistan surrendered the entire territory and Bangladesh was declared a sovereign nation. At the beginning of 1972 it faced many new problems, but the familiar one of rice self-sufficiency was still there.

Complete uncertainty reigned in agriculture throughout 1971 and 1972 as a result of this sequence of events. Amost all senior government officials were changed more than once. What had been routine became difficult to achieve at all. The food supply picture was complicated and confused, reports of rice cultivation and harvest were quite unreliable, and massive amounts of food aid moved through the country in highly politicized channels. Food and the input requirements of HYV rice were both used as weapons by the old government in 1971 and the new government in 1972.

An order from the nationalist government in exile of Mujibur Rahman in India prevented most guerrillas from attacking the rice research operations. But some guerrillas did not know of the order and thus threatened people who continued with business as usual. In July 1971 EPARRI's director was transferred elsewhere and not replaced. A senior rice production specialist and field worker were killed, and one senior soil scientist left the country as a refugee.[10] The military also was suspicious of people who moved around the countryside so much in the course of their work. During this year the IRRI liaison scientist Rufus Walker moved in and out of the country because of the fluctuating risk of remaining on the job (he was an ex-USAID official with experience in Latin America and Vietnam). Walker was predicting 5.65 million hectares soon to be sown to HYV rice, if fertilizer, seed and pesticide were available, but the ADC's seed multiplication scheme was weak and USAID, the usual source of chemicals, was preoccupied with food aid and politics. Few of the 1,800 tonnes imported the previous year were actually replanted for seed.

The Ford Foundation was thus in a curious position. Fully identified with the American presence and thus with the government of Yahya

Khan, the foundation was nevertheless nurturing projects like EPARRI and the Institute of Development Economics which were moved hastily to East Pakistan in 1970. Neither foundation officials nor expert researchers were unquestionably loyal to the government and after the March 1971 crackdown were caught in the bind of trying to protect and encourage their projects without merely reinforcing the government.[11] The foundation's representative thus described the precarious work in rice research, three months before the dissolution of the country:

One cannot forget certain primitive truths: political problems will probably disrupt the life of East Bengal for the next decade. Internal [food] production will still have to be the chief weapon against famine. Incidentally, were the times normal, the progress in identifying an increasingly complex spectrum of varieties would argue strongly for redoubled efforts to strengthen EPRRI in production agronomy and production economics... Our strategy in East Pakistan for the next few months should be quiet concentration on rice research and production at EPRRI, retaining unobtrusive ties to the Secretary of Agriculture, who for as long as he remains, will provide the administrative leadership and protection the project requires.... I would like, if you concur, to keep official calls to a minimum.... It is a good time to avoid symbolism whose use we cannot control.[12]

When in 1972 the new Bangladesh government invited IRRI and the foundation to begin work under its auspices, the rice research institute, renamed the Bangladesh Rice Research Institute (BRRI), was going through severe difficulties. When the new Minister of Agriculture and Secretary of Agriculture visited the institute, they were surrounded and gheraoed (trapped) by workers and low-paid staff, agitating threateningly for pay increases. The office of the acting director was therefore removed to Dhaka, thus 'leaving the scientists to proceed with their field work as best they could without readily accessible supervision or support'.[13] A director was appointed and then transferred in three months. Several key positions became and remained vacant; broken or inoperative equipment was not repaired or replaced, although the budget allowed for that; the construction project remained at a standstill during 1972; new equipment remained trapped in the customs bureaucracy in the port, duty unpaid; and foreigners assigned to the institute experienced extraordinary delays in their clearance. There was no effective security for equipment or rice. In most of this, the institute experienced the same indifferent treatment from the government as other organizations were getting: the difference was that senior officials, including the Director of IRRI and the most senior officers of the Ford Foundation, came and held intensive discussions at the institute during 1972 in order to resolve the situation. Reports by these people were circulated to their head offices and led to establishing certain conditions intended to solve the problems listed above which had to be met before further support was approved.[14]

Robert Chandler learned all of this on his official visits in 1972 and 1973; although a record of his reaction is unavailable we do know that he had to face the fact that after years of effort, local rice varieties still predominated in the boro crop of early 1973 and were more popular and more widely available than the HYVs from IRRI. When Chandler advised the government on the route to self-sufficiency in 1972 and 1973 and weighed the multiplication of seed against the scarcity of fertilizer, etc., he had to acknowledge the weakness of rice research in his calculations. It was from rice researchers in Dhaka that the foundation's representative Robert Edwards got the idea of developing an 'increasingly complex spectrum of varieties', but not for two more years was the organization of research capable of addressing it. Meanwhile the government was under strong internal pressure to declare a programme for sufficiency and was making major decisions without consulting the rice researchers. Two years later, in 1975, the rice research institute's director admitted, with some frustration, that 'there are people who want self-sufficiency in one year'. Such influential people were in turn frustrated with the slow progress toward realizing sufficiency.

But how was self-sufficiency being measured? We turn now to examine the first attempt to measure self-sufficiency in Bangladesh. The first calculation of the gap between food production and self-sufficiency in Bangladesh was carried out in 1972 by Robert Chandler, former director of IRRI, heading a special UN Mission to estimate yield and production of 1972–3 rice crops. This mission, conceived in the immediate post-war food crisis, was used to impress on the government of Bangladesh that the only strategy towards self-sufficiency was the HYV path. It was also used to determine levels of imported food. The food crisis came to a head during August and September 1972 while the Prime Minister was abroad for a medical operation. The annual rise in rice prices was at its peak at the beginning of August but abnormally showed no sign of levelling by the third week of August even though the aus crop had been harvested. This food problem had been correctly forecast by the government and the United Nations Relief Operation Dhaka (UNROD). For example, in the first week of August the oil tanker *Manhattan* arrived and anchored eight hours' sailing time off Chittagong, carrying 67,000 tons of wheat for UNROD. The head of UNROD in New York, Sir Robert Jackson, cabled the UN representative in Dhaka that the '*Manhattan* is of critical importance and has already become a political issue in Washington. It is essential that she should be used successfully and we must all break our backs to ensure that this happens for political reasons as well as practical'.[15] It was the season of US elections; the voyage had been personally launched by US Secretary of Agriculture Butz. The *Manhattan*'s purpose was to act as a floating warehouse on to which other big vessels would unload their grain and to which lightering vessels would come from Chittagong eight hours away. But only an additional 42,200 tons was transhipped this way until 23 November, at which point the *Manhattan* withdrew. The operation was dangerous, and absurdly expensive, costing $19,000 daily, or $4 million overall, paid by USAID to UNROD on behalf of the government of Bangladesh.[16] It was a political stunt to mask the fact that the United States was among the slowest countries to give diplomatic recognition to Bangladesh.[17]

The arrival of large food shipments did not lower the abnormally high price of rice. On 21 August the Cabinet Subcommittee on Foods and Essentials announced it would review the price of foods on every alternate day, and the Minister of Food gave two public speeches urging everyone to 'keep food above politics' and to help bring down the price of rice. The following day, on 22 August, an Emergency Operation Room was opened (with telephone numbers to call) within the Prime Minister's Office 'to deal exclusively with food prices and the law and order situation in the whole country'.[18] On the same day the Revolutionary Student Union joined the Communist Party (Leninbad) to press for a united front against the food crisis, and five days later an all-party opposition Food Action Committee was formed by Maulana Bashani of NAP (National Awami Party) and seventeen other opposition parties and groups. People in Dhaka and the countryside were now nervous about Mujib's prolonged medical absence. The Cabinet met on Sunday 27 August, under the Acting Prime Minister, and decided to import 25,400 tonnes of rice immediately from India (an unpopular source of supply) and to purchase Tk 50 million worth of inexpensive cloth for immediate distribution to the poor. At the same time requests for additional special commitments from foreign countries, which could be announced immediately, were sought by the Cabinet, and UNROD was asked to pressure donors for additional financial assistance for food purchase and for additional shipments. The Secretary-General of the United Nations suggested at this time that the survey of the aman crop should be conducted by a mission under the leadership of an expert of international authority: the government accepted this proposal in principle during August (and formal agreement was only reached in October).[19]

Four days later, on 31 August, editorials appeared in all Bengali and English newspapers on the importance of IRRI rice varieties: such editorials usually followed government directives and policies very closely. One example (*Morning News* editorial 'IRRI Rice') argued that if the two fertilizer plants worked to capacity, if fertilizer imports were increased, and the pesticide factory worked to capacity, if more deep tubewells and low-lift pumps were imported, then 'Bangladesh can in a short time become a rice exporting country'. Considering the budget, 'with this huge amount' (Tk 1 billion), earmarked for agricultural development in the current financial year, 'and further amounts to come, the impediments on the road to large scale cultivation of IRRI rice are sure to be removed'.[20]

During September, when Mujib had returned from convalescing in Switzerland, the Cabinet apparently learned of discussion in UNROD of the desire to conclude the relief operations, including emergency food shipments, on 31 December 1972. During September the largest voluntary relief organization in Bangladesh the (Catholic) Christian Organization for Relief and Rehabilitation (CORR), with a budget of $400 million, announced that no new relief projects would be undertaken and that in January 1973 it would transfer its operation to a Bangladeshi-run development agency. It was widely known that the Prime Minister opposed any such withdrawals and asked for high-level meetings between himself and UNROD; the visit of senior UN official Sir Robert Jackson from New York in October was announced in September as a result of these requests. But it was still not a foregone conclusion that UNROD would survive until the end of 1972 as the Prime Minister wanted. Indeed, as early as June the Inspector of Foreign Assistance of the US State Department had recommended that UNROD should move quickly away from an operational role to an advisory role. Simultaneously, there was new evidence of UNROD's willingness to help Bangladesh: on 23 August when the USSR announced that it would not consider a contract to salvage and clear Chalna port (near Khulna) of sunken ships, UNROD immediately employed Captain Searle, a retired US Navy officer and expert on salvage, to prepare a feasibility study by September.[21] Food shipments were reaching the port of Chittagong, but the half of the country lying east of the Padma River could get food more easily from the port of Chalna. This fact was of vital strategic importance, given the government's fear of Indian military reintervention in Bangladesh (in spite of 'good relations' between Prime Ministers Gandhi and Rahman). This port salvaging was the sort of UN activity Mujib was at pains to praise before the Jackson visit to Dhaka on 14 October. From that meeting the government of Bangladesh requested that the Secretary-General allow UNROD to continue emergency food shipments until the end of March 1973, until the boro crop was ready for harvest. The Chandler mission was supposed to estimate reliably for the United Nations what the shortfall in production would really be and therefore what food would have to be imported, whether bought or donated.

The Chandler mission arrived a week after Jackson's October meeting with the Prime Minister in Dhaka and two weeks after Kurt Waldheim announced the mission's formation and purpose. It was to evaluate the 1972 aman rice crop, estimate the 1973 boro and aus rice crops, and calculate food grain import requirements for 1973. Chandler had by this time just retired from IRRI and was about to become Director of the Asian Vegetable Research and Development Center, Taiwan. He was accompanied by a Burmese agriculturalist from the World Bank, an FAO agronomist, and an agricultural economist from Washington. Three Bangladesh officials were attached to the Chandler mission; the Director of Agricultural Research and Training (who controlled BRRI and other research institutes), Chief of the Agricultural Projects Section of the Planning Commission, and Director of Procurement in the Ministry of Food and Civil Supplies. Chandler had been visiting Dhaka for seventeen years and was well known to senior officials and to some politicians. He held regular discussions during the mission with important people in relation to the calculation of self-sufficiency: the Deputy Chairman, Member No. 1, and Agricultural Consultant of the Planning Commission; the Secretaries of the Ministries of Agriculture and of Food and Civil Supplies, and the Director of the Bangladesh Agricultural Development Corporation (BADC).[22] He also visited the rice research institute.

The mission's strategy of evaluating the aman crop would be fourfold: to observe rice fields all over the country, travelling by helicopter, light plane, and car on about one hundred trips; to interview the senior district officers responsible, the Deputy Commisioner and the District Agricultural Officer; to record and weigh all the estimates made by farmers themselves in relation to their perception of a 'normal crop'; and to send out four drought damage survey teams (in October 1972 rainfall was 61 per cent below normal for the country as a whole). (See evaluation of production in the aman crop in Table 8.1.) The mission assumed there would be average aman yields of 1,040 kg/ha compared to 1,060 kg/ha in 1969–70.

The two main causes of the aman production decline were ranked by the mission as (1) the severe drought which resulted in reduced area planted to rice (10 per cent less in broadcast seeding, 15 per cent less in transplanted rice), and (2) reduced fertilizer distribution which was also

TABLE 8.1. *Estimates of 1972–1973 aman rice crop, Bangladesh* (million tonnes)

1969–70 Official record[a]	6.95
1972–3 Official record	5.59
1972–3 Mission estimate	5.57
1972–3 Gov't estimate	5.27

Note: [a] Most recent crop with reliable reporting prior to 1970–1 conflict.

Sources: UNROD, *Report of the Special Agricultural Mission*, Dhaka, 1972; and Ministry of Agriculture, Government of Bangladesh, 1973.

carried out too slowly because of the disruption by war of transportation networks in the country. In particular 1.7 million acres planted to IR 20 was spoiled by the fertilizer shortage. Only in the Chittagong area, near the port, was there normal distribution of nitrogen and phosphorus fertilizers. Chandler wrote in the report:

It has been shown time and time again that when fertilizer resources are scarce, and when alluvial soils, as in Bangladesh, are involved, the most important deficient mineral is nitrogen, the next is phosphorus and the least is potassium.... However, the Mission was told that approximately half of the transplanted rice areas responded to phosphorus. The Mission believes, however, that it would be worthwhile to obtain more data on this point by widespread field trials.[23]

The mission's estimates differed by only 300,000 tons from the Ministry of Agriculture's aman calculation, 'within the expected margin of error'; most of the difference lay in the Rajshashi Division which the mission did not think had been as hard hit by drought as did the teams from the ministry.

Calculation of both the boro and aus rice crops were as follows: 1973 boro, mission estimate—2.1 million tons, official record—2.07 million; 1973 aus, mission estimate—2.50 million tons, official record—2.27 million. The mission assumed that the HYV yields in the boro season would be 2,460 lb./acre of clean rice, but acknowledged there would be more local rice varieties planted than HYVs so boro yields would be lower. The predictions by BADC of the number of low-lift pumps which would be actually operating was questioned by experts, and so the mission lowered irrigated area estimates from 1.1 million acres to 1.0 million acres in the boro season. The supply of fertilizer on an adequate and timely basis was also questioned in districts far from ports. In making its calculations of food import requirements, the mission reminded the government of Bangladesh that there was a chronic problem that had to be 'solved', that is, that the yield per acre of both local and HYV varieties was far too low: this had been IRRI's and Chandler's message for years. But the mission also said that the government's goal of raising average yields for HYV varieties from 2,700 lb./acre to 3,300 lb./acre, and for local varieties from 1,230 lb./acre to 1,480 lb./acre was 'a rather ambitious program', particularly if people wished immediate results.[24] Nevertheless Chandler and IRRI had been forecasting immediate results for many years, but this is how they were giving contradictory messages.

The Chandler mission was used by UNROD and other foreign donors as an independent check on the estimates of the Ministries of Food and Civil Supplies and of Agriculture because government estimates were not trusted in Dhaka, Geneva, New York, or Washington, according to most sources. How the mission calculated the need for food-grain imports in 1973 of 2.564 million tonnes is shown in Table 8.1. Actual imports in 1973 totalled 2.78 million tonnes, the greatest amount imported into the country in history. While the difference between calculations of the mission and of the government was small, its importance lies in the effect it had on the next plan for self-sufficiency. The ministries had learned how an international group did its calculations and how much the country had to depend on the judgement of foreign experts. Chandler returned in March 1973 (by agreement) to evaluate the boro crop, but that report is not available.[25] The 2.56 million tonne shortfall in 1973 was confirmed in Chandler's second visit, though it was a lower deficit than what was ultimately reported.

The Ministry of Agriculture was already calculating a 'scheme to attain food sufficiency' before Chandler returned a second time. In the scheme, released in April 1973, imports were to decline to an insignificant amount in 1978, when self-sufficiency would be declared.[26] Big increases in production of fertilizer and pesticides and in irrigated area were planned during the same period, the period of the first five-year plan (1973–8). Most important, official sources said that while there was little scope for extension of area under food grains, 'the scope for intensive cultivation to raise per acre yield by modernizing agriculture is very vast'. Yield 'could be increased by 10–15 times if farmers could be encouraged to adopt improved and modern methods of cultivation through cooperative farming which offered the best opportunities'.[27] The scheme optimistically planned reduced imports of food, and the projections are contrasted, in Table 8.2, with the actual record of imports of food grains.

In the light of this optimistic schedule, the Minister of Agriculture and

TABLE 8.2. *Food sufficiency plan and actual grain imports, Bangladesh* (million tonnes)

Year	Planned imports	Official record	Imports as % of production
1973	1.63	2.78	31.0
1974	0.72	1.64	16.0
1975	0.33	2.26	25.0
1976	0.13	1.47	13.0
1977	0.02	0.78	7.4
1978	0.01	1.78	14.0

Sources: Various reports, Ministry of Agriculture, Ministry of Food, government of Bangladesh, 1973–80.

the Cabinet reconfirmed the strategy for self-sufficiency during 1973, then, as later, placing the HYV seeds and associated technologies at the centre of the drama. Working from these initial calculations, the government set out to import at least 2.5 million tonnes in 1973. In June, after UNROD had been closed, Sir Robert Jackson visited Dhaka again and announced that the UN predicted that only 2.3 million tonnes would have to be imported.[28] It was during this period that the extraordinary rise in oil prices and in food-grain prices occurred; Bangladesh had to enter the world food market precisely when food and oil were getting more costly. From the beginning in 1972, two-thirds of the foreign exchange expenditure on imports has been for petroleum oil, food, and fertilizer. Mujib's administration was thus continually in trouble over the price of rice and other essentials; the unsuccessful 'Second Revolution' was intended to control or regulate these prices. But scarce foreign exchange had also to be used to buy the requirements of the HYV strategy, so the strategy's planners had to compete for this money with those demanding other imports.

The dialogue on self-sufficiency had tended to focus on how much food should be imported because that is what interested major international development agencies and the rest of the government. The advocates of the HYV strategy always extrapolated upward to what was considered ideally possible, seen largely as a mathematical optimum. But then the strategy for food self-sufficiency was also intended to reduce dependence on foreign food sources and supplies and to build stronger links from the central to local levels of society. It is therefore natural that there would be an enthusiastic and continuous advocacy of 'the possible' in this profoundly political strategy. It was clear from the beginning in 1972 that 'sufficiency' was a calculation which simply divided the population's per capita need into the total grain available for consumption. The high levels of malnutrition in many districts (studied by UNROD survey teams) in 1972 and 1973 were not included in the calculations; that is, actual grain intake was not included in the official definition of self-sufficiency or the formula for its calculations. In fact per capita availability had been constantly declining. Mujibur Rahman hoped that food relief delivered to recipients by his party's rural workers would mask some consumption inequalities and sustain some political linkages between hungry and well-fed households. The very same linkages between poor and rich were necessary to increase production through the HYV strategy as it was conceived. With all these contradictions, one begins to see why the HYV strategy absorbed so much economic and political energy but delivered so little, by 1980, in agricultural terms. This was partly because the flow of imported food ultimately guaranteed to the government that the inadequacies of the strategy could be covered over. Nowhere in these calculations did the role of rice researchers appear. They were relegated to the distance and to the sidelines.

8.2 Changing the Forces of Rice Production: More Seeds and Their Requirements

In the countryside there had been, by 1980, a major push to irrigate rice fields artificially, to plant HYV seeds and fertilize them heavily, to protect fields with chemical pesticides, to carry out some activities mechanically, to reduce crop losses at or near harvest, and to improve milling and storing techniques. While all of this activity over twenty years had effects on the cultivation of other rice varieties, it was aimed at spreading the adoption of HYVs more and more widely. In this section we will examine the progress of each specific technical component of the HYV strategy, omitting one crucial factor—credit. In effect this entire pattern of promoting the HYV strategy by importing the HYV requirements favoured the mercantile structure of the business class. These wholesale agents, procurement specialists, import–exporters, were part of a whole mentality of doing business by trading but not by manufacturing or producing anything. This activity thrived, inside and outside the government, along with the age-old patterns of overinvoicing and underinvoicing which were the source of extra income for bureaucrats and private contractors alike.

New Seeds

Because rice cultivation in Bengal is ancient, and rice varieties have become so differentiated by season, region, and quality, it is reasonable that cultivators would exercise care in choice of seeds, and would be interested in the merit of new seeds but sensitive to their limitations. In any village or rural tea shop one can hear lengthy and learned debates on seed types and their ecological niches. It is in fact a mark of wisdom to be known as an expert on seeds, and there is an accessible 'public' memory and rural communication system about the performance of a seed variety under optimum or stress conditions. Observations among settled cultivators, corroborated by other researchers, show that cultivator-assessment of rice varieties is systematic over the seasons. Discerning about the purity and potential of their seeds, cultivators recognize an inherent inequality of endowments not just in land and water availability or in soil type, but also in the quality of seed with which they commence the season. This entire body of knowledge was disregarded in the promotion of the HYV strategy.

Seed distribution, planting, and multiplication were perceived, from the beginning of the HYV strategy, to be the best indications of success or failure of the programme. In the crucial period, 1967–77, imports of HYV rice seed (shown in Table 8.3) was over 17,000 tonnes. When compared with 214.3 tonnes imported to Sri Lanka and 107.7 tonnes to India for the same period, these imports show a weaker domestic system of seed multiplication than in those countries. It also indicates the scale of public resources allocated to these imports, some of which were rush shipments by air.

The government's distribution of rice seed was not from imported sources alone but also from BADC seed multiplication farms which produced both for free allocation and for sale. HYV rice varieties constituted the major portion of seed multiplied and distributed and sold through the government system, though in a few districts (Mymensingh, Rangpur, Comilla, Khulna, and Patuakhali) 'local aman' varieties were sold by the government. This system was also used to replace seed stocks lost in districts affected by flood and cyclone; thus new seeds became part of the widespread government relief system. We have made district-by-district calculations of the allocation of all rice seed between 1966 and 1976, including both HYV and local varieties; districts with higher government procurement of rice in the north-west (Dinajpur and Rangpur) received favourable seed allocation. Faridpur District received consistently low allocations of seed in the 1970s (and actually received more wheat seed than rice seed in the period); it ranks last among districts with an overall allocation of less than 1 per cent in the whole period (1966–76). Barisal District, which ranked number two in receipt of seed for the whole period, received virtually no HYV rice seed during the 1960s and a great deal between 1972 and 1975. The records show an erratic decline and rise of seed allocation for certain

TABLE 8.3. *Import of HYV rice seed to Bangladesh* (tonnes)

Crop Year	Amount	Source	Variety
1966–7	10	Philippines	IR 8
1967–8	1,500	Philippines	IR 8
1969–70	4	Philippines	IR 8
1970–1	1,800	Philippines	IR 20
1971–2	701	India	IR 20 and Jaya
1972–3	7,000	Philippines	IR 8
1973–4	5,200	Philippines	IR 20
1975–6	1,100	India	IR 20
TOTAL	17,315		

Source: Calculated from Dalrymple, *Development and Spread of High Yielding Varieties*, p. 72.

districts and a dramatic fluctuation of the total seed distributed to all districts, for example a 100 per cent drop in national distribution between 1974 and 1975. Farmers could hardly be expected to rely on this system of distributing HYV seeds: a few developed private sources and others stopped planting HYVs. This is not to say that the public seed distribution system was irrelevant in the spread of the HYVs or the increase of rice production. But it helps to explain why HYV cultivation spread so slowly in the 1970s and in some places declined. A 100 per cent drop was certainly not absorbed by unofficial seed distribution channels. Although district-wise receipt of seed does not indicate planting of HYV rice in a straightforward way (owing to the existence of other sources), it confirms how unreliable the official subsidized system was from the farmer's point of view.

To rationalize the HYV seed system, the National Economic Council approved in May 1976 a scheme to develop procurement, processing, and distribution of seeds between 1976 and 1979. The seeds were to be certified by BRRI, and the bulk of the reproduction would occur on seventy seed farms managed by BADC. The World Bank announced that it would finance the cost of production of the 1977–8 target of 4,369 tonnes of HYV seed on the twenty seed farms which were then actually operated by BADC. If this was the optimistic target in 1977 of domestic government production of seed, one can see the great extent to which the government system had previously been dependent on seed imports. At the time of the World Bank agreement in June 1977, it was calculated that the country as a whole, if it was to be completely planted to HYV rice, required 750,000 tonnes of HYV seed. The domestic target of 4,369 tonnes on twenty seed farms was thus an insignificant fraction of the national requirement, but the government was presumably relying on a great multiplier effect in the private sector. Every year there were reliable reports of a shortage of HYV seeds, particularly in northern districts. When plantings of HYV rice declined, so did HYV rice seed availability in following seasons.

The gap between plans offered to international investors like the World Bank and the actual system of seed reproduction and HYV plantings was heightened by BADC's announcement in September 1977 that there had been a 60 per cent decline in the demand for HYV seed between 1974 and 1977, consistent with the revised area estimates presented by the HYV task forces.[29] Imports of HYV seeds were supposed to be discontinued in 1976, yet domestic production was too slow to replace them. There is also clear evidence at the farm level that the number of 'IRRI Blocks', or HYV pump groups using government pumps with fuel permits, had declined; accordingly, this suggests that the decline in the demand for government seed was not simply transferred to the private sector, but that there really was a reduction in planting of HYV rice, and the task force 'discovered' this. These declines were not uniform; indeed, a few thanas close to sources of the necessary inputs expanded their HYV cultivation up to 60 per cent of their acreage.

There was a strong national expectation that seeds of rice varieties developed by BRRI and IRRI grown within the country would come to replace imported seeds. BRRI varieties were popular and performed very well in international yield trials. By 1977 BRRI had released seven varieties, all of which were photoperiod-insensitive; by 1980 it had released eleven varieties, all using IRRI genetic materials (IR 5, IR 20, IR 22, IR 272, IR 506, etc.). There does not seem to have been any lag on the part of the cultivators to seek out new seed varieties even for trial purposes; the lag was really in the system of mass seed reproduction. For example BR 3 and BR 4, released with fanfare in the early 1970s, were increased slowly. Seed sales by BADC were: 1975–6, 125 tonnes (BR 3); 1976–7, 333.5 tonnes (BR 3) and 79 tonnes (BR 4); 1977–8, 105.7 tonnes (BR 4).[30] These figures show that new seed varieties from Bangladesh were of interest, but distributed on a small scale. The HYV strategy thus relied both on massive imports of IRRI variety seed and on state-financed and controlled seed multiplication—both alien to the indigenous seed system.

The cultivators' explanations for reduced HYV plantings and reduced HYV seed purchases, contrary to the expectations of the HYV strategy planners, was that HYV rice varieties had become susceptible to diseases and pests, required too much water in drought years (as in 1972, 1974, 1976, or 1979), entailed much higher production costs, and produced a coarse-quality rice which did not obtain as good a price in the market and which the government was sometimes reluctant to procure. Farmers who continued to plant HYV rice seem to have decided to plant local-preferred varieties for their household use and to sell the coarse HYV rice in the market where landless labourers and smaller farmers had to purchase up to 50 per cent of their needs. Some farmers and some officials also stressed that the HYV seeds purchased through BADC were not always of the 'best quality' and did not always germinate properly. This widespread evidence contrasted with reports of spectacular aman yields of IR 8 paddy per hectare cut in front of astonished government officials in 1978.[31] Such achievements, although reported every year, were exceptional for farmers, but point to the phenomenon of lateral linkage where select seed was passed through chains of farmers who did not rely on official seed unit multiplication. For example, some farmers without land squatted on government land beside the highways and cultivated HYV seed 'gardens', preparing rice seedlings for the market where they could be sold and transplanted to regular fields owned by others. There was also unofficial diffusion of the 'China-IRRI' and *indica–japonica* cross 'Pajam', both of which were tested by farmers along with other new varieties.

Given the complexity of local rice variety cultivation, in Bangladesh it may be early by 1980 to speak of decline in genetic diversity, though it was true in some areas of the country. BRRI's director warned in 1978 of 'loss of genetic diversity'; when BRRI announced its intention to produce a steady stream of new varieties, four each year for the next thirty years, the maintenance of diversity was on their minds.[32] In a few thanas 60 per cent of the rice cultivated was HYV rice by 1980, but in the cultivating system as a whole there remained a residual diversity.

BRRI's problem with seeds lay mainly in its lack of control, which confirmed its marginal role. Imports and multiplication were carried out by BADC, which had not, in BRRI's view, sufficient scientific expertise to fulfil all the expectations surrounding the HYV seeds. It will be recalled that the first (and all subsequent) calculations of the path towards self-sufficiency were predicated on very rapid seed multiplication. Such rapid development had still not occurred by 1980 either in IRRI-bred or BRRI-bred seeds. In fact the two most widely cultivated types of rice were Nigershail and Pajam, both improved varieties which were developed before the 1960s. Of course a great deal of private, unofficial HYV seed multiplication took place, but taken all together these accounted for less than 20 per cent of total national area. In 1980 expert and official government estimates still showed only 16 per cent of cultivated area under HYV rice.[33]

Irrigation

Water control and the extension of regular mechanical irrigation in Bangladesh was considered the key to increasing food production. The Trilateral Commission estimated in 1978 that rice production in South and South-East Asia could be doubled in fifteen years if irrigation investments were doubled.[34] The cultivation of HYV rice in the boro season was limited to irrigated areas because HYV varieties did not tolerate water shortages and required water at precisely timed intervals. HYV varieties were also grown on a limited area in the rainy aus and aman seasons along with rain-fed rice.

Complete control of water was an attitude which was largely unknown in rural Bangladesh and was considered 'foreign' by villagers. The rural people had a sophisticated understanding and use of surface and subsurface water, as well as phenomena like tidal action where they occur, and were engaged in a constant shaping of the earth to channel and store water on the local level; but they saw full control of the major river systems as an odd, if not arrogant, attitude to a powerful natural force which could remove whole villages and their land in a single night, destroying crops, drowning and killing people, cattle, etc. The differences of experience in each district meant that there was no uniform evaluation of water and its use, only that water was essential for agriculture. And there was often too little or too much water. For example, in 1976 floods affected crops on over 347,000 hectares of land in the districts of Sylhet, Comilla, and Chittagong alone, thereby qualifying them for government relief and delay of tax payments. In the same year drought affected crops in the districts of Faridpur, Kushtia, Pabna, and Rajshahi, also qualifying them for some government relief. This pattern of drought and flood was repeated over and over.

Dams, barrages, and embankments were the ancient technologies invoked for flood control, and many have been built within the past fifty years. These are usually not likely to protect residents of areas in which probability of flooding is over 66 per cent: in such areas people are not against flood control; they just do not believe the rivers can be controlled completely. Their culture helps them to adapt to natural forces, not to seek to control them. A flood is simply called 'unwanted water' in Bengali/Bangla.

Given the perceived importance of water, the conflicts over water use, and the long history of adaptation to the excesses of the water regime, what forms of irrigation technology and organization existed which could contribute to the HYV strategy for food self-sufficiency? There were two main types of irrigation in Bangladesh, one using large-scale fixed installations and the other using smaller and (sometimes) portable machines in specific seasons. The first type is, of course, very old and includes dams and canals, with gates which can be opened and closed according to need. Classic modern examples are the Curzon Canal built in Comilla at the turn of the century or the Khulna Canal which from the 1930s has joined the Padma River to Khulna through central Faridpur District. These systems usually had branch or feeder canals along which irrigated agriculture develops. Other fixed installations for rice cultivation include the ubiquitous large tank, very large in some places, like a small lake, and coastal embankments intended to limit the incursion of saline water. These installations were always undertaken with state funds and control and eventually came under the jurisdiction of the Department of Irrigation. Small tanks and ditches, however, were the work of households which had land available and which could afford the labour; they were evidence of wealth as well as its source. Repair and maintenance of these earthworks was a sign of continuing power and influence because they required massive gangs of labourers.

The second type of irrigation is distinctly modern, involving pumps and wells driven by machines, which may be fixed or portable. Earthwork in this case becomes a subsidiary matter. Labour is not so important as capital to purchase, run, and repair the machines. In this case the scale is usually smaller than in areas affected by canals or large tanks; though economists consider such irrigation a 'lumpy input' which must be shared, they agree that machine use need only be reached among a small and limited number of cultivators who nevertheless paid dearly for use of the technology. (It is this type of irrigation, by pumps and canals, which transformed rice cultivation in Texas, Louisiana, and Arkansas in the late nineteenth century.) In 1976 3.009 million acres were reported under irrigated rice cultivation, of which 1.05 million hectares were planted in the boro season alone. Another 60,750 hectares were irrigated for wheat, cotton, and sugar-cane, so that in 1976 the irrigated total amounted to about 1.27 million hectares of Bangladesh's total cultivable area of 8.71 million hectares, that is, 14 per cent. The BRRI estimate of total irrigated rice area four years later, in 1980, was 16 per cent of the total area cultivated; fully one-third of this irrigation was still achieved with traditional (unmechanized) irrigation technologies.[35]

There were striking differences in district irrigation capacity: Sylhet, Mymensingh, Comilla, Dhaka, and Chittagong ranked highest, and Patuakhali, Chittagong Hill Tracts, Jessore, and Kushtia ranked the lowest. Decisions to undertake large-scale irrigation and water control projects were taken at the highest level. The central agency involved was the Water Development Board (WDB, formerly WAPDA), an engineering agency which obtained agreement for projects from the Ministry of Water and Power, the Ministry of Finance, and the Planning Commission. The WDB was a large and relatively inflexible institution built up by foreign loans and grants and by foreign consulting agencies, including those which produced the master plan in 1964. Following the coups of 1975 the chairman of the WDB announced that 32 projects were being dropped from its list of 182 projects because of their 'political nature'; that is, they were both costly and unlikely to succeed owing to their design.[36] Many large-scale irrigation projects were not only 'political' and costly, but very slow: the Ganges–Kodabak project was begun in 1954 and in 1977 was still only 70 per cent complete. It was planned to have a 'command area' of 48,500 acres west and south of the main stream of the Padma–Ganges River, but by 1975 its command area had to be reduced to 21,000 hectares. The work advanced only as quickly as the supply of food aid which was the payment for most of the labourers who dug the earth in the project. The United Nation's World Food Program and the USAID Food for Work Program were combined in this project in order to complete it.

But the claims made for wide area projects are very great. For example, the World Bank loaned $46 million for a canal and tide action irrigation project in Barisal district in 1974–5. The project appraisal document calculated that rice production would increase 240 per cent in the four thanas within five years after the completion of the project in 1984–5.[37] Reports showed that, like the Barisal project, many large irrigation projects which had major investments were underutilized and improperly managed. Most observers believed that the scale of these projects, their design, and their subsequent cost was of greater interest to the WBD and the consultants, donors, or investors than the eventual cultivator-use of the irrigation, if any, provided. If true, such priorities help to explain why the HYV strategy had not achieved self-sufficiency by 1980.

Difficulties with large-scale irrigation projects, and their limited utility to the HYV strategy for food self-sufficiency, prompted interest in the smaller pump technologies which were available from the 1950s, and in the diesel-powered low-lift pump.[38] This pump, and the electric or diesel-powered deep tubewell became the main sources of irrigation for the boro HYV rice crop; both were considered 'lumpy inputs' by the government and needed to be shared in some way. Though the government prescription for their use was to create a co-operative, these technologies to irrigate HYV fields were easily captured by the few interest groups which obtained all the other HYV associated inputs. Thus they were ideal for the values and practices of local rural élites, in contrast to large-scale systems which outsider officials can more easily control. It was this capture which led to the regular complaints from small HYV rice cultivators that one or two people dominated the use of the irrigation system, that the manager of the system was corrupt, or that maintenance was poor and fuel was adulterated.

Supplying low-lift irrigation power pumps was very complicated. There were annual shortages of pumps, spare parts, and adequate diesel or petrol fuel; each year was a 'crisis year', even in years in which the actual planting of HYV rice was reduced. Almost all pumps were imported. In 1976, for example, parts had to be airlifted from London for the government's power pumps, of which 29,000 were already in the field and 11,000 were awaiting repair. The BADC, which controlled power pumps, also needed 150 million litres of oil for these pumps, 100 million litres of diesel, and 50 million litres of lubricating oil—all of which had to be imported. BADC had 100 trucks and 11 oil barges for movement of this fuel, all imported. It needed fifty new oil trucks and complained that the oil companies did not pay BADC a commission for selling their products to the farmers. In 1975–6 BADC expected to field 40,000 low-lift power pumps, but fielded only 36,000. In the following year, 1976–7, only 28,000 power pumps were fielded against the planned 40,000 pumps. Though the demand seemed to exceed the supply, because subsidies were a strong incentive, the suppliers could not organize their logistical system to meet it with pumps already available. This information was one of the central explanations offered by rice cultivators as to why the adoption of HYV rice was not greater. It was impossible, they said universally, to rely on the BADC system of irrigation technology supply.

Deep tubewells, on the other hand, once installed did not have to be returned to the government and then fielded another year. But although private contractors and big farmers enjoyed them as fixed installations, tubewells in Bangladesh disappointed the original investors like the World Bank and donor governments. They often lay idle and were under-used or 'improperly' used against the terms of their installation, and they always irrigated smaller areas than intended. While deep tubewells have more cubic feet per second (cusec) capacity than the low-lift power pumps and consequently can irrigate larger areas of rice, they require a greater degree of co-operation in their use or a greater degree of coercion by the households which capture their use. The results of observation and enquiries in the late 1970s recall and confirm the results of enquiries into the use of tubewell irrigation in Comilla in the late 1960s, that is, that co-operatives formed in order to use these lumpy inputs were captured by the larger farmers. Co-operative subsidies and loans were used by these people for their own purposes, including money lending to other members of the same co-operative. These same people controlled or regulated the use of the water and ensured that they obtained disproportionate benefits from a resource which is publicly subsidized and provided. This capture seems easier when the technology is in the form of deep tubewells and low-lift power pumps than when it is a canal with sluice gates which is operated by a bureaucracy.

An example of the combination of large-scale irrigation planning using deep tubewells as the unit of technology is the Thakurgaon project. The chairman of the WDB announced in December 1975 that the Thakurgaon project, which was planned to have 318 deep wells, had actually irrigated 28,000 hectares in 1970; 2,585 hectares in 1974; and 8,484 hectares in 1975.[39] While 1974 was a year of severe drought, the drop between 1970 and 1975 shows serious organizational problems with the tubewells. Stepanek calculated that the ultimate cost of this project was $28 million for the 318 wells or $410 per acre for the proposed acreage (higher per acre cost if only the actual acreage irrigated is calculated). There were at most times about 100 wells idle in the project out of 300. The cost for all deep tubewell pumps combined was calculated at $75,000 per pump on each deep tubewell.[40] It should be remembered in all these projects that there were additional high and often unrecorded transaction costs involved in securing the international loans and contracts. Clearly only the state and its wealthiest clients could afford the up-front cost in these projects.

In another large irrigation project, the World Bank and the Water

Development Board planned to irrigate 57,000 out of 59,400 cultivable hectares in the Barisal Irrigation Project. Finally 8,000 hectares were irrigated at the project's completion, achieving only 20 per cent of the projected food production increase. This scheme was planned entirely without significant contribution from BRRI's substation in Barisal and was begun before BRRI's substation was rehabilitated after years of neglect. By 1986 the Bank conceded that the project was not nearly delivering the expected rate of return, despite the earlier expectations and announcements of success.[41]

Difficulties with these technologies, with their continued importation, costly fuel consumption, and political unpopularity among small farmer classes, in addition to the reputation of small farmers for efficiency in using other modern inputs, prompted interest in hand-pump tubewells, bamboo tubewells, pedal-operated water lifts, and other small-scale technologies. Hand-operated tubewells were well known in most parts of the country as sources of drinking water. Efforts to design and fabricate in Bangladesh a pump from indigenous materials or of more durable design, needing fewer repairs, did not overcome the role of the cast iron and steel pump, parts of which were imported, including the raw material required if it was made in a foundry in Bangladesh. In 1977 USAID announced a $14 million loan for 240,000 hand pumps for irrigation purposes which would be channelled through the IRDP and the Krishi (Farmers) Bank. In the same year BRRI announced a prototype of an efficient double diaphragm hand pump which would lift water up to 4.5 m, and could be manufactured and repaired by village blacksmiths. These pumps would permit small farmers to irrigate HYV rice fields and other crops, without the problem of continuous interaction with large numbers of their neighbours.

It is possible that new forms of co-operation and productivity might have arisen from these small-scale technologies. Such possibilities encouraged Bangladeshi designers to make a half cusec capacity pump (diesel or petrol fuelled); the Bangladesh Machine Tools Factory began to produce 5,000 of them in 1978. It seems that only by 1980 did the problems of scale and underutilization of irrigation investments reach sufficient proportions to cause the government, which ultimately controlled water use and irrigation technology, to search for a small mechanized irrigation system. Even World Bank and government plans made in 1974-5 for the Barisal irrigation project in four thanas included 100 huge pumps of 25 cusec capacity, and 2,500 pumps of 2.5 cusec capacity. The movement to the smaller-scale and hand-operated technologies must also be interpreted in a more critical light; the attempt to share the use of large-scale technologies was carried out without any philosophy of co-operation, without government discipline and purpose, and without protection for the small farmer members of the groups. These co-operatives were imposed, formed around technologies suited only to large-scale organization. Only bigger farmers, backed by strong households with political networks, could make these irrigation systems work within the existing competitive agrarian conditions.

BRRI had little influence on this key variable in the HYV strategy. Its engineering department was relatively weak and not in a position to make a comprehensive study of the problems in the field, even within the project area (or development block, as it was later called). Again the BADC, part of the same Ministry of Agriculture, was in command. Again the pressure for immediate results led to imports before local development. BRRI scientists pointed repeatedly to irrigation as a major obstacle in the HYV strategy, saying the issue was so complex that no one agency or institution could resolve the bottle-necks in irrigation. Most previous attempts had failed to address adequately the actual practices and values of cultivators in regard to water, and so the vast sums poured into irrigation had provided little return on investment. Even major supporters, for example USAID, of government schemes were dissatisfied by 1980. In view of the fact that only 16 per cent of cultivated area was irrigated, a 1982 USAID report lamented that 'the government's twenty-year "crash program" has not produced a well functioning system of community canals from which farmers obtain water with Low Lift Pumps, and the Low Lift Pump and Deep Tube Well program itself is grossly inefficient with the average acreage irrigated per pump or well at less than half of the equipment's potential'.[42]

Fertilizer

Improving soil quality and structure and increasing soil fertility and the nutrients available to rice plants have been traditional goals of cultivators in Bangladesh. While households have clearly varied in their treatment of their land, they have always mixed organic wastes with the soil. This old technique is limited by the ceiling on availability of organic wastes, and its effectiveness is thus limited in the HYV strategy for self-sufficiency. This strategy is based on using rice plants which respond to greatly increased fertilization, particularly nitrogen applications. Although much traditional material was used in fertilization of rice plants, except for experiments in 1950 with human sewage from Dhaka Municipality as fertilizer, human wastes have not been systematically studied in agriculture in Bangladesh. Random deposition made in fields or ponds can increase nutrients available to rice plants, but it is not in concentrated form and does as much to communicate disease as to increase soil fertility. Cattle dung has been used more as a cooking fuel, along with rice straw and jute stalks, than as a systematic fertilizer. Pigs have not been kept in rural areas, and goats and chickens provided most of the animal waste, along with about one-third of the cattle dung not used as fuel. An otherwise noxious weed, water hyacinth, is used in many districts as organic compost to deposit nutrients and improve the texture and structure of the soil, but it does not provide the range of chemicals and trace elements required by the HYV rices. Legumes planted in the dry season have been used to fix nitrogen in the soil, but planting of these crops has decreased considerably in the past twenty years: on some land they have been displaced by HYV rice in the boro season.

Tyers estimated from work in 1974–5 that the greatest source of energy for Bangladesh agriculture was composted crop residues and that cattle dung and chemical fertilizers (and pesticides) were the least important source for the whole system. The proportions in terms of kilocalories were: crop residues, 90.3 kcal; cattle dung, 9.5 kcal; and chemical fertilizers, 4.1 kcal.[43]

Crop residues were thus contributing ten times more than cattle dung to cultivation. Cattle dung contributed double the amount supplied from industrial sources, yet it was the inelastic supply of such a traditional source of fertilizer in the cultivation of HYV rice which was inviting replacement by chemical fertilizers from industrial sources. The traditional fertilizers were not marketed or transported and could not be allocated by government agencies; these technologies were not organized into a form which could be marketed and integrated with the HYV strategy. Cultivators who did use fertilizer seldom applied it to either HYV or traditional rice varieties in the recommended amounts posted by the Ministry of Agriculture, for a number of reasons: its cost, difficulties in obtaining it, or a personal belief that the recommended amounts are posted to simply increase the use, purchase, and therefore production of fertilizers. Yet cultivators considering adopting the HYV rices knew the fertilizer demands of these plants, and all other farmers knew of the utility even of applying fertilizer to non-HYV varieties like the popular introductions Pajam and Nigershail. Thus the government and many cultivators perceived a gap in the fulfilment of the fertilizer need from the traditional sources such as dung, a gap which naturally was a great obstacle in the efforts to spread the HYV rices. This explains why one rice breeder at BRRI said in 1975 that 'we are totally infatuated with chemical fertilizers'.[44] The entire government was infatuated with chemical fertilizers.

Fertilizer was sold to farmers at a price fixed by the size of a subsidy. Fertilizer for the HYV rice planted in the boro season accounted for 80 to 90 per cent of all fertilizer used. This season was preferred because in other seasons rains could disperse the fertilizer too quickly, making it unavailable for plant uptake. Expanded planting of HYV rice to the aus and aman seasons increased fertilizer use outside the boro season, and the annual allocation or sale increased steadily.[45] (See Table 8.4.)

The district-wise allocation of fertilizer between 1966 and 1975 showed a great variation in receipt or sale of fertilizer, all of which was controlled by BADC and its agents. Chittagong District, presumably because of its proximity to the port, received almost 16 per cent of the total, while Faridpur, Patualkhali, and Chittagong Hill Tracts received less than 1.5 per cent. This pattern is consistent with allocation of other inputs necessary for the cultivation of HYV rice; allocation was greater in the half of the country on the eastern side of the Padma River, most of it served by trucks and daily freight trains (Chittagong, Noakhali, Comilla, Dhaka, Mymensingh, etc.) which originated at the port in Chittagong. Dhaka, Comilla, and Sylhet moreover were sites of the country's own fertilizer factories and the districts with the greatest area planted consistently to HYV rice.

Throughout the 1970s about 55 per cent of the fertilizer used in Bangladesh was produced in the country and 45 per cent obtained through imports. In 1974 Norman Borlang visited Bangladesh and urged the government to use its deposits of natural gas to produce anhydrous ammonia and urea.[46] These, he argued, would not only make Bangladesh self-sufficient, but put it in a position to be a major exporter within a few years. In 1975 it was announced that fertilizer self-sufficiency would be reached in 1979, but this objective proved elusive, even in urea, the main fertilizer produced in Bangladesh. Through the period two-thirds of the urea (which is 46 per cent nitrogen) sold in Bangladesh was procured in the international market because fertilizer obtained that way (on concessional loans) was comparatively cheaper than the cost and risk of new production or new plants. It is important in comparing the two sources of fertilizer, domestic and imported, to consider the problems of supply and distribution to cultivators separately from those of production.

TABLE 8.4. *Annual allocation/sale of chemical fertilizer, all sources ('000 tonnes)*

Year	Amount	Year	Amount
1966	106	1974	372
1967	160	1975	285
1968	211	1976	n.a.
1969	225	1977	521
1970	277	1978	731
1971	337	1979	745
1972	242	1980	855
1973	384		

Source: Ministry of Agriculture (Fertilizer Procurement), various reports.

Domestic Production of Fertilizer

The first fertilizer factory in the country was under constuction by 1962, as a result of the recommendations of the Food and Agriculture Commission in 1960. This was the urea plant at Fenchuganj in Sylhet District, where natural gas fields had been discovered. The original plant was built completely by Kobe Steel Company of Japan and was modernized and repaired by the same company in the 1970s under a $12 million commodity loan from the fourth yen credit to Bangladesh. Though it was originally designed to produce 100,000 tonnes of urea annually, production fell to 36,000 tonnes in the mid-1970s. After repairs were completed production increased again.

The second fertilizer plant, manufacturing triple super phosphate (TSP), was begun in 1970 at Gorashal on the border of Dhaka and Comilla districts. It began production in 1974. A third plant was under construction during the 1970s at Ashuganj, near Gorashal. It was being constructed for the Fenchuganj Fertilizer and Chemical Company by the international firm of Foster Wheeler. The plant had contracted with the Titas Gas Company of Bangladesh for the supply of natural gas. The World Bank, the Asian Development Bank, the governments of the UK, West Germany, and Iran loaned $142 million in 1975 for the construction; it was scheduled for completion in 1979 but actually began operations only in 1981. It was designed to produce 580,000 tonnes of urea and 300,000 tonnes of ammonia annually, though big factories and plants in Bangladesh seldom operate at more than 50 per cent capacity. In 1979 total domestic production was 55 per cent and in 1980 64 per cent of installed capacity.[47]

Each of these plants is an example of large-scale industry which proved difficult to operate and maintain in Bangladesh. They were built as turnkey projects by international firms. Thus Bangladeshi engineers did not learn to design or construct them, but only to operate them; much of the production process has become automatic. The plants occupied large areas of land which had to be fenced. Legal acquisition of sufficient land was a continuous problem. Large groups of non-local labourers had to be maintained in the rural area, far from amenities. These big plants required extensive electrical and rail connections, and water supplies. The entire supply of fertilizer originated at three sites and was vulnerable to transport delays and to site-specific natural hazards. In addition there was another vulnerability: in 1975 the Gorashal TSP factory was seriously damaged from sabotage by antigovernment groups. These plants have tied the agricultural economy of Bangladesh more closely to the international industrial system. The equipment was imported intact and assembled, parts and chemicals had to be bought on the international market (except natural gas), and major repairs or changes required foreign expertise. The cost cannot be borne by the country, and so loans for these plants contribute to the country's debt service charges. Nevertheless this integration with the international system was a key feature of the HYV strategy.

Fertilizer Imports

Fertilizer has been imported to Bangladesh since the early 1960s at least, and imports have accounted for almost half the amount used in the 1970s. At first these imports were in the form of grants and easy concessional loans, but there was a gradual shift toward commercial sales in the 1970s. All countries which export fertilizer seemed to begin their relationship with Bangladesh in terms of grants, hoping that purchases would eventually be made or that fertilizer factory contracts for design or construction would be signed.[48] The most regular supplier was the USA. In 1976 imports from the US were 73,000 tonnes of TSP.[49] In 1977–8 USAID renewed its line of credit arrangement to allow Bangladesh to import 50,000 tonnes of urea and 80,000 tonnes of TSP, worth $27.4 million. In 1978 USAID financed about 22 per cent of total imports, and in 1979 about 29 per cent of total imports. Other countries extended similar annual lines of credit: Canada for potash, the Federal Republic of Germany for urea, etc. One of the largest transactions occurred in 1977–8 when Saudi Arabia sold (on concessional terms) 214,000 tonnes of urea.[50]

In 1980 a bizarre phenomenon occurred: Bangladesh planned to export 152,000 tonnes of its domestic urea to its neighbours. 'US AID mission and other donors protested the exports which were apparently cancelled except for 40,000 tonnes contracted with Pakistan and Sri Lanka'.[51] Supplies had greatly exceeded market demand, largely because of a reduction of the fertilizer subsidy resulting from the urging of the World Bank and donors like USAID. Counting on imports, and the start-up of the Ashuganj factory, the government decided to export domestic urea despite a 'need' for fertilizer in the country. In this instance the demand for export earnings and antisubsidy ideology had triumphed over the official support for the HYV strategy. It was a sign that some people felt that the strategy was consuming too much money and producing poor results.

As in the case of other inputs for HYV rice cultivation, importation of fertilizers remained the monopoly of government agencies. The sole importing and distributing agent in the country was BADC; proceeds from the sale of the fertilizer were realized by BADC, and the difference between these proceeds and the international price was made up by a government subsidy. The extent of the subsidy virtually determined the extent of HYV rice plantings in the past. When the proposed power subsidy was announced, typically some politically important IRRI cultivators threatened government officials that they would not plant HYV rice unless the subsidy was increased.[52] Sometimes rice procurement prices were increased to adjust for fertilizer subsidy declines. There was normally a crisis in distribution of the fertilizer, whether imported or not, because most farmers wished to buy it at the fixed, subsidized price. Dealers and agents, on the other hand, preferred to show sales at subsidized prices according to regulations and to make their actual sales on the open market at higher black market rates. The crisis usually occurred at the districts more distant from Dhaka, the fertilizer plants, or Chittagong port.[53] Beyond problems of supply for these regulated sales, there were chronic complaints of poor storage, neglect, and spoilage of fertilizer bags, of favouritism in allocation within a thana, and even jurisdictional disputes. For example, in 1978, 1,300 tonnes (32,000 bags) of fertilizer were left lying for weeks in the rain on the river bank near Madaripur town in Faridpur District because the transporter, government official, and selling agent could not agree on who should move it and store it.[54] At Kishoreganj the railway officials and BADC officials quarrelled because a large fertilizer shipment had arrived short by 100 bags; neither would take responsibility to report and account for the loss.[55] Finally, there was regular smuggling of imported fertilizer to India.

Though these vertical supply and distribution problems were chronic, fertilizer in the end reached fields on which HYV rice was planted. The farmers who eventually purchased the fertilizer for use were those who had money and influence, and who could also obtain pumps and fuel, or pesticides, or better quality seed at the subsidized price. Most estimates are that fertilizer applications were actually about one-fifth the amount recommended. The people who actually applied fertilizer to the fields were usually tenants, daily labourers, or sharecroppers. There were, however, a few small farmers able to buy a little on the black market and apply fertilizer to their own small plots, thereby enhancing their yields (unless they could get enough organic wastes more cheaply). Cultivators were thus drawn into vertically integrated sources of supply, into deeper dependence on unpredictable government agencies or exploitive private traders, and into demands for credit with which to purchase these inputs for their HYV varieties of rice.

Although BRRI was consulted in certain aspects of fertilizer production and importing, only in the late 1970s did BRRI begin to examine critically the fertilized plant response and discovered that common fertilizers did not address some deficiencies in the soils of parts of Bangladesh. In the case of sulphur deficiency and zinc deficiency, detected in 1978, an estimated 3,232 million hectares (8 million acres) was affected by one or both these deficiencies. Corrective tests showed increases in yield up to 1,000 kg. of paddy per hectare: if these deficiencies were corrected, said the 1980 Review Mission, 'and the increases in yield estimated to be only a modest quarter ton per acre, an additional 2 million tons of paddy would result'.[56] But between 1960 and 1980, unfortunately, this question had not been studied in a truly comprehensive manner by BRRI or any other institutions. And for years IRRI scientists and other advisers like Robert Chandler had presumed that the correct fertilizers to use in Bangladesh had already been identified. BRRI's lack of independence is reflected in the fact that this presumption was largely accepted for so long.

Pesticides

Estimates of rice crop losses due to pests vary from a conservative 15 per cent of the total annual production of the country to an astonishing 30 per cent of the potential rice harvest estimated by the 1978 BRRI Review Mission. Farmers have all reported that the number of attacks on their rice crops have increased in the past ten years and that the range of pests has also increased, with many 'new' insects, in their opinion. Chemicals used against insect pests which reduce the yield or production of a specific plant (pesticides) are often highly toxic to other organisms and plants and may have a long life and accumulate in the food chain. In some cases they do not break down and remain chemically active for many years. Pesticides also affect non-target organisms; for example the parasites that keep an insect population in check may be killed by a chemical applied for other reasons. When its parasites decline, a non-target insect population may expand very rapidly and compete for other plants with man (cotton, sugar-cane, jute, etc.).

It is unclear when major crop pesticides were first applied in the country, but systematic importing began in 1962. There was evidence of the existence of three or four biotypes of the brown plant hopper. There were also attacks from stem borers, ufra nematodes, and greenleaf hoppers, which transmit the Tungro virus disease. In addition, birds and rodents have become aggressive competitors with man for the same food sources. While the pesticides were used on all the insects, no effective controls were available in the 1970s for bacterial diseases in Bangladesh. It will be remembered that the first doubt expressed among IRRI and Ford Foundation experts about quick self-sufficiency in 1969–70 concerned the complexity of the disease and pest regime in Bangladesh and its profound effects on IRRI varieties. Though it was widely believed in the Ministry of Agriculture that pesticides were the sole means of reducing pests, voices had been raised for other forms of pest management. Since at least 1975 BRRI questioned the economic utility and agronomic effects of pesticides.[57] The pesticide emphasis of the HYV strategy, however, was much stronger than these questioning voices. The Ministry of Agriculture was sensitive that it had presided over the importation of new crops and plants which had brought along with them new, unknown insects. Even before they were fully identified these pests were being attacked with chemicals as the only way to control them and get yields up.

The Directorate of Plant Protection in the Ministry of Agriculture established almost absolute control of importation and application of pesticides by 1974. There were at least two domestic sources of supply, CIBA-Geigy of Switzerland in Bangladesh and Insecticide Enterprises of Dhaka (parent company unknown). These companies formulated and labelled their own brands (for example Malathion for Cyanamid USA). Imports were bought under concessionary loans and grants from many countries and involved numerous chemical companies. One government report in 1976 estimated that 150 brands of pesticide were in use in Bangladesh, but the quality and performance of only a few brands were known by thana officials who advised farmers about its use.[58]

American pesticides were normally imported under a line of credit arrangement in a USAID loan, largely indented and administered by the Directorate of Plant Protection. USAID 'vigorously promoted pesticide use' in developing countries and exported about $500 million of chemicals since 1957, according to a meeting in June 1979 at the State Department.[59] About 30 per cent of US annual production has been exported (270 million kg out of 720 million kg), including all types of pesticide banned for use in the US by the Environmental Protection Agency at the end of the 1970s. The US Department of Commerce proposed a bill to remove any restrictions on the export of banned pesticide in order to improve the US balance of payments. The majority of these chemicals were used in countries like Bangladesh, where misuse and overuse were regular. For example, in 1976, 2,900 persons were poisoned by Malathion in rural Pakistan, and 1,000 buffaloes used in rice cultivation in Egypt died from poisoning by Leptophos pesticide.[60] There was little official analysis of this risk in Bangladesh in the 1970s owing to the dependence of the HYV strategy on pesticides.

The Directorate of Plant Protection had no research function and few personnel trained in relevant disciplines; it was more like a logistical bureaucracy concerned with price concessions, shipping contracts, storage, etc. Until BRRI completed a major survey of rice pests in the country there was little prolonged study of the effects of pesticide use in Bangladesh. Indeed, little is known on the general question of the accumulation of agricultural chemicals in the environment in Asia in the 1970s, according to a USAID Report to the US Congress.[61] Thus interviews and newspaper accounts have been the source of information presented here, in addition to the district reports of the directorate. The Directorate of Plant Protection was jealous of its autonomy within the Ministry of Agriculture (obtained in order to speed the importation and dispatch of pesticides), and officials were believed to be sensitive to any reduction in their privileges or in their unofficial incomes resulting from their (autonomous) importing privileges.

District-wide allocation records are available for 1973, 1974, and 1977. Total amount of pesticides allocated in these years was as follows: 7,000 tonnes in 1973; 4,980 tonnes in 1974; and 3,164 tonnes in 1977.[62] Some districts consistently received little pesticide; for example, Faridpur ranked 15th in 1972 and 18th in 1973 in area covered with pesticide; in amount it ranked 16th in 1973, 17th in 1974, and 13th in 1977. Districts whose pesticide allocation was near the mean were the surplus rice districts of Dinajpur, Rajshahi, Bogra, Pabna, Tangail, and Patuakhali. In area and amount of pesticide the top-ranking districts in 1972, 1973, 1974, and 1977 were Dhaka, Mymensingh, Chittagong, Comilla, and Barisal. Higher pesticide allocation was to districts in which HYV cultivation had consistently exceeded 20 per cent of total boro area, so that in 1977 the first-ranking district in allocation, Comilla, received 510 tonnes of pesticide. Importation was calculated in terms of the HYV area, according to observers and a well-informed letter to the editor in the *Bangladesh Times* on 1 April 1976 by A. S. M. Huq. According to Huq, procurement of pesticide was made by calculating the total IRRI rice crop area and the commercially recommended dosage of different pesticides: 'procurement of a huge quantity of copper oxychloride bears testimony to this'. There was apparently no emphasis on correct utilization, and so old and new products were 'pushed into the Bangladesh market'. The Trading Corporation of Bangladesh (disbanded after the 1975 coups) imported large amounts of pesticide which were later reported unused. About 4,000 tonnes of granular pesticide were formulated annually in Chittagong, at the CIBA-Geigy plant under government supervision, and only 'a small fraction of this huge stock has been used'.[63] The Ministry of Agriculture agreed that before the 100 per cent subsidy on pesticide cost was gradually removed (to reduce government expenditures) about 6,000 to 7,000 tonnes were 'sold' annually in Bangladesh. After the subsidy was reduced to 50 per cent, sale of pesticide dropped in 1975–6 to 3,000 tonnes, according to a ministry source.

In 1976 a newspaper enquiry revealed that there were about 11,000 tons of accumulated pesticide in Bangladesh stored in inadequate sheds or simply lying outside.[64] Stockpiled since 1962, much of it was declared unusable. Stockpiles had increased, admitted the Ministry of Agriculture, after the loss of the 100 per cent subsidy on pesticides and spray applicators because farmers no longer had an incentive to use them.

Field reports in March 1976 from Thana Agricultural Officers described large quantities of pesticide in their warehouses or stored under eaves of buildings. The agricultural offices did not have authority to dispose of pesticide stockpiles and were keeping them for purposes of audit and inspection in order to avoid malpractice charges. Actual control of care and distribution of pesticides was usually assigned to a low-status *moqaddam* (record-keeper) in the office of the Thana Agricultural Officer, a person not likely to be able to read complicated chemical labels in English. It is inferred that officers and private agents acting for the government held on to pesticide supplies, in fact may have stockpiled them, in anticipation of a dramatic price rise following removal of subsidies. Large amounts certainly leaked into the ground water. The dramatic reduction in pesticide use may have caught these people by surprise. It was estimated by foreign firms that safe destruction of these stockpiles would cost Tk 5,000 per tonne, or Tk 55 million for the whole stock.[65] In September 1976 an editorial disclosed that 700 tonnes had been destroyed at a cost of Tk 2 million (about Tk 2,900 per tonne), and that over 8,000 tonnes were still awaiting safe disposal.[66] Shortly afterward government pesticides were reported on sale in the black market, and the Ministry of Agriculture then announced that 2,235 tonnes of new pesticide were ready for use in the forthcoming boro season. A study by the Institute of Nuclear Agriculture confirmed the existence of these stockpiles in 1977–8.

The effect of these pesticides, whether stored or applied, was largely unstudied. Reliable reports from Sylhet, where pesticide was sprayed by CIBA planes on to rice crops in 1973–4, state that these flights resulted in widespread death of livestock (chickens, cattle, goats) and fish in ponds. There was no study of the cumulative effects in Bangladesh of these pesticides in the food chain through milk, fish, fruit, etc. A case of the effects of Malathion disposed in a deep well in the industrial area of Dhaka came to light in 1978; an expert committee composed of representatives of BADC and the Plant Protection Directorate had instructed a private pesticide firm (Insecticide Enterprises) to dump Malathion down the company well in 1975.[67] In 1976 the public water supply from a nearby deep tubewell developed an unpleasant odour, and public water supplies were stopped after 1976 because decomposed malathion had percolated into the underground water system.

Among rice farmers there was a willingness to use pesticides even when there was little or no information on its effects, roughly in the way they used strong medicines for illness in their households. Pesticide and medicine are called by the same name in Bengali. While concerned about prices and subsidies, farmers said the big problem was that pesticide applied in one field was useless because the pest moved to another field or regenerated in another field and reinfested the sprayed field when the chemical effect wore off, thus making an HYV rice field even more vulnerable to attack than fields with more resistant rice varieties. There was wide public understanding of the loss of HYV rice from pest attacks. Bigger farmers who adopted HYV rice could afford to rent sprayers and purchase pesticides if they wished to. A minority of farmers, usually those who had direct contact with the overuse or misuse of pesticide (for example in Sylhet), were reportedly cautious about the effect of pesticides in the ecological system on which they depended. Those who also drew cash income or their diet from milk, egg, fish, and other commodity production were more likely than rice farmers to question pesticide use. Nevertheless, actual reductions in the use of pesticide seemed to be related more to the decline of HYV rice plantings in the period 1974–5 and 1977–8 than to concern about pesticide use. Further reduction occurred in 1979 and 1980. Interest in biological control and other forms of pest management was clearly stimulated among officials when the farmers reduced their use of pesticides following removal of the 100 per cent subsidy.

Although BRRI created a Task Force on Pest Management in the late 1970s, it was still not studying pesticides comprehensively by 1980. While the phenomenon of pest resurgence was recognized at BRRI—for example that serious brown plant hopper outbreaks could be traced to the use of certain insecticides—the 1980 Review Mission urged a more comprehensive approach: BRRI should 'include all insecticides now available to farmers, as well as those proposed for distribution, so that they may be carefully and thoroughly evaluated for their resurgence capacity and that recommendations be made, on the basis of such tests, for withdrawal from the market or cancellation of license'.[68] It is not difficult to see that BRRI's recommendations on this subject would, if pressed towards withrawal or cancellation, run directly against the work of the Directorate of Plant Protection of the Ministry. Their autonomy would presumably be useful on such an occasion. The fact remains that this was the sort of systematic pesticide testing which should have been conducted by BRRI from 1962 and was not.

Mechanization

While one of the promises of the HYV strategy was always that more and more mechanization would inevitably occur, by 1980 Bangladesh remained quite resistant to these promises. In three fields—milling, storage, and cultivation machinery—changes in practice were very slight and BRRI's study of them was marginal. In the field of modern rice-milling, Harriss reviewed eighteen of the twenty-five project proposals made since 1966, including the performance of the only fully automatic rice mill built before 1978. This stream of project proposals (which showed no evidence of BRRI's own analysis) contained the reiteration of unexamined assertions about greater efficiency and reduced costs of high-volume mills manufactured in Japan, Taiwan, Italy, Great Britain, and India. The review was carried out in the context of a special FAO Mission in 1977–8 to promote the sale of modern rice mills in Bangladesh.[69] Composed of a former Arkansas rice miller, the engineer who became IRRI team leader in Sri Lanka (Moomaw), two Burmese officials of FAO, and a British economist, this mission's work included accompanying Bangladesh contract-signing officials to Malaysia and the Philippines where automatic rice mills were in operation. The problems of the existing Comilla automatic mill (then for sale with no buyers) were avoided by the mission, as was the profoundly negative impact which modern rice mills would have on the common method of milling rice, that is, by the foot-operated *dheki* used by women. Female employment, and the efficiency of the home-based method in the Bangladesh context, is the reason why such mills had not displaced the *dheki* by 1980.[70] The fact is that only 10 to 15 per cent of the rice crop was traded at any one time, and much was stored and sold (if at all) in very small amounts. Big mills were not economic in such a context, but they were being advocated and built, usually at government expense.

The storage of rice affects the amount of grain available for consumption. Rodents, insects, birds, and mould take their toll of rice improperly stored. Means of dealing with these problems are ancient, including big cane baskets, clay pots (to hold seed rice), and wooden boxes; and modern, including jute sacks, oil drums, and metal bins. Farmers had in the past used leaves from the neem tree to deter insects, but by 1975 some were prepared to use a dehydrating agent like calcium oxide. Others fumigated the paddy with Malathion dust.[71] Researchers have tested the sun-drying and artificial drying of rice in the monsoon, the problem of rodent-proofing storage spaces, and the value of parboiling rice as a preserving technique. Because the amounts stored by most farmers were small, research on large-scale storage problems would have only benefited large farmers, co-operatives, and the Food Department of the government—which stored thousands of tons of rice in the public grain-distribution system in its huge warehouses. Twice in the 1970s there was a storage crisis for rice: bumper crops in some areas coupled with transport bottle-necks left piles of rice sitting on the ground in the rain, and CIDA was requested to create many temporary storage structures. This could have been avoided with smaller storage systems. The general pattern, however, was of shortage of rice, not surplus. By 1980 BRRI had not systematically measured grain losses in the post-harvest system and had not taken any means to avoid these losses, despite their evident importance at every level.[72]

Finally there was a great expectation at IRRI that mechanization would transform Bangladesh rice cultivation, just as it was in the Philippines. At a BRRI board of governors meeting in 1973, IRRI's Director Brady said that 'contrary to common belief, ... farm mechanization would increase ... job opportunities instead of throwing people out of farming operation'.[73] Mentioning the tractor for quick ploughing and the mechanical drier run by burning paddy husks in the monsoon, Brady said mechanization would help increase production and prevent post-harvest losses. By the 1970s BRRI became involved in the prototype construction of irrigation pumps, driers, threshers, and a two-wheeled tractor. But most of its expertise was preoccupied with repairing and maintaining BRRI's own mechanical equipment and building up the BRRI workshop. A pump and drier were deemed ready for farmers' experiments in 1980, but the adaptation and production of IRRI-designed power-tillers and threshers did not occur despite high expectations. The 1980 Review Mission said this failure was due to the absence of firm orders from BADC, banks, or other agencies in a position to stimulate factory production.[74] Any vision of a mechanized rice agriculture in Bangladesh remained unrealized, because cheap labour was widely available. Labour's constant availability, its low cost, and its flexibility and replaceability were all crucial. Like land and credit, labour was an element of rice cultivation over which BRRI had very little influence and this contributed to its ambiguous status—applauded yet disregarded.

The forces of production described above (irrigation, etc.), and the other key to the Green Revolution—credit—were clearly under state control. Ignoring their world price, the government and its foreign supporters offered these inputs for sale at subsidized prices as part of the attempt to induce farmers to cultivate the HYV rice varieties and also to reward those who had previously adopted the HYVs and remained committed to them. State control was exercised largely through one corporation, BADC, which during the 1970s directed 60 per cent of development investment in the agricultural budget. Most of the foreign exchange component of the cost of these inputs came from grants and loans from foreign donors, whose own corporations were normally the suppliers, thus creating a pronounced appetite for these inputs within the government and among donors who had aid budgets to disburse.

It was this very appetite, among other factors, which led to the exaggeration of the reports of the spread of HYV rice cultivation. The planning process required that expenditures and budget be seen to be tied to the statistical progress of the programme which they were promoting. Thus input purchase and distribution has tended to depend on the statistical evidence of the spread of the HYVs from the previous season. Although the Ministry of Agriculture produced these exaggerated reports in the 1970s, the Ministry of Finance, the World Bank, the Planning Commission, and the President's Cabinet needed and used them. As a sign of their importance, Dalrymple summarized the results of the three official Task Forces (see Table 8.5).[75] In the boro season, the original target of the HYV strategy, the task force judged that official records were correct within 10 per cent when they showed a decline of HYV plantings from 94.4 per cent in 1968–9 to 40 per cent of the boro area in 1976–7. Even with this decline the HYV boro crop was said, by 1980, to contribute about 30 per cent of the country's total rice production. In any case, the task forces acknowledged a decline in HYV

TABLE 8.5. *Percentage of rice area planted to HYV, aman season (%)*

	Earlier official record	Task force decision
1974–5	34	11.5
1975–6	36	18.0
1976–7	31	20.0

Source: Calculated from Dalrymple, *Development and Spread of High Yielding Varieties*.

plantings in the key boro season. In the aus season, reports of area planted to HYV rice were judged to have been inflated between 50 per cent and 100 per cent. Thus the 27.5 per cent of the area reported under HYVs for the 1976–7 aus season were revised to between 14 per cent and 18 per cent.[76]

The Division of Agricultural Economics of BRRI said in 1978 that the official records might still be somewhat high and suggested an overall national figure of 12 per cent for the rice acreage planted to HYV rices combining all three seasons. But why were statistics unreliable at the lowest level? A clear answer was offered in 1978 by a thana-level officer of the Integrated Rural Development Program, which regulated the flow of modern agricultural inputs to the villages.

We sit in this office and fill these forms full of statistics. But we know these are inflated or deflated according to need of the government. At the first level of reporting, in the contact with farmers, our people are deceived. The farmer is 'chalak' [shrewd, cunning] and he under-reports. The lowest level official is just a sort of clerk and he estimates production, more or less. We send their figures upward, and they are used to conjure up a false impression if necessary.

I mean the government wanted the price of jute to fall so they calculated the total production of jute to be very large and offered a low procurement price. But there was not nearly so much jute as they announced, and now there is scarcity, so prices are 50% higher than the government's procurement price. The result is that the government cannot buy jute. This happened in the aman rice crop; in hopes of keeping the price down the government predicted a much larger crop than there was. We junior officials meet each other, we talk about the use of statistics, we know what is going on. Last week I attempted to explain my concern to a senior man visiting from Dhaka. He would not hear of it. Politely he told me it was none of my business.

Explanations offered by the Task Forces for overestimation were that HYV rice was conspicuous and often in the foreground near roads and paths and river or canal banks, thus enjoying an exaggerated visibility to busy officials, that figures were arbitrarily altered at higher levels, and that district offices inflated figures to demonstrate sufficient improvement to guarantee increased input allocations next season.

The official pressure for overestimation stands in contrast with farmers' underestimation. The state was delighted to think that its efforts bore good fruit; the farmers learned how to avoid both the states' interest (say, in tax) and their neighbours' *hingsha* (envy or jealousy). Even agencies of the government differed: if estimates of the volume of the rice crop were low, then the Ministry of Food asked for more food aid from donors, and could demand more money from the Ministry of Finance for food purchases. If estimates of the spread of HYVs were high, then the Ministry of Agriculture could insist on more money for procuring enough HYV inputs to meet an increased demand curve next year. And with this increased demand went an expansion of the bureaucracy to meet the demand.

In order to see how reports of the spread of HYV rice might correlate with allocation of the forces of production, the district receipt of inputs was analysed for the decade of the 1970s. There was strong distribution in Dhaka District (around the capital), Comilla District (where the programme began), and Chittagong District (where the inputs arrived at the port). There was a definite attempt to deliver inputs to the rice surplus districts of northern Bangladesh, the districts from which the government consistently procured rice and wheat for the public grain system. The extreme variation in allocation levels from year to year suggests that there was no logical connection between HYV input distribution and the agro-ecological zones or districts of Bangladesh. That said, there was a pattern in which some districts, such as Faridpur, consistently received the lowest volume of inputs. (No reports were made for agro-ecological zones, only for districts.)

Input allocation might also have been used as a form of reward (or punishment) for political behaviour, for support or rejection of the government in power in the elections held in the 1970s. While there are a few districts where such a hypothesis might hold, in others it does not, at least not with the data available.[77] What is probably more likely is that within districts, within thanas, allocation of HYV inputs to various farmers, households and *paras* (neighbourhoods) might have correlated with political behaviour. The fine tuning of reward and punishment, of opportunity and advantage was possible at this level and could be observed empirically in many places.

Given the capital-intensive nature of the HYV inputs it is not surprising that the richer farmers predominated in the Green Revolution. The planners tried to mask these inequities by forcing these inputs through a 'co-operative' IRRI-block form of organization. These cooperative blocks were mostly dominated by richer farmers who used their money and political influence to guarantee supplies of inputs, credit, fuel, parts, etc. Many of such blocks dominated by the powerful farmers did not repay the loans. But ten years after the beginning of the public promotion of the HYVs, understanding of the relationship of power, the co-operative IRRI-blocks, and the Green Revolution was filtering into the public consciousness. This critical description of the relation between the strategy for self-sufficiency and the bigger farmers was widely held by farmers who had not benefited by the strategy; in urban and official circles it was slowly receiving approval. A particularly lucid expression of this view even appeared in a 1977 editorial of an influential daily newspaper:

The myth of 'Green Revolution' now stands shattered and the system of property relations, exploitative insitutions and debt payment keep defeating all rural development efforts. Both by the compulsion of resource constraints and by the superiority of the alternative, our rural development strategy must take co-operatives as the fundamental structure. But how these co-operatives can also become an extension of the exploitative arm of the vested class we have seen in the recent past. As a result all governmental efforts to extend credit, bullocks, power pumps or other implements only strengthened the hand of rich farmers.[78]

Research work had to be inserted into this complex of economic and political conditions established by the HYV strategy. Just as these conditions defeated rural development efforts, they defeated researchers' efforts by isolating them and rendering them irrelevant.

NOTES

1. On the theory of entrapment, see Joel Brockner and Jeffrey Rubin, *Entrapment in Escalating Conflicts*, New York: Springer-Verlag, 1985.
2. Ziaur Rahman's speech to the National Economic Council as reported in *Bangladesh Times*, 14 May 1979.

3. A. R. Serniabat's statement reported in *Bangladesh Observer*, 23 December 1972.
4. For analysis of events leading to the two 1975 coups, and development strategies immediately following, see Robert S. Anderson, 'Impressions of Bangladesh: The Rule of Arms and the Politics of Exhortation', *Pacific Affairs*, Fall 1976. Also, Robert S. Anderson, 'Stop Everything in Bangladesh: Communication, Martial Law and National Strikes', *Canadian Journal of Communication*, 1988.
5. Government of Bengal, 1938, p. 33. See also Amartya Sen, *Poverty and Famines*, Oxford: Clarendon Press, 1981.
6. Quoted in *Christian Science Monitor*, 18 March 1975.
7. This contest of values was revealed in the work of Theodore W. Schultz (ed.), *Distortions of Agricultural Incentives*, Bloomington: Indiana University Press, 1978, in which R. Barker speaks of government 'Barriers to Efficient Capital Investment in Agriculture', and Crawford's comments on Abel's 'Hard Policy Choices in Improving Incentives for Farmers' reveals that it is profoundly difficult to answer this question—distortions from what? At best, deviation from a mathematically-modelled optimum appears to be understood as distortion.
8. Tajuddin Ahmed's statement reported in *Morning News*, 26 September 1972.
9. Ziaur Rahman's statement reported in *Bangladesh Times*, 5 July 1977.
10. D. E. Bell to McG. Bundy, 'Recommended Grant Action (for Bangladesh Rice Research Institute), 16 May 1973', p. 6, FFA 73-455. Note that this document anachronistically refers to BRRI even though at this stage it was still EPARRI. Legal change of names took more than a year after surrender of the Pakistan army in December 1971.
11. George Rosen, *Western Economists and Eastern Societies*, Baltimore, Johns Hopkins University Press, 1985, see ch. 7 on the split in the Institute of Development Economics on the Nationalist Question, and the role of the Ford Foundation.
12. R. H. Edwards to E. S. Staples, 2 October 1971, FFA 66-125.
13. Bell to Bundy, 'Recommended Grant Action, Background, 16 May 1973', p. 6, FFA 73-455.
14. Ibid. 7.
15. T. W. Oliver, *The United Nations in Bangladesh*, Princeton: Princeton University Press, 1978, p. 145.
16. Ibid. 146.
17. In February 1975 the Canadian International Development Agency shipped 127,000 tonnes of 'utility wheat' and 6,100 tonnes of rape seed on board the *Amoco Cairo*, a new oil tanker built by Mitsubishi. The ship's first voyage was to test this world's largest shipment of grain, there being a surplus of oil tankers with too little oil to carry. The grain and the ship charter cost $25 million, but then it was stuck in Vancouver during a strike, costing an additional $1.2 million in demurrage charges. Like the *Manhattan* before it the *Amoco Cairo* anchored far off Chittagong because of its size, and small boats ferried the grain back to port. Automatic vacuums to pump the grain failed frequently, forcing the government to put much of the grain by hand

into bags. Like American food aid, the Canadian wheat was sold through the ration system to support the government's main clients, but there were so many criticisms of the distribution that Canada's minister in charge of CIDA had to go to Bangladesh to investigate a few months later.
18. *Morning News*, 22 August 1972.
19. Oliver, *United Nations in Bangladesh*, p. 145.
20. *Morning News*, 31 August 1972.
21. Oliver, *United Nations in Bangladesh*, p. 141.
22. UNROD, *Report of the Special Agricultural Mission*, Dhaka, December 1972.
23. Ibid. 10. Chandler did not mention the mission or its political importance in his *Rice in the Tropics: A Guide to the Development of National Programs*, Boulder, Colo.: Westview Press, 1979.
24. UNROD, *Report*, p. 11.
25. Chandler also returned from Taiwan in 1974 to advise the Ministry of Agriculture again. See Robert Chandler, *Vegetable Research and Development In Bangladesh and Suggestions Regarding Co-operation with the Asian Vegetable Research and Development Centre*, Dhaka: The Ford Foundation, 1974.
26. *Morning News*, 29 April 1973.
27. Ibid.
28. *Morning News*, 10 June 1973, 'Jackson Optimistic about Food Position in Bangladesh'.
29. *Bangladesh Times*, 5 September 1977.
30. Ibid. quoting 'a BADC source'.
31. *Bangladesh Times*, 22 and 29 June 1978.
32. Personal communication, December 1978.
33. Reporting on area planted to HYV rice and other varieties was done regularly by field officers of the Ministry of Agriculture.
34. U. Columbo, D. G. Johnson, and Toshio Shishido, *Reducing Malnutrition in Developing Countries: Increasing Rice Production in South and Southeast Asia*, New York: The Trilateral Commission, 1978: 'The report proposes a fifteen year international program for the doubling of rice production in South and Southeast Asia focussed on irrigation improvement as the leading factor in generating production increases. The emphasis is basically on farm ditch construction neglected in the past.... The total capital cost of this program is estimated at $52.6 billion in 1975 prices' (p. xiii).
35. For a good review of the subject, see Stephen Biggs and Jon Griffith, 'Irrigation in Bangladesh', in Frances Stewart (ed.), *Macro-Policies for Appropriate Technology in Developing Countries*, Boulder, Colo.: Westview Press, 1987.
36. *Bangladesh Times*, 24 December 1975: 'WDB Abandons 32 Projects'. Interview with WDB Chairman Asafuddaula.
37. World Bank, *Appraisal of the Barisal Irrigation Project*, Dhaka, 1974.
38. This idea was stressed by Heydayatullah in his presentation to the International Rice Commission in 1952.
39. WDB Chairman Asafuddaula, *Bangladesh Times*, 24 December 1975.
40. J. F. Stepanek, *Bangladesh: Equitable Growth?*, New York: Pergamon Press, 1979, p. 125.

41. A. B. M. Shafiqur Rahman, 'The Matrix of Institutional Communication in Development Projects: A Case Study of the Barisal Irrigation Project', Ph.D. dissertation, Simon Fraser University, 1986.
42. *Project Paper on Water Management Systems*, Dhaka, USAID (draft, 24 May 1982), p. 2. See also *Bangladesh Minor Irrigation Sector: A Joint Review by Government and the World Bank*, Dhaka, 1982.
43. Rodney Tyers, 'Optimal Resource Allocation in Transitional Agriculture: Case Studies in Bangladesh', unpublished Ph.D. thesis, Harvard University, 1978.
44. Quoted in Peter Muncie, 'New Rice and New Hope', *Christian Science Monitor*, 18 March 1975; statement of M. A. Choudhury, plant breeder trained at Texas A & M University.
45. Note that not all fertilizer produced or imported in Bangladesh was consumed there. From some districts large amounts were smuggled to India, depending on price and currency differentials.
46. N. E. Borlaug to G. Zeidenstein, 2 March 1974, p. 4, FFA 73–455.
47. Comptroller-General of the United States, *Poor Planning and Management Hamper Effectiveness of AID's Program to Increase Fertilizer Use in Bangladesh*, Washington, DC, 31 March 1981, p. 14.
48. Discussion, visiting FAO fertilizer consultant, Dhaka, January 1978.
49. *Bangladesh Times*, 1 September 1977.
50. Ibid., 29 June 1977.
51. Comptroller-General, *Poor Planning*, pp. 9, 10.
52. Discussions, IRRI-block managers, Madaripur, January 1976.
53. Discussion, visiting FAO fertilizer consultant, Dhaka, January 1978.
54. *Bangladesh Times*, March 1978. See also ibid., 8 and 23 June 1978.
55. Ibid., 5 April 1978.
56. 'Report of the BRRI Review Mission, Dhaka, 1980', p. 12, FFA 76–109.
57. Various discussions, including BRRI scientists, Dhaka, January 1978.
58. *Bangladesh Times*, 3 March 1976. This leaked report was not given a title.
59. John Walsh, 'United States Beginning to Act on Banned Pesticides', *Science*, 29 June 1979, 1391.
60. Ibid. 1393.
61. USAID, *Environmental and Natural Resources Management in Developing Countries*, Washington, DC: Department of State, 1979.
62. Data from Directorate of Plant Protection, Dhaka, 1978.
63. Note that a member of the board of directors of CIBA-Geigy, Victor Umbricht, was UNROD Chief of Mission 1972–3. Porter-Pilatus crop-spraying planes owned by CIBA and on contract to the Directorate of Plant Protection were placed on UNROD duty during 1972–3.
64. *Bangladesh Times*, 3 March 1976.
65. Ibid., 10 March and 21 October 1976.
66. Ibid., 4 September 1976. Plans were reported to re-export this unused pesticide.
67. Ibid., 20 June 1978: 'Dumped Pesticide Pollutes Water in Tejgaon Area'.
68. 'Report . . . Review Mission, 1980', p. 14.
69. Barbara Harriss, *Rice Processing Projects in Bangladesh: An Appraisal of a Decade of Proposals*, Dhaka: Bangladesh Agricultural Research Council, 1978 (draft).

70. Barbara Harriss, *Post Harvest Rice Processing Systems in Rural Bangladesh: Technology, Economics, and Employed*, Dhaka: Bangladesh Agricultural Research Council, 1978 (draft).
71. R. Boxall, M. Greenley, J. Neelakanta, *Report on Visit to Bangladesh by IDS/IGSI Grain Storage Team*, Dhaka: Bangladesh Agricultural Research Council, 1976.
72. *BRRI and IRRI Cooperative Project Annual Report 1979*, Dhaka, 1980, p. 13, FFA: PA 76–109.
73. *Proceedings . . . Eighth Meeting Board of Governors BRRI*, Dhaka, 15 December 1973, p. 7, FFA 73–455.
74. 'Report . . . Review Mission, 1980', p. 17.
75. Dana Dalrymple, *Development and Spread of High Yielding Varieties of Wheat and Rice in the Less Developed Countries*, Washington, DC: USDA Foreign Agriculture Report 95 (6th edn.), 1978, p. 40.
76. The Task Forces were concentrating their attention on the planting of HYV rice in the aman season, in preparation for the new Intensive Transplanted Aman Program (ITAP), which was announced by the government in 1976, a programme which would give special benefits to farmers planting HYV rice in the aman season. For a historical review see M. Alauddin and M. K. Mujeri, 'Changes in the Crop Sector of Bangladesh', in F. Vivekananda (ed.), *Bangladesh Economy: Some Selected Issues*, Stockholm: Bethany Books, 1988.
77. Analysis was done with election returns for 1973 and 1978.
78. 'Rural Development', *Bangladesh Times*, 28 June 1977.

9

The Limitations of BRRI's Objectives, Organization, and Donors

> serious problems of management [in BRRI] ... must be corrected. If we [IRRI] have not leverage ... we are powerless to bring about the needed changes.
>
> Nyle Brady, 1975
>
> BRRI should 'begin soon to develop the necessary institutional capacity to self-manage'.
>
> BRRI Review Mission, 1980

DURING the 1970s BRRI consolidated its headquarters, but hardly developed the extension of its work to cultivators. The enormous energy and considerable funds absorbed by Bengalis and foreigners establishing the institute at Joydevpur limited the resources available for its outreach work and its substations. The headquarters did indeed require great attention, as will be demonstrated, but this requirement distracted attention from the bigger picture, that is the contribution of rice researchers to rice cultivators. What developed can be called the headquarters syndrome: the genuine attempt to make a scientifically competent centre in a poor country which is itself largely indifferent to the requirements of researchers. This proved such a difficult task that very little energy was left over for BRRI's other undertakings and responsibilities. Centralizing tendencies contributed to this syndrome, reinforcing the skewed allocation of resources and narrowing objectives when earlier ambitions were not realized. What follows is first an account of how the headquarters was established and second an assessment of the relation between rice cultivators and BRRI, particularly through its system of ecologically specific substations. Evidence is offered here which suggests why BRRI's effective engagement in the HYV strategy, at least until 1980, was marginal at best.

Reconstructing BRRI after the 1970–1 chaos took much longer than expected. No proper legal mandate was obtained until 1973, and even then BRRI's board perceived some conflict between its Presidential Ordinance and its Act of Parliament, both of which seemed to address

BRRI's autonomy.[1] Labour unrest in 1973 led to the removal of administrative functions back to the city from Joydevpur. Hiring at BRRI was subject to the Public Service Commission's rules and thus many vacancies were not being filled because the Commission allocated positions to other ministries. Although the World Bank was prepared to grant $1 million to BRRI (which was in BRRI's budgetary terms a very large sum), BRRI's board of governors felt in 1973 that the Planning Commission and the National Economic Council were holding up formal approval of BRRI's plans for expansion despite the Ministry of Agriculture's approval.[2] Government expenditure in agriculture in 1973–4 was certainly not focused on research; in percentage terms water control accounted for 44 per cent, HYVs and their inputs for 33 per cent, the Rural Works Program for 9 per cent, forestry–fisheries–livestock for 6 per cent, miscellaneous for 3 per cent, the Integrated Rural Development Program for 2 per cent, and research and development for 1 per cent.[3] In 1973 BRRI was not permitted even to name its own latest rice variety (BRRI 3) as 'Biplab' (meaning 'revolution') because, according to board minutes, while the Secretary of Agriculture was in favour of the name the Prime Minister curiously did not approve of it. This word was in common use at the time, including by the Prime Minister. Infants were commonly named Biplab in 1972 and 1973. This rejection added to the feeling that BRRI was not being taken seriously by decision-making bodies in Bangladesh.

A year later the pace of building construction was judged too slow by IRRI and the FF and they sought to introduce a range of new practices. Hours of actual research work were still limited to thirty per week owing to time travel lost in to and from Dhaka. Security was almost non-existent and equipment and buildings were very poorly maintained; electricians, mechanics, and plumbers were brought all the way from IRRI to make repairs. Preparing for a major grant review, the Ford Foundation and BRRI brought in three experienced American analysts (Russell, Minehart, and Freeman) who presented a long-term plan for BRRI. They proposed to create an efficient bureaucracy which did not require scientists (who might otherwise be interested in insects) to run it. This proved hard to implement because basic management capabilities in BRRI were so weak: the long sought-for autonomy was still unrealized, directly affecting such matters as the salaries which could be paid to repairmen, the permission to obtain its quota of cement for construction, and even the relative prestige of its director in comparison to other officials in the Ministry of Agriculture.[4] Construction costs had tripled between 1970 and 1974, absorbing most of the budget. This was the year, 1974, in which Bangladesh experienced its worst drought and famine of the decade: thousands died in the north or were roaming the countryside looking for food.

In the following year complete autonomy was formally secured, but headquarters construction was still incomplete. The report of the 1976 Review Mission was concerned that scientists were dissipating 'their energies in unproductive work' (for example one Ph.D. scientist was managing the transport pool in the 1970s) and said that unless they gained mechanisms to communicate the importance of their work to the public, research staff '... will become inward looking and frustrated'.[5] Lower down the hierarchy, uncompetitive salaries were affecting recruitment of junior scientists and technicians. Labourers on the research farm went on strike for a week. There was one report that poor land preparation on the research farm had damaged soil fertility.

Slowly the apartments at Joydevpur were completed, and slowly they were filled. Families of senior scientists tended to be the last to move in. Reluctantly they occupied the subsidized housing, eliminated hours of travelling, and prepared to brave the rigours of country life. By 1980 more than 120 families lived at Joydevpur, so security and maintenance were improved with a larger community there. Despite occasional labour disputes—there was even a full-scale strike of Ministry of Agriculture staff workers in 1978 which split scientific ranks—by 1980 BRRI had considerably matured as an organization.

Its scientific capabilities in terms of producing 'a stream of new varieties' were now much enhanced, for all the disciplines relating to producing a new rice variety were installed at headquarters. Other disciplines, however, were weak: for six years, for example, BRRI attempted to create a strong agricultural economics group but had not succeeded by 1980. The size of the overall staff, nevertheless, had doubled. The problem was basically to keep employees: technicians and drivers etc. were all attracted to the high-wage boom in the Middle East. Young scientists were recruited to BRRI on the promise of prompt consideration for advanced training, whether in the Philippines or USA. But on return they often found their services in demand in institutions prepared to pay higher salaries (parastatal organizations, banks, etc.). Some left, some stayed. Each review mission commented on the adverse effect of this; for example senior scientists had to spend much extra time in personnel selection and training. Scientific productivity therefore was limited at all times, despite BRRI's gradual evolution, by structural difficulties over which scientists had little control.[6]

Three structural features of BRRI's organization further limited scientific enquiry and responsibility: the availability and distribution of staff in various divisions, the training of staff, and the communication of research information. By 1977 BRRI announced to its donors that only the breeding division 'had almost reached stable strength'; other divisions were 'understaffed'.[7] In 1975 eleven BRRI officers were being trained in the United States or the Philippines, and the number of trainees remained the same in 1980. The majority of researchers were young and relatively inexperienced: many were waiting for long-promised permission to travel abroad to obtain doctoral degrees. In some cases they sought employment at BRRI only because it offered prospects of foreign study. Meanwhile employment in banks and agricorporations in Bangladesh was more attractive. As some staff moved away, others had to be interviewed, recruited, and trained on the job. In only three divisions was there more than one senior scientist. Scientists confirmed the adverse effect of administrative responsibility on their scientific work. BRRI began to conduct field days, seminars, and annual reviews among its staff in order to overcome some of these organizational problems.

All BRRI's training and career advancement had been paid out of foreign currency accounts provided by the major donor-investors in development in Bangladesh. The 1977 BRRI proposal to foreign donors admitted that 'BRRI is beginning to experience difficulty attracting the desired quality of applicants', and suggested that providing air-conditioned offices, daily special transport for senior scientists, travel funds for overseas contacts, and improved library resources seemed to be the only ways to retain trained scientists.[8] (Indeed, although few offices were air-conditioned, some of these suggestions were implemented.) Of thirty-five BRRI scientists in 1976 twenty had basic degrees from Bangladesh Agricultural University at Mymensingh and twelve from Dhaka University.[9] Thirteen scientists had taken additional degrees in foreign universities, none of them in the United States, and the other four in conjunction with IRRI. Both BRRI and its donors acknowledged the problems of a weak commitment to research and to the institute's goals among junior staff. Arranging foreign study seemed to be the only way to attract talented and ambitious researchers, but BRRI found it difficult to hold them once they obtained it. It should be noted, of course, that such contradictions were not unique to BRRI.

In addition to these staff difficulties and imbalances, and the dependence on foreign sources to provide the cost and site of training, the communication of research results to the public had little impact in Bangladesh. Though much of the daily work at BRRI was conducted in Bengali, the scientific language of the institute was English, which no rice farmer spoke. Publications from the institute appeared in English: most scientific work was discipline-oriented (for example, entomology, agronomy), and most was published abroad. Fourteen BRRI scientists reporting in 1976 had published ninety-two papers in their careers: 43 per cent were in Bangladesh and Pakistan journals, 22 per cent in IRRI publications, 21 per cent in American journals, and the rest in Japanese, European, Sri Lankan, or Indian ones.[10] This pattern is consistent with Hargrove's finding that the rice breeders of Asia reported in 1975 that most of their communication was with IRRI, and that IRRI scientists published more of their work in American journals than anywhere else. Like other scientific institutions in Bangladesh BRRI did not engage in much public reporting of its work; it had neither the skill nor the resources to do so. In particular it did not seek to communicate widely with its ultimate clients, the farmers. It should now be easier to understand why senior scientists spent little time with farmers or at far-away substations, and why those substations grew so little, as will be shown below. Basically there was little energy left in the research system which could be turned towards the cultivators themselves. The objectives had become fixed and, arguably, wrong.

9.1 BRRI's Objectives and Organization

By the mid-1970s BRRI was a complex of low, gleaming white buildings twenty miles outside Dhaka, beyond the military academy and the telephone factory. Travelling two hours a day, BRRI scientists and staff moved between homes in the city and their work-place in the country. Except for occasional visits to plots or outbuildings, most scientists remained inside for most of the day, while staff moved up and down long corridors carrying files or tea. After the library and cafeteria were constructed, scientists were waiting, until 1978, for completion of living quarters on site. The homes were modern and subsidized, but most scientists and staff were city people and few really wished to live behind the brick walls that defined the institute's boundary. Outside the wall buses and trucks lurched past a few little shops on the road. Beyond the road farmers had everything under cultivation as far as the eye could see. Farmers seldom came inside the wall; scientists seldom went out to meet them. The vehicles, the salaries, the books, the flower gardens, and the technical equipment set the institute well apart from farmers.

BRRI was established to introduce and to promote the spread of IRRI-bred rice varieties; it was one of the few national rice research institutes to be founded in Asia *after* IRRI was in operation. Breeding and testing were its primary functions from its very beginning: BRRI described itself as 'the agency responsible for developing more productive and pest-resistant varieties, and designing and testing and recommending appropriate packages for production technology'.[11] In announcing that seven rice varieties had been bred there since its beginning, BRRI predicted in 1980 'a stream of new varieties' in the future, perhaps four per year, because a collection of thirty or more varieties 'may be required to secure high levels of farmer adoption'.[12] These rices should not be of the shortest type, because of widespread flooding; they should be resistant to pests; and they must tolerate or resist drought, cold, salinity, and diseases, while being responsive to fertilizers. BRRI in 1977 called these objectives 'the biological engineering of the rice plant'. BRRI's director said that work had to begin with high-priority areas, like increases in production. The breeding programme was begun first because there were at least three experienced breeders and because breeding is a cheap technique requiring only a pencil and forceps. The laboratories were used, he said, for diagnostic purposes; the research plots were the true laboratories for BRRI.

BRRI's other objectives had been minor in terms of budgets and personnel; of ninety-five researchers by 1977 fifty-six (60 per cent) were engaged in the 'classic cluster' of disciplines considered essential to produce new plant varieties—breeders, entomologists, pathologists, soil chemists, agronomists, physiologists. This is the same cluster of diciplines at IRRI or at any American rice research station where the objective was biological engineering. BRRI researchers regarded their objectives in 1975 as being fully consistent with the goals of other breeders in ten Asian countries.[13] That is, regarding 161 genetic crosses selected for study by Hargrove, Asian breeders said 86 per cent were made for yield potential, 85 per cent for fertilizer response, 75 per cent for non-lodging, and 73 per cent for grain quality. Other objectives such as growth duration, disease, and insect resistance were subordinated.[14] BRRI's budget and staff assignments at this time conform to this pattern.

However, among BRRI researchers in 1978, there were signs of a broadening of objectives and perhaps a re-emergence of the 'improvement philosophy' which would lead to incremental (less than 50 per cent) yield increases with existing varieties. Driven out in the 1950s and 1960s by the strategy of promoting brand-new varieties and transferring technology, the objective of selective improvement of indigenous varieties now seemed more effective and less risky. Researchers in East Bengal and East Pakistan had investigated both improvement and variety-transfer techniques until the late 1950s, when they were swept up in the search for the new hybrid varieties which would be transferable anywhere and would yield well anytime.

Officially, scientists were still concentrating at the end of our study period on the biological engineering of the rice plant. A 'stream of new varieties' would, in BRRI's judgement, be the key to meeting 'the sustained annual increase of 3.5 per cent in rice output' required by Bangladesh:

> The population growth trends and other factors point to the need for this sustained annual high rate of increase in output well into the next century until population growth stabilizes at zero level. This is a most sobering projection when one realizes that in the history of agricultural development seldom has such a high compounded growth rate in rice output been sustained over a fifty year period.[15]

The Limitations of BRRI

BRRI's work was more on the production side of the equation than on studying losses from the field and the kitchen or of the consequences of earlier technology change recommended by scientists. By 1980 BRRI had better working relations with the Bangladesh Agricultural Research Council, Jute Research Institute, Water Development Board, and Agricultural University at Mymensingh.[16] Somewhat closer relations were also achieved with the Extension Department in the Department of Agriculture. These were mostly for the purpose of increasing production, not for assessing the consequences of previous activities. BRRI's production orientation was summed up in the task of producing four new varieties every year, a task that could best be done, it was decided, in a centralized institution. This centralization, so typical of other 'mass-oriented' projects of the state apparatus in Bangladesh (for example, family planning or malaria control), made it convenient for busy government officials, BRRI's foreign donors, and those IRRI scientists who did research at BRRI. These people became the real clients of BRRI's work, more than the great mass of small farmers. But were strategies convenient for officials, donors, and scientists also the most effective for feeding a growing population or even for achieving 'self-sufficiency'?

It is clear that almost everyone wished the projection of a 3.5 per cent annual increase in rice production would come true. When matched with the historical trends of slow growth studied by Islam and Ahmed, the requirement of a 3.5 per cent growth rate in production was a most formidable objective indeed. Nevertheless, BRRI said that the 3.5 per cent had to be a 'sustained annual increase' and offered publicly the explanation that this increase would be possible 'because the Comilla region experienced a growth rate of 4.5% between 1961 and 1971'.[17] What was not said was that the domestic and foreign investment in rice production in Comilla between 1961 and 1971 could hardly be duplicated in the rest of the country during the 1970s.

Rice researchers and expert observers were not confident that this 3.5 per cent target could even be reached, let alone sustained. Many BRRI scientists were privately not convinced that the HYV strategy as it was pursued would come near this goal. They pointed to the fierce competition for small amounts of land, to rates of landlessness being about 35 per cent of the rural population and increasing, to high rural unemployment, to the large percentage of the population under the age of fourteen, to the chronic indebtedness of rural households and the high interest rates for credit from moneylenders, and to the prevalence of land sales among those with less than one hectare. They knew that the increase in rice production largely in the summer season had been paralleled by a decline in per capita grain consumption.[18] They were aware that Bangladesh's population growth might not stabilize within fifty years. Indeed, all these factors which made Bangladesh an international *cause célèbre* were very well known to BRRI scientists. Yet publicly these subjects were not raised.

Some BRRI researchers acknowledged that previously BRRI had, like IRRI, neglected the study of adverse cultivating environments, and they worried about the decline of genetic variation which was resulting from the replacement of indigenous rice with HYV rice. The official exaggeration of the actual spread of the HYV rice, revealed and revised by the government's HYV Task Force, was also on their minds. These were issues in which they had direct involvement. But the other factors which affected the utilization of their work were beyond their control, they said privately: land fragmentation, landlessness, unemployment—all of which affect the HYV strategy—made them feel powerless because these crucial features of the agricultural revolution were regarded as being outside the scope of their scientific enquiry. Such limitations thus made them helpless in the study of the real cultivating conditions of millions of small farmers, limiting the contribution they could make to solutions to the food problem. This contrast between the stated objectives of research and the researchers' private lack of conviction came, we think, from their general, uneasy conclusion that a major shift in their practice was called for. Though they were unsure which major shift in research would succeed, they concurred that no major shift was likely under present circumstances because of obstructions, as they saw it, 'beyond their control'. What researchers and planners did not seem to see clearly is that the organization and conception of scientific research is itself a major variable in the outcome and is clearly something over which scientists have a degree of control and for which they have a responsibility.

During the years from Bangladesh's beginning in 1971 to the end of the decade a gradual maturing of the institute's capabilities took place, as well as a consolidation of its organization and staff. Its relationship with IRRI changed slightly, and dependence was decreased a little: new foreign donors entered the picture, and one or two grants were made at the end of the decade directly to BRRI without going through IRRI. The views of foreign funding agencies remained the major determinant of BRRI's specific objectives, however. As staff began actually to live at Joydevpur, the headquarters syndrome was complete; the relative weakness of its substations and of relations to farmers became, if anything, more of a contradiction. What follows is an analysis of the evolution of the institute's objectives and the role of donors in this process.

The Project Area

One potential source of scientific knowledge of consequences could have been the institute's fledgeling applied research programme, but its limitations were difficult to overcome. Some applied research was conducted on land called the 'project area': a 148 square kilometre zone drawn arbitrarily on the map of the territory surrounding BRRI. Though it contained 'diverse land forms and soil associations' it was a generally upland area seldom flooded and thus unrepresentative in drainage.[19] It was semi-urban, cut through by paved roads serving the industrial estates nearby. Villages near BRRI were often electrified, and communication and commerce with Dhaka regular. Some families combined farm and urban incomes. It naturally had many progressive farmers with whom BRRI could co-operate. The director of BRRI in 1973 apparently drew the area's boundaries across other administrative boundaries; thus an independent check on the accuracy of agricultural reporting generated by government officials was next to impossible, even though this check was potentially one of the most important contributions of off-station research and was required for beginning a thorough study of consequences. BRRI thus had to rely for social and economic information almost entirely on records generated for administrative purposes, not scientific purposes. The exaggeration by ministry officials of the HYV area in the early 1970s in which BRRI scientists were passively co-opted, is an example of this.[20] One can see that the critical scientific function that BRRI should have played was sometimes minimized by the nature of the project area.

In addition to applied research which identified 'progressive farmers' in the project area, BRRI also studied the factors which inhibited other farmers from cultivating HYV rice. The questionnaire prepared by IRRI for the study of constraints was administered in a short version because neither the junior field assistants nor the farmers wished to spend much time on it, according to discussions in the field. There also were case studies (limited by the small size of the staff) of the influence of moneylenders and fertilizer dealers on the adoption of new varieties, of cultivators' response to a crop insurance proposal, of the government's massive (pre-election) agricultural loan programme, and of the shift away from grain and pulse crops to more profitable vegetable and fruit crops on land irrigated by government tubewells. Though considered interesting by other scientists, the results of these studies were not aligned with the ongoing work of other departments and task forces. BRRI's research commitment to these complex and important issues in terms of staff and budget was minimal.

In BRRI's 'project area' applied research was conducted by identifying 'progressive farmers' who had adopted HYV cultivation, then offering them training, encouragement, advice, and more seed. To increase the multiplication of HYV seed BRRI linked these 'progressive farmers' in production chains which were supervised by junior staff under instruction. It is difficult to see whether this applied research has resulted in much insight to scientists. Relations between farmers and scientists were symbolized by an official BRRI photograph of researchers talking to progressive farmers who were wearing white shirts, and watches on their wrists.[21] Progressive farmers had access to tubewell irrigation and all the inputs required for HYV cultivation. Some had non-farm income from jobs, shops along the road or in town. Because of the special location of the project area and the identification of a specially privileged class of farmers for these relationships, BRRI naturally learned more from it about the cultivating conditions of 'progressive farmers' than of other farmers in Bangladesh. By 1980, said BRRI's director, HYV rice was grown on 80 per cent of the cultivated land in the project area.

This absence of responsible relations with groups representing the great majority of farmers resulted from and reinforced a conception of science as merely the generation of knowledge, not the communication of knowledge. The underlying assumption was that a single-source scientific assembly line probably could, if modelled on other single sources like IRRI and if regulated by controlled methodologies, produce four new rice varieties annually. How relevant this generation of knowledge was to farmers, what was actually done with these varieties, and what was learned from the consequences of that action, were questions that were subordinated to the objectives of biological engineering.

This weakness was a corollary of the preoccupation with bioengineering and varietal manipulation, a natural weakness of that conception of scientific enquiry and responsibility that includes the following formula: (1) knowledge springs from interaction with the scientific literature, recent experiments in controlled environments, discussions with other scientists, conferences, and so on, and not from understanding the consequences in agrarian systems of the actual use of research; and (2) scientists' responsibility is more to ministry officials, IRRI scientists, world scientific disciplines, and foreign investors, and less to the people who actually experience the consequences of induced technical change. To overcome these limits on its contribution BRRI would have had to identify new potentials in the cultivating environment and to respond to the conditions outside the insitute walls for which the institute was erected. The substations would have had to be much more important.

Substation Case Study: Barisal

The underdevelopment of BRRI's five substations led to a failure to study the agro-ecological variations of the country. This was all a result of BRRI's centralized organization which had concentrated almost all of its resources at its headquarters outside Dhaka. Although the importance of agro-ecological variations was recognized in the 1930s when some of the substations were founded, there was a decrease in their use beginning about 1960, partially caused by the preoccupation with a search for a universal solution to rice cultivation problems through the HYV rices (and in some cases for reasons of economy). Consequently, the determining role played by district and intradistrict variations in cultivating conditions, and therefore in production potential, began to be ignored just when they were most important. BRRI's five substations were physically and organizationally isolated from its Dhaka headquarters. This does not mean they were not noticed, but that as the headquarters syndrome set in the substations were progressively underdeveloped. Some headquarters scientists regretted this, and review missions commented on it. The Ford Foundation Review Mission in 1974 said substation development was too slow; 'in the various ecological zones of the country' said the 1976 Review Mission Report, there should be substations 'to do, under special local constraints, those things which cannot be done at Headquarters'.[22] Recommended again by the 1978 and 1980 review missions, this issue was subordinated to the more pressing organizational problems at headquarters. The substation was not a high priority for BRRI—its 1977 proposal for foreign financial support discussed substations twice in passing—but it was always a source of complication. What was needed was a system by which the isolation of the substations from headquarters was turned into a source of decentralized advantage for BRRI as a whole, but the social forces against this were there to stay.

On a visit to the Barisal substation which is responsible for research on rice submerged by tides, it was possible to compare personal observations based on this case with reports from the other substations.[23] Barisal was listed in 1938 as a rice surplus district, and while it lost that status around 1964, the only senior scientist at the substation said that the district could probably produce a surplus again by making incremental increases in the yield of the aman and boro crops. Eighty per cent of the district is subject to daily tidal submergence, salinity, regular flooding, and high winds or cyclones. The district contains the usual proportion of very small farms, but also has coastal islands composed of farms of hundreds of hectares with absentee landlords. The population density (excluding water surface) in 1974 of 740 people per square kilometre was above the national mean. Between 1966 and 1976, Barisal ranked second in total area planted to rice, but fifth in total production. Rice yields were below the mean in each year except 1973: in the overall period 1966–76 the district ranked fifteenth in yields in all three seasons; in yields for the new summer HYV rice crop, however, it ranked sixth. Cropping intensity declined in the period from 151 per cent in 1969 to 146 per cent in 1976, and Barisal ranked from tenth to nineteenth among districts in this period. The picture emerges of a large rice crop grown in very difficult conditions using lower yielding well-adapted varieties to provide for subsistence.

BRRI headquarters determined the main work of the Barisal substation, which in 1978 was to carry out twenty-three separate tests on new international and national rice varieties. The test methods were standardized by IRRI and headquarters for research control reasons. But these methods could not always be transferred to local cultivators. For example, the experimental fixed-distance planting technique is judged inferior to the local transplanting technique: because of high tillering the resulting plant stands are too thick and give undesirable mutual shading. The senior scientist therefore encouraged the local method on district farms but followed the experimental method on the substation, as instructed. The fertilizer doses prescribed by BRRI for its rices had proved to be too high for Barisal soils and water conditions. Although the senior scientists at the substation suggested that farmers reduce the prescribed dose by 20 per cent, he conducted experiments at the prescribed level, as instructed. The large number of experiments with rigid specifications, he feared, caused the creation of peculiar soil and drainage regimes on the substation's 8 hectares. Unless more staff were assigned, enabling BRRI to do research on farmers' fields, artificial conditions on the substation, caused by international methodology, would detract from the relevance of the results to the district.

Given the shortage of staff, the large amount of mandatory work required of the senior scientist left him little time to increase local rice yields incrementally or to deal with urgent district research. The scientist was, however, in his spare time, also crossing local rices which tolerate tidal submergence with better yielding BRRI varieties which tolerate common pests like the brown plant hopper. He was hoping to increase yields in the transplanted winter crop, grown without irrigation or fertilizer, from 1,400 to 2,100 kg per hectare, a 50 per cent increase. The senior scientist understood local conditions only because he used time left after completing headquarters tests to meet farmers around the district. In 1978, however, there were many places where he still had not been; because there were no other scientific staff and because he had no boat (until 1980–1), he could not stay away from the office very long. The senior scientist laboured on mostly unassisted: he was thirty years old, did not have a Ph.D. in 1978 but had done a year's training at IRRI. Although there were three long-vacant positions for junior scientific officers at the substation, his request for a complete staff had drawn no results by 1978. He continued to work alone with three field assistants and one clerk.

In 1977 and 1978 a few commercial farmers came to the substation looking for rice seed. Though a net rice importer, Barisal still exported to other districts fine rice (Balam) which has very high commercial value. A new BRRI variety released in 1977 (BRRI 7) was in demand in the market because it was expected to be a fine exportable rice, and so bigger commercial farmers sought this seed at the substation. The yield potential of all the HYV rices was well known among these farmers. On a 1-hectare field adjacent to the substation, BRRI-bred HYV rice planted by a Catholic educational institution yielded double the district average for the 1977 winter crop, that is 2.85 tonnes per hectare, and yielded 4.72 tonnes per hectare in the 1977 summer season. This institution could easily afford the hand tractor, seed, fertilizer, and irrigation it used for this field, and it got excellent scientific advice from the substation. These results were well known to farmers. But the objective of the substation should not only have been to devote attention to surplus farmers wishing to grow fine rice for export, but also to conduct research on subsistence farmers' fields in adverse conditions.

The substation was visited by most of the principal scientific officers of BRRI since 1972, most of the IRRI team in Bangladesh, a regional director of the Ministry of Agriculture, and a member of the planning commission. The problems of the substation were known, but remedial action was slow in coming. For example, requests for a portable irrigation pump to be used in trials of IRRI rice took three years to fulfil. Many problems had to be resolved at the local level; for example, the substation had no security retaining wall around it, so livestock wandered into the plots, and rice plants were stolen. BRRI headquarters was prepared to pay the substation to build the wall but with materials obtained at fixed prices through local government resources. Making arrangements with district government officers and securing scarce construction materials stretched over two years and took up a considerable amount of the senior scientist's time. District government officers had spoken of the importance of the substation to the district, but although good relations were established with some government officials who eventually visited the substation, they were often transferred away; the contact then had to be re-established with someone new. Attempts to place the substation in a tidal area more representative of the district, on a piece of land thrown up when the river changed course, also required lengthy and delicate negotiations at the local level, in which BRRI headquarters gave little assistance.

Underdevelopment of the substations kept the links between them and international research insignificant; no foreign scientist stayed at the station for more than a day or two in the 1970s, and no research collaborations were formed. Though an FAO/World Bank team visited the substation in 1972 'to see the role the substation could play in an internationally financed agricultural development program in the district', the 1974 bank appraisal of this irrigation project mentioned the existence of the substation only once, and in the next four years not one contractual link was made between BRRI and this $46 million project to increase rice production in four project thanas of the district.[24] In fact the World Bank is reported by BRRI to have interfered in substation development plans (having allocated the funds), and in particular opposed the development of the Barisal substation as a tidal submergence research centre because of its anomalous location. The Bank probably underestimated how difficult it would be to acquire new land to which to move it. In 1981 it was reported that the World Bank had finally stopped opposing the funding of the Barisal substation for tidal zone research. This resulted in fifteen additional staff positions being authorized during 1980–5. BRRI's director said that a speedboat had just been attached to the Bank's Barisal irrigation project and could be borrowed by BRRI personnel. By 1986 a block of apartments had been constructed for staff using Bank funds. On these questions of major long-term funding, of course, working scientists have only minor influence in comparison to senior ministry officials and foreign financiers like the Bank. Thus national research policy is made by people with no experience in research.

BRRI's substations could have provided a wealth of contacts with rice farmers all over the country and detailed knowledge of specific agro-ecological conditions which called for technical change. But these contacts and knowledge were not available to BRRI's headquarters. Of the attention BRRI scientists did pay to farmers, a disproportionate share naturally went to conditions immediately around them. While it is useful to test national and international varieties at substations, an equally great potential lies in crossing local varieties to achieve modest yield or resistance changes. A 1980 BRRI report shows such crossing was beginning. And there were other issues equally important, and location-specific, like saline- and tide submergence-tolerant rices, control of the brown plant hopper, and so on. If any research could contribute to the self-sufficiency of rice production in Barisal, most of it probably should have been done in Barisal rather than in Dhaka.

There were important exceptions. One area of Barisal was found to have a zinc deficiency so severe that farmers could not obtain a rice crop in the late 1970s. BRRI applied $ZnSO_4$ to their fields, 'and they were astonished to see its magical effect', so that later they were clamouring for zinc to rectify their deficiencies.[25] But the substations remained an undistinguished and small part of BRRI's activities. Thus while the substations were more intimate with the environment which they were intended to understand, because of constraints on their work, this intimacy was not the source of knowledge and creativity which it should have been. Thus another opportunity for a linkage with farmers was missed, and a crucial avenue for scientific enquiry blocked.

The Study of Consequences

The weakness of the study of consequences of earlier research and technological change reflected scientists' preoccupation with on-station research in controlled environments and with the next phase of research rather than the effects of previous work. It is striking that there was no communication loop from past recipients of BRRI research to researchers, no cumulative tradition except in the memories of senior scientists. While such memory enriched judgement, there was no corpus of evidence to which new researchers could be introduced. Except for one history (up to 1960), scientists did not collect and analyse these crucial records—indeed, some important parts of the record, like chemical practices, might be missing. What study there was of consequences of the HYV strategy was haphazard and removed from current practice, as well as occurring in an unhistorical framework.

The research conducted at BRRI off the station at Joydevpur was thus not the important source of information about rice cultivation that it should have been in the 1970s, whether in the project area nearby or in the five substations around the country. It was widely believed at BRRI that the previous director in the early 1970s did not like research done off-station: one senior scientist said 'he wanted everything to be done in front of him, to see it and control it'. Consequently, research in the project area was given low priority except when donors were reviewing the programme. (Whether donors actually considered off-station research important is not clear; what is clear is that officials at BRRI attributed this view to the donors.) Most research at BRRI was directed at maintaining test plots and laboratories. One division head was reported in 1978 to have kept all his work on-station because he did not want his division's prestige and scientific rigour diluted on farmers' fields. A few BRRI scientists said they were reluctant to encourage off-station research because they feared theft of seed and other inputs which were hard to obtain. They also doubted the reliability of results produced by inexperienced junior field staff who were 'hard to supervise' in the fifty-seven square mile project area. 'Do I really know if people actually go to the field or if they guess and fabricate the data?' asked one senior scientist. They were no doubt correct about the methodological difficulties of off-station research, but did not seem to recognize the extreme limitations of on-station research. The discovery, by BRRI researchers in the late 1970s, of the continuous depletion of organic matter in the soil, because some farmers thought fertilizers supplanted organic matter (and so stopped applying it) and of serious zinc deficiencies in the soil of Barisal shows that research can trace the consequences of previous activities and can have a direct bearing on the advice given by government workers to farmers. BRRI's director's view was that although the on-farm trial was a good mechanism, the interpretation of this evidence was rendered 'difficult by multiple variables and lack of suitable statistical designs to control the confounding of main effects and interactions'.[26] Moreover short on-farm trials needed to be supplemented by long-term tests, he said, in order to acquaint researchers with consequences of a technology in farmers' fields before they advocate it.

If substation work was recommended, if these are 'good mechanisms', why were they underdeveloped? The 1980 Review Mission visited the Comilla substation and found conditions similar to Barisal—work assigned by headquarters dominated the substation's activities and so was skewed away from local conditions and farmers' problems.[27] Two years later, in 1982, Zillinsky reported that the substations still remained underdeveloped and their potential thus unrealized.[28] Recommendations for substations, for enlarging the BRRI system, were made anew because of the clear limitations upon adoption of new rice technologies and because of the donors' desire that the investment in BRRI should finally pay off in adoption by farmers. But throughout this period the forces against this enlargement were strong. BRRI was administratively very weak: autonomous, but barely able to manage the headquarters activities. Basically scientists were becoming bureaucrats in order to manage it. Staff were reluctant, it will be remembered, to move out to Joydevpur—twenty-two miles from Dhaka on a good road—never mind substations. Such an assignment might be, in effect, a lateral demotion, or a condition from which it would be hard to return as a senior scientist. In addition middle-class families of scientists wanted the best schools for children, medical facilities, good shops, and better than rural housing. Few had chosen a scientific career (or married a scientist) to live in marginal discomfort or exile in the countryside. Junior staff would stay there unwillingly for two or three years, but all the intrigue and politics related to working conditions occurred in Dhaka; briefer periods away from headquarters would allow one to keep abreast of these crucial circumstances. Furthermore, giving the substations full automony would perhaps undermine or dilute the political authority of directors of divisions and task forces. The fear of loss of control was a central principle in all social organizations in Bangladesh; full growth of substations would excite precisely that fear. Recognized widely by members of review missions and BRRI alike, these social forces militated against substations and were, in effect, part of the headquarters syndrome. Even when the concept of new rice varieties for each agro-ecological zone was accepted in the 1970s, the system resisted change. It was centralized even when it did not wish to be.

9.2 Relations with Rice Cultivators

Rice farmers in Bangladesh knew where new rice varieties were developed, even if they did not cultivate them. They called HYV rices, including the Chinese Purbachi introduced in 1968, by the name 'IRRI'—as in 'China-IRRI' or 'IRRI 8'. Most farmers also knew where BRRI was and knew BRRI seeds. In fact the neighbouring farmers, whose lands had been expropriated for BRRI, were invited by scientists in 1972 to name two promising new rice varieties. They called them Chandina and Mala and were growing them on about 60 per cent of their land as a rain-fed crop by 1980. But in general they believed that science had little to say about rice-farming beyond propagating new seeds. They believed that other important factors in cultivation—particularly the supply of irrigation, fertilizer, pesticides, disease control, fuel, rice mills, transport, and storage—had little to do with researchers at BRRI. Researchers might study these things but could not influence their availability. Enquiries with people in the immediate neighbourhood of BRRI's headquarters in 1978 showed that they thought the question of their relation with BRRI to be a strange one, perhaps even a joke. They asked why scientists would want to know about their cultivating conditions. 'They do not need to talk to us or see anything here to know our conditions in order to do their work', one said, 'because they are educated and we are not [ashikita]'.

But rice farmers—not just the big commercial farmers but also smaller farmers under real pressure to make the most of a complex cropping system on a hectare of land—were continually carrying out their own experiments. They tried new varieties and new techniques. One case of cultivator's practices will illustrate a historic cropping experiment and reveal farmers' attitudes towards the HYV technology. Hundreds of land-poor Hindu rice farmers in a swampy area of Rajoir thana (and adjacent thanas) in central Faridpur District had made a practice of going north by boat to harvest rice in a swampy area of Sylhet District. Oral history among these farmers suggests that the practice began at least before the partition of Bengal in 1905; it is probably much older. Payment was a fixed proportion of the rice they harvested; they brought some of this rice 300 km. home to Rajoir. About 1966 some of them brought a variety of rice grown in the summer season in Sylhet back to their farms: apparently no one knew if this variety (which they call 'Lakhidigi') would grow at home. As one group of farmers said, 'It was just an experiment [parikar], the same way people were wanting to see about IRRI rice.'

The similarity of ecological conditions of the two areas and the interest of these poor farmers brought about the planting of this rice at a place 200 miles away from its place of origin, at about the same time as the introduction of the HYVs in Rajoir.[29] The first 'IRRI-block' of HYV rice was planted in Rajoir thana in 1966–7 in the summer season on the land of a very rich merchant at the same time as the Lakhidigi transfer. Ten years later, in 1976–7, the Lakhidigi variety was being planted, also in the summer season, on 1,000 hectares, compared to 560 hectares of HYV rice. Average yields in this season for Lakhidigi in this thana were 2.19 tonnes, and average yields for HYV rice (IR 8) were 3.65 tonnes a hectare. The Lakhidigi rice required no fertilizer, was pest-resistant, needed little irrigation because it was planted in the excess swamp water, but was tall and thus vulnerable to unseasonable hail or rain. In 1978 it was selling in the local markets at the same price as HYV rice though it was thought to be better-tasting. Some farmers planted it in places where they expected yields of only 1.43 tonnes per hectare because there was so little cost associated with its production. Above all it was a variety which entailed little risk.

There was an additional reason behind the adoption of this variety from Sylhet. These farmers were from the Hindu minority and lived deep in a swampy area which was inaccessible except by boat or long indirect paths. The indigenous variety did not require interaction with, or the goodwill of, distant government officials. Minority status, isolation, and poverty meant that adoption of this rice conferred on these farmers an advantage which they would not otherwise have had. The decision to plant the HYV seed, which required costly inputs, was similarly based on complex reasoning, partly related to the supply of technical requirements of HYV cultivation and the need to be in active relation to government programmes. These same cultivators had earlier adopted the exotic variety Nigershail (in the 1950s), and now treated it, and the newer Pajam (Mahsuri) as something akin to 'traditional' varieties quite separate from the HYV rices. Both were, of course, modern and improved varieties.

In a classic separation of labour, BRRI scientists had to relegate most relationships with farmers to partially trained extension workers. Though BRRI tried to retrain them (in the face of resistance), extension workers did not conduct research and convey questions to scientists until well after 1980. BRRI thus learned little from farmers or from extension workers; indeed, researchers from the classic cluster of disciplines, expected to produce a stream of new varieties, have, until recently, been isolated from this essential contact. Nevertheless, donors and supporters of BRRI remained hopeful, asking for expansion of substations and encouraging BRRI's training of extension workers. Near the end of the 1978 Review Mission Report one finds a classic statement of the proper role for agricultural research, a public relations statement which could have been lifted from a Louisiana rice brochure:

The bottom line is, 'How much more production is possible and how much more profit is to be earned by the suggested changes in this cropping system?' Also, the additional question, 'What are the additional risks and extra capital needed to make the suggested changes?' Farming is a gamble. The physical scientists, the agronomist, the entomologist, the plant breeder, the pathologist, outline the opportunities in the bet. The economist, if he is any good, sets the odds. If the odds are 2 to 1 or better, then the extension agent can make his recommendations with some assurance of success.[30]

Only someone wholly unfamiliar with cultivating conditions in Bangladesh and with the nature of BRRI's research on cultivating conditions could subscribe to this view of 'the odds'. It is, of course, a gamble to farm rice, but only the farmers in their localities really knew the odds, the risks, and the chances of success. Certainly the experts mentioned in this statement did not really know the odds. Until recently the extension agent has been almost out of the equation in Bangladesh, except in a few areas. Some researchers knew 'about' cultivating conditions and odds, but whether they were economists or entomologists, they did not understand the odds. These odds were not even systematically calculated. Even drawing up district-by-district 'farm budgets', common in North America, would have required an enormous trained staff. No one in Bangladesh could presume that 'the bottom line' was so simple, or that production and profit were the only considerations of an Asian rice farmer confronted by technical change.

Although relations between researchers and extension agents sometimes succeeded on an individual basis, structurally research and extension were unrelated and unfriendly activities. Indeed the attempt by BRRI to train extension agents in 1977 and to test them regarding what they learned before awarding a certificate resulted in a strike of extension agents against BRRI which the director was unable to resolve. The Minister of Agriculture had to come in to negotiate and settle the matter. Some years later the humiliating status and salary differences between research and extension began to be removed. But the situation in the countryside was actually far from the 1978 Review Mission's statement about the scientists outlining the 'opportunities in the bet'.

9.3 The Responsibilities of BRRI's Donors

Half the cost of BRRI's operations were met, at the end of the decade, by grants from the governments of Canada, the US, and Australia, and from the Ford Foundation, covering most aspects of career advancement, foreign travel, and scientific conferences, and much of the new capital costs for buildings and a library during the 1970s.[31] BRRI's physical and organizational expansion was completely dependent upon foreign investment (see Table 9.1). When the Ford Foundation began investing in rice research in East Pakistan in 1966 there was no mechanism for

TABLE 9.1. *Aspects of BRRI's finances*

A. Total expenditures for salaries, research, administration (in million taka and current US dollars):

1974–5	1975–6	1976–7
Tk19.20	Tk25.50	Tk27.10
$1.28m	$1.70m	$1.81m

B. Foreign funding approved by donors for IRRI–BRRI collaborative research project (in current US dollars):

1978	1979	1980	TOTAL
673,800	738,850	802,550	2,215,200

C. Foreign funds released to and expended by IRRI–BRRI collaborative research project (in current US dollars), 1978, 1979, 1980:

Ford Foundation	250,000
Australian government	543,285
Canadian government	1,404,730
Total	2,198,015

Note: No figures were available to allow comparison of total BRRI expenditures (see A) and foreign funds released to and expended by BRRI (see C).

Sources: *Foreign Financial Support Grant Proposal*, BRRI, 1977 pp. 6, 57.
IRRI–BRRI Collaborative Research Project Terminal Report, BRRI, 3 August 1981.

attracting other donors to share the cost: the Ford Foundation continued alone until 1974, except for small USAID training grants. But in 1972 the Consultative Group on International Agricultural Research (CGIAR) was established, and potential donor-investors began to meet regularly to decide on priorities for international research centres like IRRI and relations among international and national institutes like BRRI. This is how new donors were found. From the beginning, foreign funds for BRRI were channelled through IRRI, whose director has been automatically a member of BRRI's board of governors since BRRI's inception. In 1978 there was a flow of $678,000 through IRRI for its 'collaborative relationships' with BRRI (see Table 9.1), and this sum accounted for almost all the foreign funds received by BRRI. IRRI thus had a kind of donor-status relative to BRRI in this period.

IRRI's research team in Bangladesh functioned (whether it intended to or not) as a monitor on this investment, exercising considerable influence on the report of the regular review mission sent by donors. Four IRRI staff were stationed at Dhaka by 1967. The team naturally varied in size and composition, but it always offered special assistance and training in addition to maintaining collaborative or co-operative research for IRRI with BRRI. IRRI considered the team, which by 1980 had five members, as part of its own outreach programme.

BRRI frequently relied on the IRRI team in Bangladesh and on Ford Foundation officials in Dhaka to negotiate its needs with its own Ministry of Agriculture which, like most other ministries, was more responsive to the petitions and inducements of foreign agencies than to those of its own staff. Though no longer the major donor to BRRI, the Ford Foundation was perceived to have a significant voice in the management of IRRI and thus to be crucial to BRRI's continued existence. It had experience and analytic capabilities. BRRI's board of governors was reported by observers in Dhaka to be relatively indulgent to the requests of its three major donors and to the review missions. Scientists thus traditionally looked as much to foreigners for leverage on their ministry as to officials for leverage on their donors. In 1974 a review mission sponsored by the Ford Foundation advised a major diversification of BRRI's financing. IRRI's director in the 1970s, Nyle Brady, was effective in bringing the diversification about. In the following year the Canadian and Australian governments joined the foundation as donors, so that by 1979 BRRI's foreign exchange costs were financed 75 per cent by Canada, 15 per cent by the Ford Foundation, and 10 per cent by Australia. It is interesting to note that the United States, Canada, and Australia were the sources of two-thirds of the million tonnes of wheat imported by Bangladesh at this time.

It was not foreign funding by itself which generated the criticism of BRRI's role in Bangladesh, it was the fact that most of such funding flowed through IRRI. There were exceptions to this: the British Overseas Development Authority objected to the IRRI director's request in the mid-1970s that support for deep water rice research be channelled through IRRI. Following a dispute the British funded this research directly, bypassing IRRI, and British scientists did not formally participate in top-level review missions of BRRI in 1976, 1978, and 1980. Just as important, this British-funded project did not restrict itself to a substation, such as Habibganj, but spread out in many deep-water rice situations across the country. But this exception proved the rule that most foreign research support was channelled through IRRI during this time, not just funds for IRRI's advisers. IRRI's director spelled out the reasons to CIDA in 1975. It was 'not merely a matter of pride', he said, but practical experience that IRRI needed control over funds related to its own scientists and their families to avoid their frustration; sometimes there were 'serious problems of management which must be corrected. If we have not leverage', he said, 'we are powerless to bring about the needed changes'.[32]

The historic fact that BRRI inherited the relationship between IRRI and East Pakistan's rice research, and consolidated that relationship, meant that the BRRI–IRRI relationship was still sensitive even in 1980. At once proud of the evident public respect displayed by IRRI and shy about the inevitable appearance of dependence, scientists and onlookers in Bangladesh sought to deal with this historic presence as best they could, in spite of coups, new governments, and changes of public mood. IRRI for its part became sensitive to BRRI's situation, which it knew intimately from the inside, because it had a seat on BRRI's board of governors. For fifteen years IRRI guided BRRI's fortunes, from pushing for a new institute in 1965, to its own official sitting as a member of the Review Mission in 1980. In fact that mission issued a kind of warning regarding the need for 'self-management' in relation to IRRI, something which had been voiced privately in many circles but which was not written publicly. Its report said that the withdrawal of IRRI at some future date 'could seriously jeopardize' the continuation of the work it helped get under way unless BRRI could start to manage its own affairs now.[33] BRRI should 'begin soon', said the 1980 Mission, 'to develop the necessary institutional capacity to self-manage' foreign assistance and international activities while the relation with IRRI was still close. A contradiction lay in the fact that this call for self-management came from representatives of donors who so profoundly influenced BRRI's budget. Another contradiction lay in the criticism expressed by some mission members that other members consciously and unconsciously sought to restrict the scope of the review because a more thorough (and perhaps negative) appraisal would reflect poorly on their own previous decisions to commit funds to BRRI and on the agencies which nominated them to the mission, thus causing trouble in the disbursement of aid funds.

It should be remembered that IRRI was dealing with the same agencies (as donors) which BRRI was dealing with. There was disagreement at IRRI at this time whether these outreach programmes (like work at BRRI) were to be considered part of IRRI's core budget. Donors were not agreed on whether to fund outreach programmes through IRRI (as IRRI wished), or directly and independently. These issues were not resolved during the 1970s. The advice issued by the 1980 Review Mission addressed a condition of IRRI–BRRI relations which had evolved to that stage over fifteen years. IRRI was also not the only source of outside influence on BRRI. In 1974, for example, the Ford Foundation representative in Dhaka advised the Secretary of the Ministry of Agriculture to ensure removal of certain outstanding issues in BRRI's still-dependent relationship with the ministry. A year before there had been promises that these issues would be dealt with, but in 1974 the foundation's officers decided to get tough with BRRI.[34] The Dhaka representative advised his colleagues that removal of these issues should be a necessary precondition of further support. Within a year those issues had been largely settled. Funding was renewed, although at a reduced level.

Later, when Ford financial support was eclipsed in volume by Canadian support, it was reported that one Canadian official's zeal, to demonstrate that investment from CIDA was justified had permeated and distorted the work of a review mission. In fact the difficulty in reconciling the three uncritical mission reports (1976, 1978, 1980) with the sceptical private views of observers and scientists lies in the fact that these high-level missions visiting BRRI for a week or ten days were satisfying BRRI's expectations, IRRI's expectations, and donors' previous investments. Criticism in the reports was muted. Some of the tougher judgements were delivered only orally, if at all. Each mission agreed on the highest objectives: to consolidate and expand research capability and to keep international investments in it as high as possible. There was a form of escalating entrapment here. Once they were deeply committed, donors were reluctant to retreat: all they could hope for in BRRI was leverage, and that is what they got.

From the time in 1973 and 1974 when the Ford Foundation dug in its heels regarding disorganization at BRRI there were four major ideas promoted by foreign scientists and planners. The first was the need for long-range planning, which BRRI achieved by 1980. When this was first discussed in 1973 and 1974, BRRI, like most institutions in Bangladesh, did not think longer than one year ahead. The second idea was that interdisciplinary teams or task forces were needed in order to overcome the severe bureaucratic limitations of divisional and disciplinary boundaries. These teams were scarcely visible in 1980. The third idea was that BRRI was not being taken seriously by the Ministry of Agriculture (from which it was fighting for autonomy), or the Planning Commission—which controlled the budgeting process. Some observers argued that neither the Planning Commission (as a whole) nor the World Bank (the commission's major influence outside the country) really sufficiently valued BRRI's potential: foreign scientists admitted its limited present role but felt that these major bodies had missed the potential of the role of scientific research in Bangladesh. The fourth idea was the sanctity of a rice-only approach; some foreign scientists, particularly those from IRRI, tried to counter attempts to bring BRRI into line and resisted having it formally co-ordinated in an all-crops approach. Theirs was an argument for the autonomy of rice, and an argument again for the autonomy of BRRI.

Finally, of course, there was a variety of positions on what BRRI's objectives should be. In its simplest form, the mainstream view was 'study breeding first, study consequences later'. Transcending the position that the rice crop could be reduced to a handful of well-adapted HYVs, BRRI and its foreign advisers came to believe that as many as thirty varieties might be necessary in the future, requiring continuous breeding and evaluation: in short, the biological engineering of the rice plant. This continuous breeding was new, but, in the redundancy of approach, not far from the Department of Agriculture's policy of the 1930s. What is different is the sense of urgency: BRRI spoke of a *stream* of new varieties, acknowledging constant demand.

All of these ideas were translated into the recommendations of various planning consultants and review missions, adding to the day-to-day advice of scientists from IRRI and other institutions. The 1974 Review Mission recommended that BRRI be reorganized around 'problem areas' in the manner that IRRI was then being reorganized. This recommendation led to the formation of task forces which were supposed to begin the production of new rice varieties. In 1978, however, many people conceded that the task-force approach had been very difficult to implement because of the deep entrenchment of the research divisions and the high turnover of staff. The 1978 Review Mission from the major donors recommended that the preoccupation with on-station research be changed by offering transport and per diem costs for work off-station, that fertilizer trials on-station be stopped because they led to misleading results which damaged BRRI's credibility, that BRRI be more careful with its research on pesticides, that administrative steps be taken to 'speed up the transfer of new varieties to farmers' fields', and that the task forces should be more narrowly focused.[35] The Review Mission also made a number of administrative suggestions about training and staffing which had been suggested by BRRI's administrative staff.

Though the donors had great influence over BRRI's work, they clearly could not monitor every detail of its work. Important changes were worked out by BRRI scientists. But the donors were at least partly responsible for centralizing research and neglecting the substations, for ignoring the study of consequences, and for limiting the responsible relations with rice farmers.[36] This neglect was partly due to the fact that both the donors and the Ministry of Agriculture were busy with other, bigger projects. One foreign observer in Dhaka observed, 'You cannot expect a Ministry which just received a $200 million loan from the World Bank to think about one of its institutes with a foreign budget of less than $1 million a year!' Donors like CIDA of Canada had higher-profile projects and more expensive projects in Bangladesh, although their officers felt that changes in BRRI might have important results; a development officer in the Canadian High Commission in Dhaka put it in 1978: 'How much time can I spend on $2 million over a couple of years for BRRI, when I've got to spend $12 million in the next forty-eight hours on Canadian food aid to Bangladesh?' This is why donors relied so greatly on the judgement of the IRRI team in Bangladesh and of the review missions, thus reinforcing the dependence upon exogenous foreign conceptions of the role of research in changing the rice agriculture of Bangladesh. BRRI's international relations keeps its senior scientists busy with flying visits, review teams, and missions from many foreign agencies, among them potential new donors. Briefs and submissions were in preparation continuously. In a few weeks in 1977–8, for example, BRRI was visited by the Minister of Agriculture of India, the Prime Minister of Mauritius, the President and Vice-President of the Ford Foundation, a delegation from the Rice Millers of America, a United States Senator, and the President of the United Nations World Food Council. While it was important that these people visited BRRI, protocol, explanations, and tours took up much time and resources. Middle- and junior-level staff became so important to the research process at BRRI because the senior scientists were so often busy with visitors and other things unrelated to research. The presence of the IRRI team also added an additional credibility to BRRI in the eyes of the influential visitors, thus compounding the dependence on IRRI.

BRRI's history up to 1980 may be a classic instance of the difficult relations between a poor national institute and the riches of the international agricultural system. When the authors of the Asian Development Bank's Second Asian Agricultural Survey said that the existing international centres (like IRRI) 'have hindered the development of national programs by dominating them in some instances', they could well have been referring to BRRI.[37] One Bangladeshi agricultural scientist in Dhaka described BRRI in the past as 'a thoroughly colonized institute'.[38] One or two BRRI staff said privately they felt as powerless to do anything about 'misguided' foreign donor influence as they did about agrarian conditions within the country, and that both were equally important. Most BRRI scientists saw foreign funding as the only hope for BRRI, and it was in their material interest to believe this: only foreign funding made career advancement likely either in BRRI or in other government institutes. Foreign influence pushed the interests of rice when funding of other crop research was contemplated, brought in new foreign donors, and opened the possibility of international careers or short-term international consultancies. Foreigners were well aware of the degree of their influence on BRRI, but did not agree about the extent of their responsibility for limiting the contribution scientific enquiry had made to increased rice production. No major change was likely in BRRI unless donors started to object to further centralization. But they had largely paid for this centralization and simplification and had to look for a graceful way to change their minds. Donors were not particularly successful at decentralizing elsewhere— witness IRRI's difficulties with off-station research and its very partial attempts to study the consequences of induced technical change in this period. Yet donors who underwrote the cost of BRRI's external relations and most of its internal growth had profound responsibilities because they also underwrote many of the other costs of the overall HYV strategy and of most other parts of agricultural development.

What results were evident at the end of the decade as a result of these efforts? Despite statistical uncertainties, the evidence suggests a sobering picture of a constant contest between per capita needs for food and very small increases in rice production.

Boyce has investigated the uncertainties of agricultural reporting, in sown area of different types of rice in the boro and aman seasons, in the amount of land actually available for cultivation, and in the government and BRRI estimates of rice yields. Tentatively concluding a fascinating analysis of these uncertainties, Boyce states that production was probably higher than officially reported:

Bangladesh's total rice output at the end of the decade 1970–1980 is 3 per cent higher in the revised series than in the official figures, implying that the country's annual rice production is today underestimated by about 400,000 tons.... If future studies confirm the impression that aman rice acreage is substantially under-stated, by as much as 1 million acres, then the underestimation of current rice production would be considerably greater. Such output underestimation, it should be emphasized, does not imply that the magnitude of hunger in Bangladesh is any less severe than reported in past studies. Rather, the present enquiry suggests that widespread undernutrition, the incidence of which is determined in large part by the distribution of income and hence food, has occurred within the context of higher national agricultural output than is recorded in the official figure.[39]

Acknowledging the uncertainty of the official record, the North–South Institute's widely accepted Rural Poverty study concluded that in the period between 1969 (the last year of Pakistan's normal data recording) and 1982, rates of growth in production of boro rice was 2.4 per cent, aman rice was 1.8 per cent, and aus rice was 1.6 per cent per annum. Because of wheat's very low starting-point, its rate of growth was a spectacular 22 per cent. The study said:

Yield per acre of rice has been going up gradually, increasing by 13.5 per cent between 1969–70 and 1981–1982. Despite this increase, rice yield per acre in Bangladesh is still one of the lowest in the world. The average annual growth rate of yield per acre is found to be the highest for aman rice (2.05 per cent), followed by aus rice (1.92 per cent).[40]

The Rural Poverty study concluded that compound rate of growth of food production was 1.6 per cent while rate of population growth was 2.5 per cent. Boyce, on the other hand, put the rate of growth for food production (1949–1980) at 2.03 per cent. Whether it was 1.6 or 2.03 per cent, it is obvious that the attainment of BRRI's 1977 objective of a sustained annual increase of 3.5 per cent in rice production was, in 1980, a long way off.

Moving into the 1980s, the government announced another Medium-Term Foodgrain Production Plan in 1981. Stating that 809,000 tonnes of food were imported in 1980, the Plan again pointed out that 85 per cent of the land cultivable under HYVs was 'still being sown to local varieties'.[41] Research is mentioned once, BRRI not at all. In the Paris meetings of the 'Bangladesh Aid Group' each year in 1981, 1982, 1983, and 1984, the Minister of Finance and Planning Commission staff had to analyse the national macroeconomic problems of balance of payments, debt service, imported food, and aid disbursements and commitments in terms of the constant difficulties faced in food production. These restricted documents reveal a detailed blunt analysis of almost every relevant factor in rural development and the food situation.[42] Self-sufficiency remains constantly like a ghost in the text, hovering above the tables in Paris, above the twenty donors and the Bangladesh team. Almost every relevant factor is analysed except research and development. The role of agricultural research, new knowledge, or applications of old knowledge is absent. The limitations introduced long ago had taken a firm hold. And it is striking that the BRRI Review Mission of 1984 discovered a number of weaknesses at BRRI which would tend to reinforce the government's opinion of research as of questionable value. These weaknesses were anticipated in the 1978 Review Mission, including the underdevelopment of the regional substations and the dependence on IRRI. Still in 1984, IRRI was BRRI's agent for the foreign exchange component of donor funds which made up about 40 per cent of BRRI's total budget.[43] The Review Mission found signs of progress and signs of hopeful projects which suggest that donor investment had not been in vain. But considering the long sweep of rice research in Bangladesh, the findings of the 1984 Review Mission are particularly sobering:

it can be seen that IRRI manages and administers a significant part of BRRI's accounts.... The review identified two areas where they believe the BRRI–IRRI relationship needs to change. The first is in the general area of research capabilities, methods and philosophy and the second is in the area of administration and general management.

With two rice research institutes, such as BRRI and IRRI, working on hard rice production problems, it is very difficult to determine the intellectual relationship between the two. The problem is compounded as Bangladeshis join IRRI and thus BRRI makes significant inputs to IRRI's research programs.

In the past, BRRI was the 'younger sister' in the BRRI/IRRI relationship. This is now changing and the review team felt that greater attention should be given to accelerating this change.[44]

NOTES

1. 'Proceedings of the Seventh Meeting of the Board of Governors of the Bangladesh Rice Research Institute, Dhaka, 27 August 1973', FFA 73–455.
2. Ibid. 4; and 'Proceedings of the Eighth Meeting of the Board of Governors of the Bangladesh Rice Research Institute, Dhaka, 15 December 1973', p. 4, FFA 73–455.
3. Government of Bangladesh, *Annual Plan 1974–1975*, Dhaka, 1974, p. 38.
4. D. E. Bell to McG. Bundy, 'Recommended Grant Action (for Bangladesh Rice Research Institute)', 16 May 1973, p. 4; David Catling, 'Working Paper for Grant Review', July 1974; 'Terminal Report on Grant No. 730–0455', September 1976, FFA 73–455.
5. 'Report of BRRI Donors Review Mission, Dhaka, October 1976', pp. 8, 9, FFA 73–445.
6. 'Report of BRRI Review Mission, Dhaka, March 1980'; 'BRRI and IRRI Cooperative Project Annual Report 1979, Dhaka, March 1980', FFA 76–109. On the general malaise in the Ministry of Agriculture see Abul Quasem, 'Agricultural Administration in Bangladesh—A Note', *Agricultural Administration*, 4, (1977), 159–65. Quasem particularly notes the predominance of officials without agricultural training or interests.
7. BRRI, *Foreign Financial Support Grant Proposal for Bangladesh Rice Research Institute during the Period 1 January to 31 December 1980*, Dhaka: BRRI, 1977, p. 7.
8. Ibid. 8.
9. Bangladesh Agricultural Research Council, *Directory of Agricultural Scientists: 1976*, Dhaka, 1976.
10. Calculated from Ibid. 155–61.
11. BRRI, *Foreign Financial Support*, p. 2.
12. Ibid. 4.
13. Thomas Hargrove, 'Diffusion and Adoption of Genetic Materials Rice Breeding Program in Asia', IRRI Research Paper 18, June 1978.
14. Ibid. 5. Note that each of the following objectives were mentioned for less than 10 per cent of the crosses: adaptability, adverse soils tolerance, drought resistance, shattering, photoperiod sensitivity, milling recovery, threshability, seedling vigour, alternative gene sources, and yield stability.
15. BRRI, *Foreign Financial Support*, p. 7.
16. Donors to BRRI and other institutions, worried about the weak links between various research efforts, formed a Joint Research Review Team which would show how the Bangladesh Agricultural Research Council (BARC) could actually co-ordinate the entire approach to agricultural research. As they saw it the task was to make BARC strong enough to override narrow 'crop interests' like rice. The weak links were a result of the autonomy which donors had advocated. The team had on it the (reportedly) most influential Bangladeshi civil servant in the agricultural community, K. M. Badruddoza, and people like D. S. Athwal, an Indian who had been Associate Director of IRRI, and Albert Moseman, long-time American adviser to the FF and RF. See 'Strengthening the Bangladesh Agricultural Research System, Report of the Joint Research Review Team, April 1979', FFA PA 76–109.

17. BRRI, *Foreign Financial Support*, p. 9.
18. For a review of the evidence for these processes, see F. Tomasson Januzzi and James Peach, *The Agrarian Structure of Bangladesh: An Impediment to Development*, Boulder, Colo.: Westview Press, 1980.
19. BRRI, *The Bangladesh Rice Research Insitute*, 2nd edn., Dhaka, 1976.
20. For a review of the findings of the Bangladesh HYV Task Forces on the exaggeration of the spread of the new rice, see Dana Dalrymple, *Development and Spread of High Yielding Varieties of Wheat and Rice in the Less Developed Countries*, Washington, DC: USDA Foreign Agriculture Report 95 (6th edn.), 1978, p. 63.
21. BRRI, *The Bangladesh Rice Research Institute*, 2nd edn., p. 8.
22. 'Report ... Review Mission, 1976', p. 8.
23. Anderson chose to visit this substation in 1978 because he had travelled to Barisal frequently since 1972, studying the history of the countryside forty miles to the north of Barisal town. Familiarity and convenience were thus the reasons for this choice. We are grateful for the courtesy shown us by the staff of this substation: responsibility for interpretation of conditions there at the time is principally Anderson's.
24. BRRI substation Guest Book, Barisal, January 1978; and World Bank, *Appraisal of the Barisal Irrigation Project*, Dhaka, 1974. For comparison of rice cultivation in three different Barisal thanas, and differing relations to a major rice irrigation project which was plagued with difficulties, see A. B. M. Shafiqur Rahman, 'The Matrix of Institutional Communication in Development Projects: A Case Study of the Barisal Irrigation Project', Ph.D dissertation, Simon Fraser University, 1986. The marginal role of BRRI in this development is striking.
25. Personal communication, S. M. H. Zaman, January 1981.
26. Ibid.
27. 'Report ... Review Mission, 1980', pp. 7–8.
28. F. J. Zillinsky, 'Post-Completion Evaluation Report of the Rice Research and Training Project (Phase 1)', BRRI, December 1982, pp. 12–13.
29. BRRI's director said that other such boro transfers began even earlier, probably after the great famine of 1942, spurred on by the migration of Muslim farmers driven out of Assam in 1946–7 who settled in swampy areas of southern districts. Personal communication, 1980.
30. 'Report ... Review Mission, 1978', p. 18.
31. On the entire range of aid donor behaviour, see Just Faaland (ed.), *Aid and Influence: The Case of Bangladesh*, New York: St Martin's Press, 1981. Curiously, this book does not discuss the role of the Ford Foundation. See also Roger Ehrhardt, *Canadian Development Assistance to Bangladesh*, Ottawa: North–South Institute, 1983; and Rehman Sobhan, *The Crisis of External Dependence: The Political Economy of Foreign Aid to Bangladesh*, London: Zed Publications, 1984.
32. N. C. Brady to H. G. Dion, 3 September 1975, FFA 73–455.
33. 'Report ... Review Mission, 1980', pp. 18–19, FFA 76–109. Criticism was expressed that an IRRI adviser was sitting to review BRRI in this mission.
34. S. D. Biggs to G. Zeidenstein, 25 September 1974; G. Zeidenstein to A. M. Anisuzzaman, 4 October 1974, FFA 73–455.

35. 'Report ... Review Mission, 1978', the Mission was composed of L. F. Myers (Australia), M. Ishaque (Bangladesh), J. N. Efferson (United States), W. Freeman (United States), F. L. McEwan (Canada), and chaired by H. G. Dion (Canada).
36. For an excellent account of this process of influence, by the former vice-chairman of the Planning Commission, see Nural Islam, 'Relationships to Donors', pt. 2 of J. Faaland, *Aid and Influence*, pp. 37–81.
37. Asian Development Bank, *Rural Asia: Challenge and Opportunity* (2nd Asian Agricultural Survey), New York: Praeger, 1977, p. 252.
38. Discussion, Dhaka, December 1981.
39. James Boyce, *Agrarian Impasse in Bengal*, Cambridge: Cambridge University Press, 1987, pp. 115–16. See also J. F. Stepanek, *Bangladesh: Equitable Growth?* New York: Pergamon Press, 1979; F. Tomasson Jannuzi and James Peach, *The Agrarian Structure of Bangladesh: An Impediment to Development*, Boulder, Colo.: Westview Press, 1980; Stephan de Vylder, *Agriculture in Chains: Bangladesh: A Case Study in Contradictions and Constraints*, London: Zed Press, 1982; E. Boyd Wenvergren, Charles A. Antholt, and Morris D. Whitaker, *Agricultural Development in Bangladesh*, Boulder, Colo.: Westview Press, 1984.
40. *Rural Poverty in Bangladesh: A Report to the Like-Minded Group*, Ottawa: North–South Institute, 1985, p. 96.
41. *Bangladesh Medium-Term Foodgrain Production Plan*, Government of the People's Republic of Bangladesh Planning Commission, February 1981.
42. *Reports on 8th Annual and 9th Annual Meeting of Bangladesh Aid Group 1981 and 1982*, Ministry of Finance, Government of the People's Republic of Bangladesh, May 1981, June 1982; *Memorandum for the Bangladesh Aid Group 1982–83 and 1983–84*, Ministry of Finance, Government of the People's Republic of Bangladesh, March 1983, March 1984.
43. 'Report of BRRI Review Mission, Joydevpur, Bangladesh, November 1984', p. 35.
44. Ibid. 35.

10
Conclusions and Reflections on Research Strategies

IN this penultimate chapter we gather together some of the conclusions drawn throughout the book and relate these to one of our major concerns regarding the transferability of international agricultural research and implementation strategies in the light of local complexities. Then we raise and address a fundamental problem for science and technology that is highlighted by the Green Revolution experience, namely, the means by which diverse, specialized forms of expertise can be integrated and made sensitive to the entire context of mission-oriented research. In the final section we examine some conclusions relating to the roles and responsibilities of scientists and technologists in the Green Revolution. To this end we reconsider the Underspecification and Mango theses, which were presented in the Introduction.

10.1 The Foundations' Strategy and IRRI's Strategy

As demonstrated in Chapter 2 the scientific and technological research programmes into rice production at IRRI grew out of broad social and political concerns. Nelson Rockefeller and other influential leaders who oscillated between government service and the private foundations saw underdevelopment in Asia (and elsewhere) as the main factor limiting economic growth and blocking Western investment opportunities. Underdevelopment in turn was linked to the political problems signalled by Communist advances, such as the loss of China.

The pre-eminent aspect of underdevelopment was perceived to be the food problem. Only in those regions where food was adequate could the other economic planks in the developers' platform, such as raw material procurement and extraction, increased industrialisation, and the creation of markets for consumer goods, be implemented. Work on the world food problem was judged to be 'morally unassailable'.

Given the role of rice in the Asian diet the food problem was largely understood as the rice problem. The next steps in the reductive process were far less straightforward than noting the main food crop: Weaver and Harrar's report commissioned by the Rockefeller Foundation equated the tropical rice problem to a production problem, reduced the production problem to low yield per hectare, and focused thus on the genetic make-up of the tropical rice plant. The question of the distribution of food, or the means to produce it, was avoided.

The prescription for solving these problems was to limit scientific and technological research to the rice plant, a research area itself which the Rockefeller Foundation was told was underdeveloped. Furthermore, low yield was defined as an isolable technical problem amenable to a scientific solution which could be implemented without prior social and political changes. Although the solution was said to be isolable in this technical sense, the ramified social and political effects were never lost sight of by the planners. (The scientists' and technologists' vision was necessarily more complex, as we shall discuss.) But both the political links between new rice technology and US relations with Asia, and the long-term, broad economic links between 'modernized' agricultural production and US economic interests were crucial elements of the strategy from the beginning. In the view of one observer, Harry Cleaver:

> To be sure, the Green Revolution's high yielding varieties consisted of technological packages of seeds, fertilizer, controlled irrigation, pesticides, and so forth. But these technologies were conceived by western sponsors as part of an even larger package of social reorganizing changes—changes aimed at breaking up the countryside in such a way as to stabilize it.... The Green Revolution in short did indeed achieve a large amount of social reorganization in the process of raising food output. But the key political question must remain whether it achieved its fundamental political goal of pacifying the countryside and contributing to the creation of a labor force that could be mobilized by business.[1]

For Cleaver the increased production of rice, to which IRRI geared its research, was only a tiny question in a much larger equation: those consequences which were intended were calculated in terms of their contribution to pacification and the encouragement of capitalist cultivators.

Throughout the early 1950s the Rockefeller planners debated various alternative proposals for implementing major projects in rice research. Until 1958, when the possibility of a joint RF and FF effort was mooted, the idea of a single, centralized institute was considered too costly and risky. But then very quickly the two foundations decided to proceed with the project and to locate what was to become one of the best-equipped and generously funded agricultural research facilities in the world at Los Banos in the Philippines. Much of the substantive planning for IRRI was a direct continuation of aspects of previous Rockefeller initiatives in the southern United States, later in China, and then in Mexico.

The broad outlines of IRRI's agenda were set by the time it began active research in the early 1960s. Although there had been discussion about the multiple-cropping found on most Asian farms and there would be later debates with a somewhat different outcome, IRRI's focus was almost exclusively on rice. Genetic manipulation of the new rice plant was supposed to result in greatly improved varieties and thus large rather than incremental yield increases were expected. The target farms were to be irrigated ones rather than rain-fed or upland. In these and other respects the research agenda at IRRI was patterned on the RF's previous strategy of 'building on the best'. However, in spite of long-standing RF and FF interest in social engineering and the behavioural sciences, the scientific disciplines represented were mainly drawn from the natural sciences, the exception being agricultural economics, largely because it was (erroneously) considered the only empirical social science available.

Our view that IRRI's origins and even research agendas were linked to economic and political objectives (of the foundations) is far from being universally held. For example, induced innovation theory in agricultural economics, which was embraced by practitioners at IRRI, held that appropriate technological change is induced by relative factor prices in an undistorted market. Thus IRRI's original objective of developing a strain of rice responsive to fertilizer was, according to this theory, caused by increasing man–land ratios and low fertilizer prices. In the view of some observers, the theory thus

> essentially rejects these political connections, not by assessing the issues the historical and political context creates, but rather by denying that the connections between the historical and political context and IRRI's creation make any difference. Induced innovation theory says that IRRI's establishment and performance are matters that can be explained in apolitical terms.[2]

Although we generally side with critics of induced innovation theory such as Koppel and Oasa, we do not reject the theory totally; instead, we believe that there are different levels of explanation appropriate to different levels of analysis. Thus we believe that increasing man–land ratios and low fertilizer prices did play a role in shaping IRRI's programmes. However, we think that the evidence we and others have presented overwhelmingly supports the claim that political and economic considerations played a significant causal role in IRRI's founding and continuing mission.

One of the hallmarks of the foundations' and IRRI's concept of science was that it should have 'universalizing' capacity. Embodied in the mission-oriented search for rice varieties which would yield anywhere and any time, this grand, if simplistic, concept was the basis for IRRI's prolonged aversion to local complexity. If the 'universal' was the objective, the notion of aiming research at specific, local agro-ecological zones must have appeared as provincial and parochial. In IRRI's values of research these apparent limitations of place were to be avoided. At that time rice science was an obscure specialty, and the one way to glamorize its status was to pursue universal objectives (despite the fact this meant overlooking the reality of local complexity). Such a stance was quite compatible with a simplistic definition offered by a scientific advisory committee to the RF much earlier, in 1951, namely that 'agriculture is nothing more than the application of the principles of biology and other natural sciences to the art of growing food'. While it is true that the principles of biological and physical science are involved, there is something more to agriculture than the application of scientific principles. The debates within IRRI by the late 1970s show that many researchers had understood the poverty of this earlier definition, which nevertheless had influenced thinking for almost thirty years.

When in 1965 IR 8 was deemed ready for release, IRRI had to look towards the intended recipients of its research: to the national governments, to agricultural and research programmes in specific countries and specific districts, and to the rice cultivators of tropical Asia. We have concluded that IRRI's knowledge and concern about national research and village cultivation systems were minuscule. Basically IRRI's gaze was always either inward, to other international research centres and to donors, or skipping back to the United States. In our view until well into the 1970s IRRI and its scientists exhibited little institutional curiosity about the recipients of the research (farmers) or about the research and agricultural values and traditions in Asia.

And finally, what were the consequences of this thirty-year period of activity? Were these countries pacified? Were the economies of these countries in Asia revolutionized? As we have stressed, these questions take us far beyond the scope of this book. We do issue one important caveat that stems from our analyses and bears on questions such as these: it is dangerously misleading to suggest that changes in rice cultivation are linked in a simple, causal fashion to these bigger issues.

The evidence is conclusive that the agricultural research which contributed to the Green Revolution in rice has helped create a modern agricultural sector of rice growers economically integrated into the national economies. A careful recent estimate states that 'the volume of rice traded through market channels has grown at 4 per cent per year or more over the past three decades [1950–1980]'.[3] This annual rate of growth exceeds the estimated annual increase of Asian production of 2.66 per cent per annum for the same period, which suggests that a higher proportion of the harvest is going into the market.[4] Barker and Herdt speculate that in 1950 about 30 per cent of the harvest was being marketed, the proportion growing until 1980, when close to 50 per cent was being marketed.[5]

Moreover, the consumption of synthetic fertilizer has shot up in South and South-East Asia—15 per cent per annum in South Asia and 10.8 per cent per annum in South-East Asia for the period 1950 to 1980.[6] Even acknowledging that not all of the fertilizer is being applied to rice, the application of purchased, synthetic fertilizers to rice fields has grown enormously. Less spectacular has been the growth in proportion of the rice lands that are irrigated. For South and South-East Asia combined it is estimated that 57 per cent was irrigated in 1960 while, with an expansion of area, 76 per cent was irrigated by 1980.[7] We know from national and local studies that it is the state that builds and maintains the irrigation facilities and, in the bigger projects, administers the distribution of water. Thus both through the purchase of fertilizer and as irrigation water users, rice cultivators are increasingly integrated into the national economy. Not solely because of HYV seeds, other changes were taking place, such as the building of roads and other basic infrastructures which made possible the greater penetration of both state authority and the urban products into the countryside.

And finally, were these countries pacified by the technology? Bangladesh was not pacified by 1980 and was soon to experience a series of violent conflicts through the 1980s which left the country politically exhausted and under martial law once again. Sri Lanka was not pacified; on the contrary, it was on its way to civil war, and to occupation by large numbers of foreign troops. In 1980, after twenty years of IRRI's work, the Philippines was not pacified, was under martial law and heading for a dramatic collapse of the rapacious and corrupt Marcos regime. The major US military bases remained a contested but continuing presence. Even in central Luzon, where IRRI sat in the heartland of the adoption of new rice varieties, significant parts of the land were not under govenment control in 1980, just as described in RF memos in the 1950s. Levels of malnutrition and poverty remained significant. Yet the government and IRRI had made enormous efforts to raise enough rice to export from Luzon (and thus to try to develop an economic infrastructure which would make the government welcome).

Were the economies of these countries revolutionized? All three were certainly open to foreign direct investment, a major objective of the foundation planners. All three countries encouraged private enterprise policies, employed large low-wage labour forces, and had a significant private sector. But in no sense could anyone say that their economies had been revolutionized in the way Nelson Rockefeller intended. Neither their poor, nor their ministers of finance, nor the World Bank could make such a claim. In fact, in each case there was significant private capital flight, massive public debt, and a major outflow of skill through constant migration. And this was true in spite of the fact that the new rice technology had contributed to the increased production of food in each country.

We are not suggesting that debilitating conditions like malnutrition and dictatorship were part of the planners' and researchers' intention or that changes in rice cultivation alone 'caused' these conditions. Nor are we suggesting that planners and researchers thought that the debilitating consequences of agricultural change would be advantageous to them, as held by the 'you too can profit from the coming crisis' school of planning.[8] The Green Revolution in rice is one of the features of these changes; links there surely are, but simple links they are not.

10.2 The National Research Strategies

The foregoing strategies were addressed to a level of action intended for, but rather removed from, the rice cultivator and agrarian society. National governments were not so far removed, and were compelled to respond to conflicting advice with much more limited resources than the planners and researchers. Their own national strategies had to adapt as well as possible to both domestic and international pressures. While the differences in strategy between Sri Lanka and Bangladesh in 1950 were not great, by 1980, through different relationships with IRRI, the differences were marked and illuminating.

The Sri Lanka Strategy

Research on rice and other plants has been continuously under way in Sri Lanka for over half a century, since the founding of the Department of Agriculture in 1921. By the early 1950s, original research on rice was being conducted at three research stations: the Agricultural Research Station at Gannoruwa, Maha Illuppallama (1950), and the Central Rice Breeding Station at Batalagoda (1952). Later adaptive research was carried out at other smaller centres located in other agro-ecological regions.

The senior research officers were well trained, with doctorates from universities in the UK, Canada, Australia, Japan, and, increasingly, the US. In at least the three major research centres there were agricultural scientists with advanced-level training in botany, genetics, entomology, plant pathology, and soil sciences by the early 1960s. These scientists with advanced training were supported by junior officers who had graduated from the Faculty of Agriculture of the University of Ceylon (now the University of Peradeniya). Their research findings and field observations were reported in the Department of Agriculture's own journal, *The Tropical Agriculturalist* (founded in 1844) as well as in the *Journal of the National Agricultural Society of Ceylon* (founded in 1963).

While few in number, the researchers had established the value of their work and were recognized internationally as a well-organized and competent group.

While the research tradition was being built, the government never accepted the position that research and plant-breeding, by themselves, could solve Sri Lanka's food problem. From 1931 to 1981, far greater resources were committed to developing or restoring large-scale irrigation in the dry zone. It was believed that the extension of area under irrigated rice would both meet the food needs of the nation and provide employment for the landless Sinhalese from the wet zone. Additional resources were committed to improving the infrastructure for the small cultivator, particularly in the Kandyan areas. Construction of roads, small dams, and electrical distribution systems had been financed by the annual appropriations for the Peasantry Rehabilitation Department. While these measures were not specifically targeted at improving the rice production in the country, they were part of a broad-fronted agricultural development policy within which the more specific rice research strategy was located. At no time did the government leadership or the Department of Agriculture abandon the irrigation development policy or the infrastructural improvement policy in favour of an exclusive seed-driven agricultural development strategy.

Within the research establishments plant-breeding held the central position, yet it was never forgotten that any new rice varieties had to flourish in diverse agro-ecological environments with distinctive topographies, soil types, and rainfall regimes. From the preparation of the first soil map by A. W. R. Joachim for the department in 1945, rice varieties were selected and bred to grow in specific settings. The concern for the cultivating environment was best articulated by C. R. Panabokke, Deputy Director of Research 1974–80, who argued that agriculture must be seen as a total system built on a foundation of soil types and hydrology. Soil and water influenced both the types of crops that could be grown and the varieties of paddy that would produce the highest yields in each specific setting.

Associated with the agro-ecologically based foundation for rice-breeding was the recognition that the locally available cultivator technology and the local social organization of rice-growing had an effect on the yields. Each distinct environment limited the crop choices, which in turn imposed patterns of activity on the cultivator within his available technology and resources. Given this perspective, it was possible to modify the local system and push it towards higher levels of productivity by offering a wider range of high yielding plant types, by encouraging the organization of multiple and varied cropping systems, and by enlarging the cultivation technology and resource base of the cultivator. The Sri Lankans had developed a sophisticated research strategy derived from a holistic perception of the relations of soil, plants, and people.

A measure of the success of this developed research tradition was that the cultivators were able to draw on the knowledge it had created and the seeds it had bred; clearly this occurred with the spread of both old seeds and new HYV seeds and the increase in yields. But since this research was still divided by environment and scientific discipline, it remained for the cultivator to make the difficult and risky decisions on how to integrate the new knowledge effectively and how to use the new seeds. In our assessment, the cultivators were left to establish the real integration of knowledge with their own specific physical environments, with the aspirations of their own households and the expectations of their neighbours. No research station's advice could replace the skill of integration required by the cultivators, nor would any outsider offset the costs of a crop failure.

The Bangladesh Strategy

In Bangladesh, between 1910 and 1950 there evolved, very slowly and with setbacks, a research strategy based on an agro-ecological approach. This strategy included establishing research stations in different cultivating zones (such as the station for deep-water rice) and the international transfer and testing of improved rice varieties from all over the world (for example, one of the most widely grown, Nigershail, was introduced from Nigeria in the 1940s). Under the colonial economic system rice remained a low research priority in comparison to jute, tea, and other exportable crops. Decapitated in 1947 by the migration of its scientific leadership to India, the rice research community remained weak through the 1950s, relying mainly on international contacts with Japan, India, and the USA through the United Nations. Attempts to create a specialized rice research institute did not succeed until the mid-1960s when the country agreed to allow the first widespread field test of new HYV varieties and accepted resident advisers from IRRI. The response to the HYV seed and its requirements was a highly centralized one: most were imported, and rice researchers were largely marginalized in the process. The focus in the new autonomous institute for rice, founded in 1969–70, was even more limited, meaning that researchers had little influence on official selection of research objectives or new factors of production to achieve self-sufficiency through the 1970s: their task was simply to produce a stream of new rice varieties. Thus, perhaps as a reaction, tension gradually developed between research that was context-specific to agro-ecological zones and the more centralized 'national' approach of BRRI's headquarters. This tension appeared, in 1980, in extended dialogue with BRRI's donors and researchers concerning continued dependence on IRRI and underdevelopment of BRRI's research outstations. Well founded and sincerely held, these concerns were addressed to contradictions resulting from the simplistic approach to rice research which had been adopted by Bangladesh and by IRRI advisers. BRRI's donors had been conscious of the limitations caused by this approach throughout the 1970s.

The position of the peasant cultivator in the strategy was subordinate or ignored. Compared to the big farmers, strategists said that the peasant was a shrewd user of new technology but lacked knowledge and capital. In the 1960s, for political and economic reasons, the strategy focused on the bigger farmers, but in the 1970s the efficiency of the small farmer was increasingly recognized. As the debate grew between those who were committed to and those who were doubtful over the HYV strategy, the 'conceptual' status of the small farmer in the system rose so that in 1980 it was almost equivalent to that of thirty years before. Nevertheless the farmer remained faced with uncertainties which researchers could only minimally address. With the status of rice researchers in the national hierarchy finally secure in 1980, BRRI finally began to turn its attention to the local complexities of agriculture which had been ignored for the previous twenty-five years at such great cost.

10.3 The Natural and Social Scientific Paradigms in Research Strategies

It is common to observe that natural science paradigms are being displaced by broader approaches including the use of social sciences.[9] This observation raises two other important questions: what are the major differences between the paradigm of the natural sciences and that of the social sciences? In the context of research policy is one paradigm to be preferred over the other? We briefly address these questions, but first a caveat about the concept of paradigms in general is necessary.

Over the last twenty years the use of the term 'paradigm' in and about science has become ubiquitous, no doubt prompted by Thomas Kuhn's employment of the term in his enormously influential work *The Structure of Scientific Revolutions*.[10] This is hardly the place to enter into an extended critique of Kuhn's concepts of paradigms—we say concepts because, as many critics have pointed out, and Kuhn has acknowledged, his notion of 'paradigm' is a cluster of concepts—but for him the focus of these concepts has always been a particular scientific discipline, or even subdiscipline, its practices and practitioners. It may seem to be a straightforward matter to extend the notion of paradigm so that we can speak of a 'natural science paradigm'. In so doing we can carry out the sociological aspects of Kuhnian paradigms fairly easily, in that a

Conclusions on Research Strategies 369

paradigm is a shared way of viewing the world, of characterizing problems, and of offering solutions. But the same cannot be said of his philosophical, principled claims about the incommensurability of paradigms and the insurmountable incommunicability among holders of different paradigms. These philosophical components of Kuhn's analysis are, in our view and in the views of many others, on very shaky ground to begin with.[11] Attempts to extend them will strain them well beyond the breaking point.

The question arises: what are the fundamental contrasts between natural science and social science paradigms? The following illustration suggests an answer:

while a natural science paradigm may be quite adequate for explaining soil erosion on a single hillslope, quite different considerations come into play at a national or international level. Many erosion scientists acknowledge the socio-political dimensions of soil conservation work...but, preoccupied with a perceived need to retain an ideological neutrality, few will accept that soil degradation itself may be a landscape manifestation of a socio-political system.[12]

What the author says is that if one views the world through a 'natural science paradigm' one sees only parameters of a restricted physical kind; one is professionally blinded, for example, to the presence, action, and importance of sociopolitical factors despite the fact that these are acknowledged. We believe that while this point is correct, more elucidation and evidence is required to support an extension to a general claim. At the least, we noticed how frequently natural scientists were conscious of crucial sociopolitical factors which nevertheless disappeared in their professional and formalized accounts of the results of their research.

The difficulties begin with the fact that there is a bewildering array of mutually incompatible entities which are candidates for being a 'social science paradigm'. But, for our purpose, more important than abstract consideration of social science paradigms is an examination of what role in practice social science paradigms have played in the research strategy. At IRRI, and at most of the other international research centres as well, the continuously active social science paradigms were few during the period before 1980. Although other types of social scientists have been employed from time to time—for example, an anthropologist served as a senior scientist from 1977 to 1981—agricultural economics was the only non-natural science department at IRRI. Throughout this period economics was defined as the only social science useful to IRRI.

It seems undeniable that calling for an increased emphasis on social science approaches raises a number of questions which have received scant attention: What sort of social science should be pursued? What should be the role of social science with respect to natural science research? Finally, we have a very basic issue to address about the suggested contrast between 'natural science' and 'social science' paradigms. It is plausible to claim that a natural scientist working, say, on soil structure or plant genetics is focusing on the tail or the tusk of the elephantine problems of food and agriculture. We certainly agree that in practice natural scientists may show little concern about either the broader causes of the problems they address or the wider implications of the solutions they attempt. However, it is unclear that social scientists have, in practice, been very good at interpreting their problems and designing their research in a way that would facilitate interchange with researchers from other disciplines. Undoubtedly there are historical, institutional, sociological, and, perhaps, psychological factors which work against an 'ideal' dialectic among cultivators, administrators, politicians, and other researchers. However, this failure is not due exclusively to a conceptually-grounded myopia on the part of natural scientists.

The solution may appear to be obvious: call for a joint approach by natural and social scientists. Although we certainly concur that a communicative reconciliation would be desirable, we also believe that the difficulty of achieving this communication is vastly underestimated. Indeed, the problem of how to draw on the understandings of the natural and social scientific analyses of the world (as well as moral and aesthetic considerations) is among the most fundamental issues at the heart of this book.

10.4 Integration: A Fundamental Problem for Science and Technology

Since the seventeenth century science has advanced by employing the two-part strategy of 'reduction' followed by divide-and-conquer. The reductive step consists in casting problems narrowly. In methodological terms the divide–and–conquer approach involves treating portions of the real world as isolated systems, systems whose 'external' connections can be ignored. In institutional terms the strategy results in the development of increasingly narrow specialities and specialists.

In spite of the undeniable successes of S and T in many respects, the strategy does have a serious drawback. The difficulty is that even if S and T is successful on its own terms (that is, by characterizing and making correct predictions about the isolated entities particular specialists have chosen to examine), there is relatively little reason to suppose that there is an overriding theory or even a set of pragmatic rules that tell us how to integrate the bits into a unified whole. This problem is particularly acute in such cases as the Green Revolution where planning and implementation are swift and where different levels of knowledge and values concerning the plant and its role in society are involved.

Conclusions on Research Strategies 371

As we saw at length in Chapter 2 the foundations and IRRI, acting on the advice of scientists, defined the food problem in Asia as simply one of yield per hectare. Even after narrowing the problem in this way the promoters of the Green Revolution were very quickly confronted by the fact that attempted solutions had to deal not only with a range of features in the physical environment but also with historical and cultural factors at the micro- and macrolevels. Social micro-infrastructure and the macrostructures of property, inheritance, marketing laws, etc. were major contributors to the local diversity, which, as we have seen in Sri Lanka and Bangladesh, were often limiting factors encountered by the HYV technology.

If the problem of communication between natural and social scientists were overcome, and even if resources at such institutions as IRRI were more evenly distributed between the natural and social sciences, it would still be difficult to achieve integrated analyses of the food problems in Asia and of the efficacy of various solutions. Under the arrangements at IRRI in Sri Lanka and Bangladesh there is no accepted way for agronomists and engineers, anthropologists and economists, geneticists and hydrologists to produce unified analyses of the food problems of a district and an integrated plan to deal with the complex of issues that would be involved. While such a plan might be possible in principle, there are still formidable practical and theoretical barriers. We lack definitive evidence for the barriers, but we find support from the history and sociology of science. The historical point has already been mentioned: the reductive and divide-and-conquer strategy is deeply ingrained. Nevertheless, more powerful ecological approaches within science have recently been accepted. But it is far too early to claim, even with powerful computers, that we have adequate ways to make predictions and explanations, especially about very large, highly interconnected systems with different loci of decision-making.

Such a conclusion about integration relates to the Kuhnian analysis of science. Kuhn focuses on the problem of succession, that is, the way in which a scientific paradigm in one discipline is replaced by a successor. In Kuhn's account this process can be revolutionary in such a way that the adherents of the old and new paradigms are so deeply divided that they have neither sufficient common language nor an adequate, shared observational base from which to judge the two paradigms.[13] According to Kuhn, those who choose one paradigm over another do so not unlike those who choose one political or religious system over another.

Now rotate Kuhn on to his side: instead of having his analysis apply solely diachronically to *successive* theories within a single discipline, consider the synchronic situation, the situation in which there are several contemporaneous theories from a number of disciplines. Here in the interdisciplinary arena difficulties of the Kuhnian sort are manifest:

differences of language, conceptual approach, historical tradition, experimental design, and interpretation are ubiquitous features of interdisciplinary endeavours. There are, of course, disparities between the diachronic and synchronic cases. In Kuhn's (diachronic) version successive paradigms are in some sense competing such that at most one can prevail. In the synchronic case all the disciplines might be able to coexist. Still, we think when Kuhn is 'rotated' the result is at least as powerful as the original Kuhn. The problems are not necessarily insurmountable, but they are prodigious. If we are correct it may well be admirable to advocate unified approaches but facile to expect them to occur quickly.[14]

10.5 The Underspecification Thesis Reconsidered

In the Introduction we said that the responsibilities of scientists and technologists are seldom adequately assessed because the conceptions of science and technology and of their activities and products are too restrictive. We suggested that the assessment of responsibility is aided by considering the applicability of two claims, the Underspecification and Mango Theses. According to the Underspecification Thesis the tasks undertaken by scientists and technologists are in principle underspecified; as a consequence, supplementary goals for the central and auxiliary technologies are provided by the S and T enterprise. The Mango Thesis concerns the nature of the products of S and T. It asserts that the product may consist of a technological core which has little sociopolitical input from scientists and technologists. The thesis maintains, however, that what is usually implemented or transferred along with the core is a set of value-impregnated strategies designed to meet the requirements generated by the core. Meeting the requirements one way or another is unavoidable. Each thesis implies a broadened responsibility for S and T: the provision of supplementary goals and the provision of strategies for implementing the core technologies are the loci of responsibility. We now consider how these two theses hold up in the circumstances of our story.

In IRRI's case, the broad characterizations of problems and solutions were provided by agencies external to the S and T enterprise, namely, by the Rockefeller and Ford foundations. Solutions to these problems were, in our view, underspecified. In their deliberations the foundations obtained advice from agricultural researchers, who in turn suggested only those goals which they deemed that the state of the art could achieve. These goals were at least acceptable to agricultural researchers and probably suggested by them, since they had previously played a role in developing HYVs for other crops and even for rice cultivation in non-tropical areas like the United States. Even granting this consultative interaction involving scientists, technologists, and planners, however, we believe that the authority and accountability for adopting these general parameters rested with the overriding sociopolitical institutions, that is, with the foundations. Of course, one cannot regard these foundations as having political accountability in this case since they have no direct input from and are not directly accountable to the main group on whom their decisions impinge, the people of Asia. In the constitutional sense they acted without responsibility.

Formal participation in the process of specification was limited to North American institutions—most notably the foundations and other donors—and members of the S and T enterprise. A possible exception at the beginning was the participation of the government of the Philippines, which was certainly party to negotiations concerning IRRI's location and physical plant. Nevertheless, the evidence shows that the selection of research objectives at IRRI was not particularly affected by Philippines priorities. Since the ultimate aim of the research was the adoption of new technology by Asian governments and farmers, clearly attention had to be paid to which technological innovations would be acceptable to those governments, particularly those represented on IRRI's board. We conclude, however, that such attention did not involve formal consultation with such governments at the early stage of setting research objectives. At the point of testing new varieties in countries like Bangladesh in the mid-1960s, and obviously during the 1970s, through the board and the CGIAR, governments regularly reviewed and commented on IRRI's research objectives. But early IRRI researchers enjoyed a great deal of autonomy in their work because of the absence of this formal consultation.

As we have seen, from the beginning IRRI's research agenda was focused on rice monoculture and irrigated crops. There has been considerable debate about the merits of these project parameters. Concerning the emphases on irrigation and rice monoculture, for example, critics have claimed that both conditions favour the better-off farmer: first, approximately 25 per cent of the rice land in Asia is irrigated, is owned by richer farmers, and irrigation certainly produces more rice; second, the poorest farmers usually grow a mixture of crops, often on unirrigated land. Our purpose here is not to enter this debate but to demonstrate that these parameters and priorities were essentially the general terms within which the foundations and IRRI's board of governors expected the scientists and technologists to work. That is, although scientists and technologists did contribute to discussions about these parameters, it was IRRI's directors and donors who decided the priorities. However, as later debates about the cropping systems programme indicated, even these major parameters could be altered under pressure from scientists and technologists. Furthermore, both the balance among the parameters and the quantitative specification of these qualitative features (such as how large an increase in yield is required) were amenable to influence by scientists and technologists. In addition there were a host of subsidiary decisions which were left mainly to the discretion of scientists and technologists (individually or collectively) and which had significant effects when the result of their work was transferred.

10.6 The Mango Thesis Reconsidered

The Mango Thesis directs our attention towards the relationships between requirements associated with implementing a core technology, such as the HYV seeds, and the values associated with methods for meeting those requirements. HYVs produce significantly higher yields only if their adoption is accompanied by changes in cultivation practices involving water, fertilizer, and pest control. While each of these requirements might demand technological factors, it is also often the case that indigenous social organizations and skills could have been partially substituted for expensive inputs. For example, there is the story told about a group of IRRI scientists who visited North Vietnam in the late 1970s and found an IRRI variety flourishing on a commune. What was even more surprising than finding it in a country that had few official contacts with IRRI was that the variety under cultivation had proven to be especially vulnerable to pest infestation, and thus its use in other parts of Asia had been almost discontinued. While the scientists at first assumed that either the pest was not present or that the commune was employing huge amounts of pesticides to control the problem, they learned that neither of these alternatives was accurate; the pest was in the area and the commune was not prosperous. When questioned, the commune leaders revealed that they had developed a simple insect trap and a highly labour-intensive observation system—in which much of the monitoring was done by children—so that application of pesticides was strictly co-ordinated with the insects' most vulnerable period. With this system, a relatively small amount of pesticide was effective in ways the IRRI visitors had not thought possible.

The point of this story is to illustrate both that social organization can be an alternative to technological measures and that such alternatives may not be seriously entertained by scientists whose priorities and approaches are shaped by involvement in their own research strategy. In North American rice-farming, for example, pest control is effected by blanket (often airborne) applications of chemicals. A labour-intensive monitoring system was simply not economically feasible. Thus Western-trained and oriented scientists' recommendations for dealing with requirements generated by the new technological core may be based on the values and priorities they hold and thereby exclude feasible alternatives. This unconscious and conscious process of exclusion has been demonstrated again and again in many countries of Asia.

Taken together the Underspecification and Mango theses give grounds for denying that implemented technology is neutral in any important sense. If the theses apply to a particular case, then in that case what is implemented should be understood as consisting of a broad spectrum of technological features plus some features so closely linked as to be attributable to the scientific and technological enterprise. We are convinced that the theses apply in the case of the Green Revolution in Asia.

On a general plane the theses shed light on the long-standing controversies about neutrality. On the one hand supporters of technological neutrality, such as the advocates of the theory of induced innovation, can be seen to be using a particularly anaemic notion of neutrality. For them the neutrality of agricultural research stems from their claim that the agenda for such research is set by market demands. Even if the latter is correct—and as discussed above, we reject the idea that the theory fully explains the phenomena—it seems obvious that in practice because of infelicities in interpreting demand and because of imperfections in the market, the products of research would almost always have enormously differential effects.[15]

On the other hand, Cleaver, Winner, and others who reject technological neutrality are invariably found wanting when defenders of neutrality point to the obvious flexibility of some embodiments of technology, which, in their view, confers neutrality on technology. In the case of the Green Revolution, for example, flexibility is demonstrated by the different agricultural and political systems in which the new seeds were grown. The Underspecification and Mango theses show how the rejectors' analysis can be strengthened: we do not accept the unduly narrow notion of technology that the defenders of neutrality employ.[16]

10.7 Responsibility of Science and Technology

Even if the zone of activities of scientists and technologists is widened in accordance with the Underspecification and Mango theses, we have neither a precise prescription for assigning responsibilities nor simple recommendations for adequately discharging them. The issues we have examined are simply too complex, too interconnected, and too varied to yield to a surgically neat reduction and isolation into easily manageable packets. What we aspire to have done is to have shone fresh light on the interface between the real world of rice cultivation and the scientists' laboratories. Among the old saws that we have tried to deny is the claim that scientists bear little responsibility for their products, especially in the case of mission-oriented research, because their tasks are set by the sponsoring agency.

Another effort to diminish responsibility has its thrust in the opposite direction: scientists and technologists have limited responsibility because by the very nature of the activity a researcher does not know the outcome in advance—if he or she did, there would be no need for research. Of course there is some plausibility to these claims, but in the case of the Green Revolution the first is undercut when the Underspecification Thesis is satisfied. As for the question of predictability of research and, for that matter, of effects of implementation, a great deal hinges on the level under examination. At the most general level, that of the planners at the donors' headquarters who were concerned about the economic and political status of entire countries or even regions, the general intentions were clear and the general results of research at IRRI were predictable. The general intention was that science and technology could be relatively uncontroversial instruments for reshaping economic and political structures, allegiances, and relationships. The general results of research—that plant genetics could produce new varieties that potentially could significantly increase yield—were essentially a foregone conclusion. In the light of the experience with wheat in Mexico and with rice in the temperate zone, it was only to be expected that IRRI scientists could in time come up with the requisite technology. That it only took a few years to develop IR8 and successive varieties was certainly a significant achievement, but the result itself was not surprising.

In contrast, at the level of the individual researcher the results of investigations were far from certain. Which crosses would result in plants of short stature? Which plants would be excessively vulnerable to local pests? Which varieties could be 'universalized'—that is, used throughout tropical Asia? The answers to these and myriad other research questions were certainly not known in advance, and thus their particular effects could not be known. What was certain was that the creations of IRRI, an institute lavishly underwritten by foundations with heavy sociopolitical agendas, would be transferred and implemented. As we saw in the discussion of the Mango Thesis, the implementation process cannot be understood as being analogous to pulling out an old circuit board and putting in a new one. In the case of the Green Revolution the 'new technology' generated requirements and effects that caused reverberations throughout the recipient cultures. Research scientists and technologists definitely bear some responsibility for these reverberations.

Conclusions on Research Strategies

How can this responsibility be discharged and what is its extent? Again there is no handy formula to answer such vexed questions. Two areas in which scientists can better discharge their responsibilities are interdisciplinary and interinstitutional dialogue. In view of our previous remarks about the difficulty of integrating the natural and social sciences it may seem paradoxical to urge interdisciplinary dialogue. There is no paradox, however. Even if barriers to integration were in principle insurmountable—and we are not convinced that they are—it would still be a good methodological strategy to attempt integration. Furthermore, international institutes such as IRRI, which are intended to be interdisciplinary, offer an opportunity, a forum for scientists and technologists to extend the critical faculties that are at the heart of their enterprise beyond narrow intradisciplinary concerns. We are not calling for a mindless holism; we are calling for the broad self-conscious exercise of critical skills.

We also suggested interinstitutional dialogue as a means for scientists to discharge their responsibilities. Here we mean 'institutional' in both concrete and abstract senses. In concrete terms we mean scientists at research facilities other than IRRI. As we have seen in the Sri Lankan case study, IRRI scientists had little interchange with their colleagues in national and district research positions. In the manner of a central authority, until 1980 IRRI considered communication from the periphery to the centre to be of negligible importance compared to communication from the centre. Most people say there was no way around the hierarchy which had been established.

In abstract terms interinstitutional dialogue refers to communication beyond the realm of scientists and technologists. For example, dialogue with government officials and with the ultimate recipients of the technology, the farmers of Asia, would be another means by which scientists and technologists could discharge their responsibilities. This extends the meaning of the term 'integration' as we described it above in terms of natural and social sciences. It demands the effort to transform the hierarchic and authoritarian relation between scientists at international centres: hierarchic because there was a system of subordinate and superior institutions with different functions, and authoritarian because the locus of expertise was expected to remain in central research establishments like IRRI. The recipients and consumers of the technology had no way of influencing directly what was being done.

In rich countries the work of science and technology institutes is regulated, however partially, with reference to the consequences of the use of its creations, but international institutions, rather like international corporations, have had a kind of sovereignty unto themselves. Because they cannot presume that governments of these poor countries will fully discharge wider responsibilities, the international institutions must accept their responsibility and urge governments to do so. Such governments may, after all, be trying to mask the effects of previous technology transfers and earlier policies and may not wish to consider the consequences of new technologies. One hopes they will. But the reluctance of governments to face this responsibility may have justified IRRI avoiding theirs.

Finally, we turn to the question of gauging the degree of responsibility attributable to scientists and technologists. For that purpose we return to the apocryphal speech in Chapter 1 in which a scientist renounces the Nobel Peace Prize on the grounds that he is responsible only for his discoveries narrowly construed and not for the uses to which those discoveries are put. We said at the time that the scientist should not hold such a position. The reasons should be clear: as is emphasized where the Underspecification and Mango theses apply, scientists and technologists are indeed responsible for as wide an ambit of the consequences of their work as is recognized by the Nobel committee. Certainly there are circumstances in which the control exercised by scientists and technologists is so slight as to be negligible. And there are circumstances in which the influence could be greater, for example, when the route between discovery and implementation is short and discernible. Even here of course the responsibility of an individual scientist is small. It would be absurd to demand that an individual geneticist must also become an anthropologist in order to practise his or her profession. However, as in the case of the Green Revolution, the role of the collective, the scientific and technological enterprise as a whole, is anything but negligible.

This is one of the main conclusions we have tried to demonstrate, but not because we believe that the Green Revolution was a disaster. If there ever could be some kind of cosmic multidimensional calculus for rating effects, surely both the Green Revolution and the instrumental role played by scientists and technologists would get mixed ratings. But such a calculus is a fantasy. Back in the real world our message is that the deliberate use of science and technology to achieve international social and political goals is more likely to increase than to decrease. There may not be another process exactly like the Green Revolution in Asia, but the cognitive authority of science and technology, their ability to reduce problems to manageable size, and then to produce tangible solutions to those self-defined problems, make science and technology irresistible. This is specially true for the institutions that are expected to underwrite major international projects. Our hopes are that future efforts will learn from the subtle as well as obvious lessons of the past and that we have helped to describe and to analyse some of the intricate interactions at the boundary between laboratories and life.

NOTES

1. Harry Cleaver, 'Technology as Political Weaponry', in Robert S. Anderson *et al.* (eds.), *Science, Politics and the Agricultural Revolution in Asia*, Boulder, Colo.: Westview Press, 1982, pp. 270-1. Although we are indebted to Cleaver's insight and research on the Green Revolution, we part company with him on some issues. For example, his analysis seems to suggest that even when the results of technological research are narrowly understood, e.g., as being the genetic make-up of a new rice variety, when transferred they will necessarily bear the indelible stamp of the political–economic system in which they are produced. We find this implausible.
2. Bruce Koppel and Edmund Oasa, 'Induced Innovation Theory and Asia's Green Revolution: A Case Study of an Ideology of Neutrality', *Development and Change* (January 1987), 42.
3. Randolph Barker, Robert W. Herdt, with Beth Rose, *The Rice Economy of Asia*, Washington, DC: Resources for the Future, 1985, p. 172.
4. Ibid., table 4.11.
5. Ibid. 173.
6. Ibid., table 6.1.
7. Calculated from ibid., tables 4.12 and 7.1.
8. We note, however, that the planners and researchers did not personally experience the brunt of these debilitating conditions and were shielded against them.
9. It occurred to us that the displacement of the 'natural science paradigm' from centre stage in the 1980s might make our examination focused on the role of S and T simply of historical interest because development strategies have changed since the first blushes of the Green Revolution. But we think that the point most relevant to our concerns would remain: over the last dozen years development strategies have changed with respect to the breadth of sociopolitical factors they *explicitly* take into account. Any changes which have occurred in no way undermine the importance of asking about the role of S and T within these new research and development strategies. At the end of extended debate about strategies, it is still important to ask what, if anything, the S and T enterprise is and should be doing in the area of development. We believe that many of the issues we raise, such as those involved in the Underspecification and Mango theses, apply to both the social and natural sciences and thus to broader strategies.
10. Thomas Kuhn, *The Structure of Scientific Revolutions*, Chicago: University of Chicago Press, 1962.
11. It is, of course, questionable whether many of those who speak of the 'natural science paradigm' intend to be making general philosophical claims at all—or are even aware that Kuhn intended to do so.
12. Personal communication, Piers M. Blaikie, 1983.
13. As noted above we do not accept the Kuhnian account as being correct as a matter of principle. We believe, for example, that adherents to different paradigms could in principle have much more common language, methodology, and observational basis than Kuhn allows. Nevertheless as a *sociological* account there is much evidence that holders of different

theoretical structures use language differently, see the world through their own paradigms, define their problems differently, and have only limited resources to recognize and reconcile their differences.
14. In the recent literature of philosophy of science there are several authors who have suggested that 'disunity' in some sense is the proper state of science. See for example I. Hacking, *Representing and Intervening*, Cambridge: Cambridge University Press, 1984, and N. Cartwright, *How the Laws of Physics Lie*, Oxford: Oxford University Press, 1983. However, ours is the first suggestion we know of to give Kuhnian grounds for 'disunity'.
15. See Koppel and Oasa, 'Induced Innovation' p. 39.
16. For an extended discussion of the issues surrounding neutrality and the Green Revolution, see Edwin Levy, 'The Responsibility of the Scientific and Technological Enterprise in Technology Transfers', in Anderson, *et al.*, *Science, Politics, and the Agricultural Revolution in Asia*, pp. 277–97. There it is shown that the addition of the Underspecification and Mango theses renders plausible even so radical a position as Winner's claim that 'artifacts have politics': see Langdon Winner, 'Do Artifacts Have Politics?', *Daedalus*, 109 (1980), 121–36.

11
IRRI in the 1980s

THE tension between a centralized agricultural research strategy, focused on a limited problematic, and the complexity of rice-growing practice in diverse localities has undermined the epistemological confidence and institutional conceit of IRRI.

In 1980, the doubts raised by this tension led to a modest reorganization in the direction of problem-oriented research programmes which cut across the entrenched disciplinary and departmental boundaries. Yet with the replacement in 1982 of Brady by M. S. Swaminathan, a well-known plant geneticist and the first non-American Director-General, the problem/programme arrangements decomposed, and a number of seasoned IRRI observers described a revival of departmental power. The Genetic Evaluation Unit, comprising many departments, fell into disuse. The multidisciplinary cropping systems programme, known for its broader problem approach, lost personnel and resources, and was disconnected from multiple-cropping studies, which ultimately disappeared. No new programme approaches were instituted until in 1989 IRRI undertook another long-range planning exercise, like the one in 1979—ten years earlier. Again a problem-oriented interdepartmental structure was recommended under the direction of a new Director-General, a German aid-administrator and agricultural engineer, K. Lampe.

Taken together, the two publications of this late-1980s reorientation show some old and some new elements. In *IRRI Toward 2000*, and the (1990–4) *Work Plan* there is the familiar focus on a single crop—rice—and there is the continuation of a plant-centred approach rather than a farmer-centred approach.[1] Once again IRRI tries to position itself somewhere between 'advanced institutions' and 'national programs', thus being a link between applied and basic research. Providing this link is called 'strategic research'. It must be remembered that this very intense planning process occurred after a period in which an institutional definition of objectives had perceptibly weakened, so there may be an understandable exaggeration of IRRI's special mission. Again there is a call to arms, saying that IRRI 'must think creatively and apply the kind of ambitious vision that has put man on the moon to generating scientific ideas that could make life better for man on earth'.[2]

What is new at the end of the 1980s? The budget for 1980 was $19 million, and the budget for 1990 is $29 million—a spectacular amount by world agricultural research standards. But what would be achieved with this $29 million? IRRI clearly decided to enlarge its conceptual apparatus by expanding from irrigated rice to work on four rice ecosystems—irrigated rice, rain-fed lowland rice, upland rice, deep water and tidal wetlands rice. Although this is not the first time IRRI espoused this broader objective (see the *LRPCR* of 1979), this time the organization of the Institute is to be renovated so that each ecosystem programme will have a programme leader with overriding authority. We think that the four ecosystems distinction constitutes an oversimplification, and that within the 'relatively homogenous' irrigated ecosystem, as the *Work Plan* called it[3] there are subtle and important variations—so that even if the water regimes are similar the human components are not.

These four ecosystems may ultimately provide the organizational basis for a more complex institutional approach in which IRRI tries to approximate the complexity of Asian rice cultivation. That approach involves Subprogram III of the proposed Cross-ecosystems Research Program, namely 'ecosystem characterization and socio-economic and environmental impact analysis'.[4] When this is done, says the *Work Plan*, there may be 'policy changes to enhance or sustain benefits or to mitigate the adverse effects of technical change', and there may be a better capacity to 'determine the transferability of new technology'.[5] But what will be created first is a comprehensive classification system for rice ecosystems, and a unified agro-ecological and geographic data base organized at several levels. Not only will there be a systematic and comparative study of how the rice market works, but also research on the consequences of new rice technologies 'among people dependent on favourable and unfavourable rice-growing environments (farmers and labourers, women and men, producers and consumers)'.[6] And characteristic of the late 1980s, the new strategy is complete with 'the integration of women's concerns into all research projects',[7] and with tests of proposed research against sustainable development criteria. At the same time there is the announcement of 'the biotechnology group' which will 'give leadership to research involving forward edge techniques' through the tissue culture laboratory, the isozyme laboratory, the plant pathology laboratory, and the biochemistry laboratory and future containment facilities. If IRRI can balance these two directions, one moving outward from the rice plant to whole farming systems in ecosystems, and the other moving inward, it may play a useful role.[8] Certainly as an information centre IRRI is probably invaluable as the need for comparative analysis increases.

But it is not certain that this more complex multidisciplinary matrix approach will succeed in overcoming disciplinary limitations, or that disciplinary groups and individuals will submerge their identities in programmes unless they have a deeper conceptual rationale for doing so. Instituting such a rationale will be a major task and may, like change in other scientific institutions, rest on new recruitment and gradual replacement of institute members whose conceptual rationales are no longer relevant. All of these plans must be set in the sociological constraints of institutionalized international science, and these simplified plans are probably part of a process of responding to the underspecified objectives of underwriters of research costs. We believe there are alternative routes to increasing and/or stabilizing rice yields and these routes are embedded in the diversity of rice-cultivating environments. These increases are to be weighed against the long-term fertility of the land and the social reproduction of viable communities which grow rice.

If research is going to make a difference to increasing and/or stabilizing yields, we believe it needs to occur in a multicentred system with a clear division of labour in which efficient and useful work must be localized in specific agro-ecological regions in rice-growing countries. The research will have to balance the human and natural sciences to a much greater extent than IRRI does in 1990. IRRI may remain the central clearing house, so that there may still be a hierarchic information system but a decentralized research system.

At the same time, like other international research centres, IRRI will have to adapt continuously to its immediate environment. This relationship is not static but changing. For example, in 1989 security at IRRI had to be increased when a demonstration blocked access to the gates: people were protesting over biotechnology research on rice pathogens which they thought were dangerous. This protest coincided with reported receipt of threats and extortion letters from both 'criminal syndicates' and the 'communist New People's Army'.[9] A military spokesperson said IRRI's 54-man security force had been integrated with the National Police Troops. It will be recalled that IRRI was originally built in the Philippines because of the the security of the location.

Finally, we agree with Paul Richards' conclusion from his work on West Africa, that the complexity of agricultural environments has been pursued on too high a level of abstraction.

Intellectuals, development agencies and governments have all pursued environmental management problems at too high a level of abstraction and generalization. Many environmental problems are, in fact, localized and specific, and require local, ecologically particular responses. The issue then becomes how to stimulate such situation-specific responses. One of the answers...is thorough mobilizing and building upon existing local skills and initiatives. Everything should be done—so the argument runs—to stimulate vigorous 'indigenous science' and 'indigenous technology.'[10]

One conclusion of our book is that scientists and institutions have to put the idea of culture back into their approach to agriculture. This means recognizing the diversity of cultivating environments not as a kind of human perversity but as following a locally grounded logic. Having reviewed the literature on agricultural research institutions, Stephen Biggs agreed with Richards that scientists in institutions like IRRI have to join forces with indigenous technologies, and learn of the complexity of indigenous systems of knowledge. This choice is open to them, as always, but it is very difficult.

In the past some of the 'best' scientists by any 'scientific' criteria of excellence in their field decided to work in 'alternative science' rather than mainstream science. On occasion the importance of their work was not recognised, because of the political, policy and institutional context at the time. The same holds today. The challenge for today's excellent scientists in such new areas as biotechnology and information sciences is to decide whether to work on ways in which these new developments can be integrated with other sources of knowlege to address the problems of poorer groups in rural areas or whether they will put their skills to work on the problems of other users. Choices have to be made.[11]

One of the choices open to scientists is to acknowlege the sophistication of the indigenous knowledge required in complex rice-cultivating environments, and to build that respect into an alternative approach. Biggs astutely recognizes that neither donors nor alternative scientists will achieve this single-handedly.

In the future if the alternative [supplemental] science model is to be sustained—rather than marginalised as during the 1960s and 1970s—there will have to be widespread political, economic, institutional and financial support for this approach. There are some clear indications that agricultural research policies and institutions at the international and national levels are changing and are supporting the alternative science approach. The challenge for social and natural scientists who are interested in this alternative approach will be how to mobilise and ensure political, economic and institutional support for the alternative model.[12]

This book began with an apocryphal speech by Norman Borlaug, with which we addressed the presumed neutrality of scientists. It is fitting that we close with his reflections after forty years' work in agricultural research. We think his advice is actually an important opportunity to redefine agricultural problems and renovate the institutions which are intended to solve them.

I have found from experience that no matter how excellent and spectacular the research is that is done in one scientific discipline, its application in isolation will have little or no positive effect in crop production. I have been forced to become an integrator across scientific disciplines in order to develop a reliable package of production technology capable of dramatically increasing yields, when it is properly applied.

I must caution that agricultural research alone cannot produce miraculous improvements in all of the more marginal production areas. Some of the biological constraints are simply too overpowering for science to currently overcome.[13]

Thus the tension between a centralized research strategy and the complexity of rice-growing practices is still with IRRI.

NOTES

1. *IRRI Toward 2000 and Beyond*, Los Banos: IRRI, 1989; *Work Plan for 1990–1994*, Los Banos: IRRI, 1989.
2. *IRRI Toward 2000*, p. 11
3. *Work Plan*, p. 8
4. Ibid. 38.
5. Ibid. 38–39.
6. Ibid. 39.
7. *IRRI Toward 2000*, p. 21.
8. For a recent review see Cornelia B. Flora, 'Farming Systems Approach to International Technical Cooperation in Agriculture and Rural Life', *Agriculture and Human Values*, Winter–Spring 1988, 24–34.
9. 'Manila tightens security at rice research institute', *Bangkok Post*, 20 April 1989.
10. Paul Richards, *Indigenous Agricultural Revolution*. Boulder, Colo.: Westview Press, 1985, p. 12. See also D. M. Warren, 'Linking Scientific and Indigenous Systems' in J. Lin Compton (ed.), *The Transformation of International Agricultural Research*, Boulder and London: Lynne Rienner Publishers, 1989.
11. Stephen Biggs, 'A Reconsideration of Institutionalized Agricultural Science in Asia', paper given at Workshop on the Future of Alternative Sciences in Asia, University of British Columbia, October 1989, p. 26. Biggs, in response to comment, suggested that 'supplemental' or 'additive' science would be a better term than 'alternative' science.
12. Ibid. 27.
13. Norman Borlang, 'Accelerating Agricultural Research and Production in the Third World: A Scientist's View', *Agriculture and Human Values* (Summer 1986), 5, 10–11. We would add to 'biological constraints' economic, political, social, and cultural constraints as well.

Glossary

aman	agricultural season in Bangladesh, July to December	Maha	agricultural season in Sri Lanka, September to March
aus	agricultural season in Bangladesh, May to September	thana	literally, police station; was until 1980 major administrative unit below district in Bangladesh: now called upa-zilla
boro	agricultural season in Bangladesh, December to April		
bund	retaining wall to contain water which floods rice fields; often defines boundary of property	Union	administrative unit smaller than a thana in Bangladesh
chena	cultivation or shifting cultivation in Sri Lanka	Yala	agricultural season in Sri Lanka, April to August
Huk	short for Hukbalahap, movement of armed resistance to Japanese and American influence in the Philippines		

Index

Abeyratne, Ernest 175, 176, 187, 194, 218
adaptability of rice varieties 66, 69, 70, 362
Agnew, Spiro 127
agricultural advisers, foreign 129
Agricultural Development Council (ADC) 31, 138, 265, 291
Agricultural Economic Department, IRRI 94–5
Agricultural Instructors (AIs) 189–90, 193, 203, 205
agriculture, rice:
　conservatism of 53
　inefficiency of 51, 206
　problems of 35–6
　restructuring relations of production in 173–4
　traditional 91–2
　variations in 7, 44–5, 152–9, 174–5
agrochemicals 191–2, 313–14, 316–20
　control of 119
　dangers of 134, 178, 316–17, 319–20
Ahmed, Tajuddin 290
Alim, A. 258, 264
All India Crop Rice Improvement Program (AICRIP) 68, 142, 143
alternative science model 384
Amoco Cairo oil tanker 326
Asian Development Bank (ADB) 138–40, 354
Athwal, 131
Australia 350
Awami League, Bangladesh 280, 283
Ayub Khan 243, 244, 246, 249, 250, 253, 254

'backward practices' 238
bacterial leaf blight (BLB) 176–7, 179, 181
bagging/transport of rice crop 201
Bahalagoda, Sri Lanka 166–7, 180, 198–206
Banda, M. D. 175, 180
Bandaranaike, S. W. R. D. 172
Bandaranaike, Sirima 172, 187
Bandung Conference (1955) 29
Bangladesh 131, 222–75, 280–325, 364
　diversity in 225–32
　forces of rice production in 300–25
　Green Revolution in 247–56
　neglect of rice research in 242–7
　politics of self-sufficiency in 284–99
　researchers' values in 256–73
　rice research in Bengal 237–42
　strategy of 367–8
　surpluses/deficits in 232–7
Bangladesh Agricultural Development Corp. (BADC) 296, 297, 301, 302, 304, 307, 310, 314–15, 322
Bangladesh Rice Research Institute (BRRI) 280, 283–4, 292, 303, 304, 309, 310, 315–17, 320, 322, 323, 330–56, 367–8
　finances of 349
　objectives/organization of 334–45
　relations with rice cultivators 346–8
　responsibilities of donors 348–56
Bank of Ceylon 200
Barisal Irrigation Project, Bangladesh 309
Barisal substation 340–3
Barker, Randolph 70, 85, 363
Barnard, Chester 35
Batalagoda, Sri Lanka 166–7, 179, 180, 198–206, 220
Beachell, Henry 65–6, 212
behavioural science 31, 123, 124
Bell, Daniel 26
Bengal 222, 233–42
Bhashani, Moulana 249
Bhutto, 249–50
Biggs, Stephen 384
biological engineering, *see* genetic manipulation
blast 166, 167, 181
Bombuwela, Sri Lanka 206–11
Borlaug, Norman 2, 12, 67, 108, 143, 258–9, 312, 384–5
Borlaug Award 143
Boyce, James 129, 355
Bradfield, Richard 42–3, 44–5, 46, 86–7, 88, 90, 268
Brady, Nyle 50, 62, 73, 74–5, 81, 84, 107, 117, 139, 188, 322, 350, 381
Brammer, H. 230
broadcast sowing 177, 207
brown plant hopper (BPH) 92–3, 181–2, 191–2, 204, 316, 320, 341
buffaloes 197, 200, 202

Burma 131, 235
Byrnes, Frank 123, 124, 125, 126, 129

calorie deficits 290
   see also malnutrition
Cameron, James 21
Canada 268, 314, 327, 350, 352, 353
canals 305
cash crops 192, 195, 210
Central Agricultural Research Institute (CARI), Sri Lanka 180, 188–92, 214
Central Rice Research Institute (CRRI), India 234
centralization 354
Centro Internacional de Agricultura Tropical (CIAT), Colombia 71, 125
Centro Internacional de Mejormiento de Maiz y Trigo (CIMMYT), Mexico 143, 264, 265
Ceylon/Sri Lanka Fertilizer Corp. 169, 172, 216
Chandler, Robert F. 42–3, 47, 49–51, 54, 62, 66, 68, 70, 71, 73, 74, 88, 109, 120–1, 128, 188, 222, 263, 266–7, 271, 272, 280, 285, 293, 295–8
Chandraratne, M. F. 166, 174
'change-agents' 126
characteristics required in rice varieties 65–6, 178–9, 181, 195, 216, 334–5
Chen, James 30
chillies 194
China 22, 24, 36, 72, 246, 265
Cleaver, Harry 361, 379
climatic variables, *see* environmental variety
coconuts 199, 201, 220
Colombo Plan 25, 45
colonization schemes 162–4
commercialized consumer societies 118
communication/promotion activities by IRRI 117–18, 122–8, 137–8, 141, 142
Communism 24, 25
consequences, study of 344–5
Consultative Group on International Agricultural Research (CGIAR) 62, 76, 83, 86, 107, 109, 128, 129, 139, 140–1, 349
   Technical Advisory Committee 88–9, 130
co-operatives 307, 308, 309–10, 325
Co-ordinated Rice Varietal Testing (CRVT) 179–80, 212
Cornell University 30, 35, 37, 50
Coromandel Fertilizers Ltd 143
Council on Foreign Relations (CFR), RF 27, 29
cowpea 194
credit, agricultural 171, 172, 252

cropping systems research 79, 82, 84–92, 135
cultivation season, *see* seasons, rice-growing
cultural diversity 7–8, 231–2, 384
cultural values 289–90
Cummings, Ralph 88, 131
cyclones 272

Dalrymple, Dana 323
Damodar Valley Project, Bengal 233–4, 241
Dasananda, S. 68, 133
De Datta, S. K. 212
debt, rural 171–2
decline in rice yields 234
deep-water rice 239–40, 247, 350
deficits, food 267, 272, 280, 284
development 238
   link with science 2
   see also politics of rice
Dhaka Farm, Bengal 237, 239–41, 246, 257, 259
diazanon 92–3
Dillon, Douglas 28
diseases of rice 67, 166, 167, 176–7, 181
distribution:
   of fertilizer 65, 201, 202–3, 296–7, 315
   of land 153, 154, 156, 158, 230–1, 241–3
   of seed 301–2
distributive questions 288–9
District Agricultural Extension Officer (DAEO) 189, 190–1, 193, 202, 205, 208, 210
divide-and-conquer approach 370
dormancy period of rice 177–8
double-cropping 168, 169, 196, 199
drought 284, 296, 305, 331
dry zone, Sri Lanka 155–7, 162, 163–4, 171, 192–8
dualism, American 56
Dulles, John Foster 27, 28, 29
dwarfing of rice 66, 72, 177

East Bengal State Acquisition and Tenancy Act (1950) 242–3
East Pakistan Accelerated Rice Research Institute (EPARRI) 261, 263, 269, 271–2, 290, 291, 292
ecological disasters from agrochemicals 134, 317, 319–20
economic rationality of rice farmers 8
ecosystems, rice 382
Edwards, Robert 293
'efficiency', search for 288–9
Eisenhower, Dwight 28
encouragement of rice-growing 161–5
environmental variety 152–4, 226–8
equity in size of holdings 101

Index

Esso 61
Evenson, Robert 129
expectations, rising 118–19
expenditure, agricultural 287
experimental farms 78, 81, 82, 84, 113, 332

Fairbank, John 24–5
family planning approach 34
famine 233, 331
farm types 97
farmers:
  experimental outlook of 346–7
  help from research agencies 190–1, 193–4, 208–10, 347–8
  problems of 168–72, 200–1
  'progressive' 338–9
  resistance by 125–6
  rich 325
  under-reporting of rice crops 100, 323–4
Feldman, Herbert 249
Fernando, H. E. 178
fertility of land 222
fertilizer 310–16
  consumption of 363–4
  costs of 100
  distribution of 65, 201, 202–3, 296–7, 315
  domestic production of 312, 313–14
  imports of 314–16
  requirements 169–71
  research 245, 246
  response of rice to 3, 39, 66, 166, 167, 212–13, 257
  subsidies for 172, 311, 314–15
  use of manure as 241, 245, 310–11
field preparation 196–7, 200–1, 202
fixed-distance planting technique 341
floods 228, 260, 305
food:
  deficits of 267, 272, 280, 284
  foreign aid with 250, 290, 293–5, 326–7, 353
  imports of 120, 187, 222, 224, 232, 235, 250, 254, 294, 298–9
  problem of 26, 38, 56, 160–2, 187, 360, 366, 371
  self-sufficiency in 119–20, 175, 224, 225, 233, 235, 251, 260, 267–8, 280, 284–99, 356
  subsidies for 151–2
  surpluses of 232
Food and Agriculture Organization, UN (FAO) 53, 119, 120, 233
Ford Foundation (FF) 1, 10, 22, 45, 47, 51, 54, 55, 56, 68, 71, 76, 122, 222, 262, 265, 268, 291–2, 331, 348–9, 352
Foreign Affairs 22, 36
foreign aid 224, 255–6
  see also food, foreign aid with
Franklin, George 28
Freeman, Wayne 129

gall midge 182
Gandhi, Indira 295
Gandhi, M. K. 21
Ganges-Kobadak Irrigation Project 244, 306
Garcia 51
genetic diversity 72, 112, 304, 337
genetic evaluation and utilization programme (GEU) 75
genetic manipulation 65, 107–8, 335, 362
genetic materials 9–10, 71, 117, 303
Germany 314
Gini coefficient 101, 103
goals, competing 10–11
Golden, William 123, 125, 129, 178, 264
Goodell, Grace 9, 134
Gray, Clarence 87
Green Revolution 1, 2–3, 15, 16–17, 56, 88, 263, 272, 325–6, 361, 363, 376, 378, 379
  in Bangladesh 247–56, 273–5
'grey market' 9
growth rate in production 335–6, 355
Gunawardena, Philip 172–3

hand weeding 177
Hanson, Haldor 264
Hargrove, Thomas 72, 335
Harrar, George 30, 37–43, 48–52, 54, 62, 121
Harrison, Anna J. 12
Harriss, Barbara 320
Harwood, R. R. 85, 89–90, 91
Hazarika, S. H. 259
Heald, Henry 48
Hector, G. P. 239
Hedayatullah, S. 244–5
Herdt, Robert W. 363
Hernandez 52
'high pay-off inputs' 6
high yielding varieties (HYVs) of rice 3, 94, 99, 103, 106, 124, 132–4, 165, 167, 182–3, 224, 252–4, 256, 259–61, 264–9, 271, 272, 280, 283, 293–5, 297, 299–304, 310, 322–5, 337, 346, 347
  see also rice, new varieties of; varieties of rice
highland crops 156, 192, 195, 197, 198, 203, 210
Hill, F. F. 48, 51, 69, 73, 123
Hoffman, Paul 48
Hopper, W. David 8, 85
Huq, A. S. M. 318
Huq, Fazlul 249
hybrid vigour 39

Index

'ideological myths', scientific 138
Ikehashi, 211
imports, food 120, 187, 222, 224, 232, 235, 250, 254, 294, 298–9
incomes 104–5, 196
India 37, 42, 44, 46, 72, 123, 129, 234, 246, 295
Indian Agricultural Research Institute 45, 46
indica type of rice 3, 39, 65, 66, 166, 176, 246
'indigenous science' 383–4
Indonesia 45, 72, 85, 93
innovation 11
  induced innovation theory 362, 375
input allocation 324
integration, scientific 370–2, 377
Intensive Agricultural Development Program (IADP) 68
interinstitutional dialogue 377
International Agricultural Development Service (IADS) 131
International Co-operation Agency (ICA) 45
International Crop Research Institute for the Semi-Arid Tropics (ICRISAT) 84, 87, 194, 198
International Development Advisory Board 22
International Development Research Center (IDRC) 85, 195
International Monetary Fund (IMF) 27
International Rice Commission (IRC) 44, 49, 119–20, 166, 245, 246
International Rice Research Institute (IRRI) 1, 3, 9, 17, 29, 188, 198, 222, 245, 246, 249, 259, 260, 262–4, 268, 285, 292, 294, 303, 331, 334, 349–53, 356, 361, 368, 376, 377
  budget of 75–7, 108–9, 382
  communication/promotion activities 117–18, 122–8, 137–8, 141, 142
  concept of research 119–21
  constitution of 54–5
  establishment of 47–50
  evolution of 62, 107–10
  methodological problems at 78–93
  mission of 63–4, 73–4, 117, 121, 136, 362–3
  in 1980s 381–5
  in Pakistan 265–73
  research priorities for 50–3
  sexism of 136
  in Sri Lanka 211–14
  values/culture of 135–6
International Rice Testing Program (IRTP) 211
investors, foreign 302, 308, 348–51, 353, 354, 364

iron in soil 216
IRRI Toward 2000 381
irrigation 44–5, 163–5, 193, 199–202, 215, 244, 245, 247, 252, 257, 283, 297, 304–10, 364, 366, 373

Jack, J. C. 230
Jackson, Sir Robert 293, 295, 299
Japan 42, 44, 235, 258
Japanese National Institute of Agricultural Sciences 45
japonica type of rice 3, 39, 66, 166, 258
Java 44, 95–107
Jayawardena, J. 208
Jennings, Peter 65, 71
Joachim, A. W. R. 366
Johnson, Lee 263, 267, 268, 269
Joydevpur, Bangladesh 331, 332, 337, 344, 345
jute 236, 324
Jyasekera, Earl 191

Kennedy, John F. 28
Khan, Akhter Hameed 244
King, John 36–7
Kobbekaduwa, Hector 187, 218
Korean War 22, 24, 25
Krushi Karma Viapthi Sevakas (KVSs) 190, 193, 203, 205
Kuhn, Thomas 368–9, 371–2, 379

Ladejinsky, Wolf 82, 122
Laguna province, Philippines 9
Lampe, K. 381
land:
  distribution of 153, 154, 156, 158, 230–1, 241–3
  landlessness 154–5, 157, 158, 230–1
  opening up of 162–4
landlords, extortionate 238, 242–3
'language of rice' 125
leaf roller 204
'lethality' in rice 39
level/standard of living 102–4
lodging in rice 176
Loevinsohn, Michael 134
Long Range Planning Committee, IRRI (LRPC) 67, 69, 72, 79, 91, 109, 117, 132, 137, 138, 143
Love 133

Ma, Dean P. C. 119
McCarthyism 22, 28, 29
McClung, A. Colin 120, 128, 131, 181, 263, 266, 267, 269, 271
macrostructure of agriculture 8
Magsaysay Award for International Understanding 68

Index

Maha Illuppallama, Sri Lanka 166–7, 192–8
Malathion 317, 319, 321
Maligaya Rice Research Farm 53–4
malnutrition 290, 299, 355, 364, 365
Manalo, Engenio B. 226, 227
Mango Thesis 15–17, 372, 374–5, 376
Manhattan oil tanker 293–4
manure, use as fertilizer 241, 245, 310–11
Mapa, P. L. 34–5, 47, 49
Marcos, Ferdinand E. 127
marketing of rice 201, 203
Masagana 99 project 81, 93, 134
Mason, Max 31
maturation period for rice 167, 168, 181, 195–6
mechanization on farms 98–9, 320–2
of irrigation 305–6, 307
Mellor, John W. 6
methodological problems 78–93
Mexico 5–6, 46, 52, 60, 67
military establishment 291
mills, rice 320–1
Minikits, IRRI 127–8
Minneriya scheme 163
misreporting for statistical surveys 323–4, 355
mission-oriented basic research 32, 79, 121
model methodologies 82–3
Mohammed, Gulam 267
monoculture, rice 373
monsoons 227
Moomaw, James 149, 178, 321
Mujib 284–5, 286, 295, 299
multidisciplinary matrix approach 382–3
Myers, William 37

national research strategies 365–8
natural science paradigms 368–9, 379
Nepal 131
neutrality, technological 15–16, 375, 384
New York Times 68, 69
Newaz, Shafi 257
North Vietnam 374
nutrition requirements v. economics 287, 288–9, 290

Oasa, Edmund 141
okra 194
OPEC oil price increase 62, 299
outreach work 123, 127–8, 130, 347–8, 351
see also communication/promotion activities by IRRI

'packages' of rice technology, transfer of 16–17
paddy, see rice

Paddy Lands Act No. 1 (1958), Sri Lanka 173, 174
Pakistan 67, 123, 245, 246–56, 259, 265
Panabokke, C. R. 176, 186, 217, 366
paradigm, notion of 368–9, 371–2
Parthasarathy, N. 68
Peradeniya Gardens, Sri Lanka 159, 160, 161, 163, 214
Peries, J. W. L. 149
Peries, Leslie 174, 178, 179, 212
pesticides 191–2, 316–20, 374
 dangers of 134, 178, 316–17, 319–20
 stockpiles of 318–19
 subsidies for 318
pests 92–3, 113, 178, 181–2, 191, 204, 209–10, 316, 319–20, 341, 374
Philippines 9, 22, 25–6, 34–5, 37–8, 44, 45, 47–51, 54, 64, 72, 76–8, 91, 97, 119, 122, 132, 134, 269, 364, 373
Philippines Seed Board 66–7
photosensitivity of rice 209
plant-breeding procedures 21
plateauing of rice production 215–18
plot-lab methodology 78
political upheaval 187, 291, 364
politics of rice 36–7, 62, 119–20, 285–90
Ponnamperuma, Felix N. 212
Pontius, Steven K. 133
Population Council 31, 34, 48
population problems/control 34, 38, 228–9, 234, 236, 336
prices, rice 149, 151–2, 236, 242, 251, 293, 294, 299, 323–4
priorities, research 50–3
problems, unanticipated, see unanticipated problems
production of rice, forces of 300–26
profits from rice 102–4, 287
'project area' for BRRI 337–9
protein deficiencies 290
publications, research 333–4
pumps, irrigation 245, 305–6, 307, 309
Punjab, India 8

Rahim, Abdur 257
Rahman, Mujibur 249, 251, 272, 284, 285, 295, 299
Rahman, Ziaur 284
rainfall 226–7
rain-fed cultivation 202
Rashid, Haroun Er 226, 227
'ration rice' 250
Ravenholt, Albert 25
reductionism, scientific 370–1
reporting bias, farmers' 100, 324
Rerkasem, B. and K. 132
research *see under individual headings, e.g.* Bangladesh; Bangladesh Rice Research

research (cont.)
    Institute (BRRI); International Rice
    Research Institute (IRRI); Sri Lanka
research strategies:
    national 365–8
    scientific paradigms in 368–70
resistance by cultivators 125–6
responsibilities:
    of BRRI's donors 348–56
    of science 375–8
restructuring relations of production
    173–4
'return above costs' 97, 99
rice:
    bagging/transport of 201
    and cultural values 289–90
    deep-water 239–40, 247, 350
    dormancy period of 177–8
    dwarfing of 66, 72, 177
    forces of rice production 300–25
    genetic diversity of 72, 112, 304, 337
    genetic materials for 9–10, 71, 117, 303
    goals for 10–11
    importance of 38, 50–1, 125
    marketing of 201, 203
    maturation period for 167, 168, 181,
        195–6
    new varieties of 64–9, 107–8, 166–9,
        176–82, 204–5, 209, 268–9, 303–4,
        334–5, 346–7; see also high-yielding
        varieties (HYVs) of rice; varieties of rice
    pests/diseases of 67, 92–3, 113, 166, 167,
        176–7, 178, 181–2, 191, 204, 209–10,
        316, 319–20, 341, 374
    politics of 36–7, 62, 119–20, 285–90
    price of 149, 151–2, 236, 242, 251, 293,
        294, 299, 324
    yields of 13, 39–40, 51–3, 99, 241, 266–7;
        see also high-yielding varieties (HYVs)
        of rice
    see also under individual headings, e.g.
        genetic manipulation; profits from rice
Richards, Paul 383
risks of rice production 347–8
Rockefeller, David 27, 28
Rockefeller, John D., Jr 26
Rockefeller, John D. III 23, 27–8, 34, 37, 38,
    43, 48, 49
Rockefeller, Nelson 22–3, 27, 56, 360
Rockefeller Foundation (RF) 1, 2, 6, 10, 11,
    22–3, 26–33, 37–9, 41, 43, 44, 47, 53,
    54, 55, 56, 60, 62, 64, 68, 76, 88, 245,
    361
Rodriguez 52
Roosevelt, Franklin D. 27
Ross, Vernon 125, 129
Rusk, Dean 27, 28–9, 33–4, 43, 53, 54

Saburo Okita 139
Saudi Arabia 314
savings, importance of 287
Schultz, Theodore 6
science/technology:
    link with development 2
    nature of 12, 15–16
    neutrality of 15–16, 375, 384
    paradigms in 368–70
    reductionism v. integration in 370–2
    responsibility of 375–8
    see also technology, new
scientists, attitudes of 108, 126–7, 259,
    260–1, 267–8, 272
seasons, rice-growing 167–9, 240, 246–7
seed:
    'grey market' in 9
    introduction of 300–4
seed-beds 169, 177
self-sufficiency, food 119–20, 175, 224,
    225, 233, 235, 251, 260, 267–8, 280,
    284–99, 355
    measurement of 293, 295–9
Senadhira, D. 204
Senanayake, Dudley S. 161–2, 163, 187
sexism in research institutes 136
Shastry 143
Silva, G. V. S. de 172
Silva, Mahinda 218
size of farms 95–107, 115–16, 154, 156, 158
social problems 80
social sciences 368, 369–70
socio-economic variations 229–31
soils 227–8
Sonora province, Mexico 5–6
Soustelle, Jacques 25
soya bean 194
specification, process of 373–4
Spender, P. C. 25
spiders 92, 93
Sri Lanka 7, 85, 123, 131, 149–83, 186–218,
    364
    beginning of rice research in 159–61
    environment/organization of rice-
        growing in 152–9
    help for rice-growers in 161–5
    and International Rice Research Institute
        (IRRI) 211–14
    new research strategies in 214–18
    research at Batalagoda 166–7, 180,
        198–206
    research at Bombuwela 206–11
    research at Central Agricultural Research
        Institute (CARI) 188–92
    research at Maha Illuppallama 166–7,
        192–8
    strategy of 265–7

successes/limitations of rice research in 165–83
staff, research 332–3, 334, 345
  see also scientists, attitudes of
statistics, falsification of 323–4, 355
stem borer 92, 181, 182
storage of rice 321
'strategic research' 381
Subramaniam, C. 112
subsidies:
  for fertilizer 172, 311, 314–15
  for food 151–2
  for pesticides 318
substations, BRRI 339–43, 345
Suhrawady, Hussein 249
surpluses, food 232
Swaminathan, M. S. 108, 112, 381

Taiwan 72, 85
Tanco, Arturo 77
tank irrigation 195–6, 201–2, 305
technology, new:
  and farm size 100–2
  transfer of 2–3, 15–17, 257–8
tenancy 154–5, 157, 158
testing methodologies 64
Thailand 45, 85, 99, 132, 133
Thakurgaon project, Bangladesh 308
threshing 177
thrips 209
tractors 197, 201, 202, 322
training:
  of extension workers 126–7
  of research staff 332–3
transfer of new technology 2–3, 15–17, 257–8
transplanting of rice 169, 177, 207, 258, 341
'trickle-down' myth 251, 288
triple-cropping 196
Truman, Harry 22, 27
tubewells, deep 308
Tyers, Rodney 311

Umali, D. L. 50
unanticipated problems 72, 168–71, 176–8, 303
underdevelopment 360
  see also development
Underspecification Thesis 13–15, 17, 372–4, 376
United Nations 294, 295–6
United Nations Relief Operation Dhaka (UNROD) 293–5
United States Agency for International Development (USAID) 6, 68, 85, 138, 252, 255–6, 266, 291, 294, 310, 314, 317

University of the Philippines at Los Banos (UPLB) 45, 49–50, 54–5, 113, 117, 128, 138
University of Tokyo 45
urbanization 229
urea 312, 314
USA 6, 56, 72
  see also Ford Foundation (FF); Rockefeller Foundation (RF)

values:
  cultural 289–90
  of International Rice Research Institute (IRRI) 135–6
  new 118
  of researchers in Bangladesh 256–73
  see also scientists, attitudes of
varieties of rice:
  Ambalantota 62–355, 204–5
  BG 11–11 179, 180, 186, 195, 212
  BG 34–6 180, 196
  BG 34–8 180–1, 196, 206
  BG 35 191
  BG 90–2 206
  BG 94–1 205, 206
  Bluestick 239
  BR 3 303
  BR 4 303
  BRRI 7 342
  Chandina 269, 346
  Dee-Geo-Woo-Gen 176
  Engatek 179
  H2 167
  H4 167, 168, 174, 176, 179, 180, 186, 205
  H5 167, 176
  H8 179
  Heenati 202
  Indrashail 246
  IR 5 264, 268, 269
  IR 6 70
  IR 8 66–70, 72, 87, 124–7, 175–9, 261, 264, 268–9, 303, 363, 376
  IR 20 69, 70, 71, 74, 264, 269
  IR 22 69, 71, 74
  IR 24 69, 70, 74
  IR 26 182
  IR 36 93
  Irrisail 269
  Lakhidigi 347
  Latishail 239, 246, 247, 257
  M. 301 165, 166
  M. 302 166, 167
  Mala 269, 346
  Mas M. 24 166
  Nigershail 239, 241, 246, 304, 311, 347, 367
  Pajam 304, 311, 347

varieties of rice (*cont.*)
  Patnai 23 239, 246
  Podiwee A 8 165
  Ptb 33 182
  Purbachi 265
  TN 1 258
  Vellai Perumal 180
vegetables 192, 194, 203
Vietnam 45, 127
Vietnam War 127
Village-AID scheme, Pakistan 243–4, 252

Walagambahu Cropping Systems Project, Sri Lanka 194–7
Waldheim, Kurt 295
Walker, Rufus 291
Wallace 30
water, attitudes to 304–5
Water and Power Development Authority (WAPDA), Pakistan 244, 252, 265
Weaver, Warren 31–2, 39, 40–3, 121
Weeraratne, Hector 166–7, 174, 176, 178

wheat 5–6, 67, 254, 258–9, 289–90, 355
women's role 275–6
Work Plan (IRRI) 381, 382
World Bank 27, 76, 188, 198, 208, 252, 302, 306, 309, 331, 342–3
World War II 41, 164
Wortman 73, 120–1, 123, 131, 263

yields of rice 13, 39–40, 51–3, 99, 241, 266–7
  see also high-yielding varieties (HYVs) of rice

Zaman, M. Q. 246, 258
Zaman, S. M. H. 272
Zandstra, Howard 90, 149
Zia 285, 286, 290
Zillinsky, F. J. 345
zinc deficiencies 343, 344

Index compiled by Peva Keane